Birkhäuser

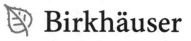

Advances in Mathematical Fluid Mechanics

Series editors

Giovanni P. Galdi, University of Pittsburgh, Pittsburgh, USA
John G. Heywood, University of British Columbia, Vancouver, Canada
Rolf Rannacher, Heidelberg University, Heidelberg, Germany

Advances in Mathematical Fluid Mechanics is a forum for the publication of high quality monographs, or collections of works, on the mathematical theory of fluid mechanics, with special regards to the Navier-Stokes equations. Its mathematical aims and scope are similar to those of the *Journal of Mathematical Fluid Mechanics*. In particular, mathematical aspects of computational methods and of applications to science and engineering are welcome as an important part of the theory. So also are works in related areas of mathematics that have a direct bearing on fluid mechanics.

More information about this series at http://www.springer.com/series/5032

Tomáš Bodnár • Giovanni P. Galdi • Šárka Nečasová
Editors

Waves in Flows

The 2018 Prague-Sum Workshop Lectures

Birkhäuser

Editors
Tomáš Bodnár
Department of Technical Mathematics
Faculty of Mechanical Engineering
Czech Technical University
Prague, Czech Republic

Giovanni P. Galdi
Department of Mechanical Engineering
University of Pittsburgh
Pittsburgh, PA, USA

Šárka Nečasová
Institute of Mathematics
Czech Academy of Sciences
Prague, Czech Republic

ISSN 2297-0320 ISSN 2297-0339 (electronic)
Advances in Mathematical Fluid Mechanics
ISBN 978-3-030-68143-2 ISBN 978-3-030-68144-9 (eBook)
https://doi.org/10.1007/978-3-030-68144-9

Mathematics Subject Classification: 35Q30, 76D05, 76B25, 76B55, 76B60, 76B03, 76D03, 76N10

This book is published under the imprint Birkhäuser, www.birkhauser-science.com by the registered company Springer Nature Switzerland AG.
The registered company address is: Gewerbestrasse 11, 6330 Cham, Switzerland

Preface

This book contains contributions from invited speakers of the workshop held as a part of the summer school "Waves in Flows," held in Prague (Czech Republic) in August 27–31, 2018, and organized by Tomáš Bodnár (Czech Technical University in Prague), Šárka Nečasová (Institute of Mathematics of the Czech Academy of Sciences), and Giovanni Paolo Galdi (University of Pittsburgh, USA). The summer school followed previous schools on various aspects of the mathematical fluid mechanics, held in Prague in the years 2011, 2012, 2014, and 2016, and it thus represented the fifth continuation of the series called the *Prague-Sum*.

The workshop offered a series of very interesting advanced lectures complementing the summer school program. Given the high professional quality of the lectures (and lecturers), it was decided to make those lectures in extended form available to public as a topical collection within the *Lecture Notes in Mathematical Fluid Mechanics*. This volume presents several contributions aiming at mathematical modeling and numerical simulations of various phenomena from mechanics of fluids and their interactions with solids.

The summer school, workshop as well as the present book was made possible due to generous direct support from Czech Academy of Sciences (special project to support research and educational activities for young people), Institute of Mathematics (institutional support Research Plan RVO 67985840), European Research Community on ercoŁftac.com Flow Turbulence and Combustion (ERCOFTAC), and Czech Science Foundation under the project No. P201-19-04243S.

Prague, Czech Republic	Tomáš Bodnár
Pittsburgh, PA, USA	Giovanni Paolo Galdi
Prague, Czech Republic	Šárka Nečasová

Contents

Chapter 1

Semigroup Theory for the Stokes Operator with Navier Boundary Condition on L^p Spaces

Chérif Amrouche (✉)
Université de Pau et des Pays de l'Adour, Pau Cedex, France
e-mail: cherif.amrouche@univ-pau.fr

Miguel Escobedo
Universidad del País Vasco, Departamento de Matemáticas, Lejona, Spain
e-mail: miguel.escobedo@ehu.es

Amrita Ghosh
Université de Pau et des Pays de l'Adour, Pau Cedex, France
e-mail: amrita.ghosh@univ-pau.fr

We consider the incompressible Navier–Stokes equations in a bounded domain with $\mathcal{C}^{1,1}$ boundary, completed with slip boundary condition. We study the general semigroup theory in L^p-spaces related to the Stokes operator with Navier boundary condition where the slip coefficient α is a non-smooth scalar function. It is shown that the strong and weak Stokes operators with Navier conditions admit analytic semigroup with bounded pure imaginary powers. We also show that for α large, the weak and strong solutions of both the linear and nonlinear systems

T. Bodnár et al. (eds.), *Waves in Flows*, Advances in Mathematical Fluid Mechanics, https://doi.org/10.1007/978-3-030-68144-9_1

are bounded uniformly with respect to α. This justifies mathematically that the solution of the Navier–Stokes problem with slip condition converges in the energy space to the solution of the Navier–Stokes with no-slip boundary condition as $\alpha \to \infty$.

1.1 Introduction

In this article we prove the existence of solutions to the following problem for the Navier–Stokes equations:

$$\begin{cases} \dfrac{\partial \boldsymbol{u}}{\partial t} - \Delta \boldsymbol{u} + (\boldsymbol{u} \cdot \nabla)\boldsymbol{u} + \nabla \pi = \boldsymbol{0} & \text{in } \Omega \times (0,T) \quad (1.1.1) \\ \operatorname{div} \boldsymbol{u} = 0 & \text{in } \Omega \times (0,T) \quad (1.1.2) \\ \boldsymbol{u} \cdot \boldsymbol{n} = 0 & \text{on } \Gamma \times (0,T) \quad (1.1.3) \\ 2[(\mathbb{D}\boldsymbol{u})\boldsymbol{n}]_{\boldsymbol{\tau}} + \alpha \boldsymbol{u}_{\boldsymbol{\tau}} = \boldsymbol{0} & \text{on } \Gamma \times (0,T) \quad (1.1.4) \\ \boldsymbol{u}(0) = \boldsymbol{u}_0 & \text{in } \Omega \quad (1.1.5) \end{cases}$$

where Ω is a bounded domain of $\mathbb{R}^3$, not necessarily simply connected, whose boundary Γ is of class $C^{1,1}$. The initial velocity $\boldsymbol{u}_0$ and the (scalar) friction coefficient α are given functions. The external unit vector on Γ is denoted by $\boldsymbol{n}$, $\mathbb{D}\boldsymbol{u} = \frac{1}{2}\left(\nabla \boldsymbol{u} + \nabla \boldsymbol{u}^T\right)$ denotes the strain tensor, and the subscript $\boldsymbol{\tau}$ denotes the tangential component i.e. $\boldsymbol{v}_{\boldsymbol{\tau}} = \boldsymbol{v} - (\boldsymbol{v} \cdot \boldsymbol{n})\boldsymbol{n}$ for any vector field $\boldsymbol{v}$. The functions $\boldsymbol{u}$ and π describe respectively the velocity and the pressure of a viscous incompressible fluid in Ω satisfying the boundary conditions (1.1.3) and (1.1.4).

The boundary condition (1.1.4) was introduced by H. Navier in [31], taking into account the molecular interactions with the boundary and is called *Navier boundary condition*. It may be deduced from kinetic theory considerations, as first described in [29] and rigorously proved under suitable conditions in [28]. It has been widely studied in recent years, because of its significance in modeling and simulations of flows and fluid–solid interaction problems (cf. [15, 22, 32] and references therein). In that context the function α is, up to some constant, the inverse of the slip length. We then impose the condition $\alpha \geq 0$ in all the remaining of this work.

The system with Dirichlet boundary condition (also called "no-slip condition") has deserved a lot of attention. In particular, a good semigroup theory has been developed in a series of work by Giga (cf. [16, 18–20]). Here we wish to establish a similar framework for $\alpha \not\equiv 0$. We will study two different types of solutions for (1.1.6), (1.1.2)–(1.1.5): *strong solutions* which belong to $L^p(0,T,\boldsymbol{L}^q(\Omega))$ type spaces and *weak solutions* (in a suitable sense) that may be written for a.e. $t > 0$, as $\boldsymbol{u}(t) = \boldsymbol{v}(t) + \nabla w(t)$ where $\boldsymbol{v} \in L^p(0,T;\boldsymbol{L}^q(\Omega))$ and $w \in L^p(0,T;L^q(\Omega))$. We will also consider different hypothesis on regularity of the function α. In particular, we collect some of the relevant results available for the Navier–Stokes problem with no-slip condition based on semigroup properties and prove them for the system (1.1.1)–(1.1.5) for the sake of completeness, so that this paper can be used as a basis for further work.

Let us give an overview of some related works. The system (1.1.1)–(1.1.5) has been studied in [13] in two-dimension where $\alpha \geq 0$ is a function in $C^2(\Gamma)$, with the main objective to analyze vanishing viscosity limit where the existence of weak and strong solutions has been established. Also in [12], the authors have studied stochastic Navier–Stokes equation with Navier boundary condition, similar to (1.1.1)–(1.1.5) where they considered same assumption that $\alpha \geq 0$ is in $C^2(\Gamma)$ and proved existence of weak solution. Beirao Da Veiga [9] has considered the same problem in 3D in $\mathcal{C}^{2,1}$ domain with $\alpha \geq 0$ constant and first showed that the Stokes operator with Navier boundary condition A is a maximal monotone, self-adjoint operator on $\boldsymbol{L}^2_{\sigma,\boldsymbol{\tau}}(\Omega)$ which generates an analytic semigroup of contraction and thus obtain strong solution of Stokes problem; Also by identifying the domain of $A^{1/2}$, he obtained the global strong solution of Navier–Stokes equation under the assumption of small data as in the no-slip boundary condition. The system (1.1.1)–(1.1.5) has also been studied by Tanaka et al. [23] in Sobolev–Slobodetskii spaces in point of view to analyze asymptotic behavior of the unsteady solution to the steady solution where they have considered Γ belongs to $W_2^{\frac{5}{2}+\ell}$ and $\alpha \in (0,1)$ is in $W_2^{\frac{1}{2}+\ell,\frac{1}{4}+\frac{\ell}{2}}((0,\infty)\times\Gamma)$ with $\ell \in (\frac{1}{2},1)$ and proved existence of local in time, strong solution and global in time, strong solution for small data. Note that in this work, α depends on both time and space variable. We want to mention further the works of [22, 27] where though the main objective is again to study viscosity limit, in the first paper, Iftimie and Sueur show existence of global in time weak solution in $C([0,\infty);\boldsymbol{L}^2_{\sigma,\boldsymbol{\tau}}(\Omega)) \cap L^2_{loc}([0,\infty);\boldsymbol{H}^1(\Omega))$ for $\boldsymbol{L}^2_{\sigma,\boldsymbol{\tau}}(\Omega)$ initial value; by classical approach: first deriving some energy estimate and then using Galerkin method. There they considered α a scalar function of class $\mathcal{C}^2$, without a sign. And in the second paper, Masmoudi and Rousset worked on anisotropic conormal Sobolev spaces considering smooth domain and $|\alpha| \leq 1$. In the paper [30], the authors aimed to prove the existence of global in time, strong solution of a similar problem assuming small data, in $\mathcal{C}^{1,1}$ domain and α non-negative, Hölder continuous in time. Also we mention the work [40] where Lagrangian Navier–Stokes problem (as a regularization system of classical Navier–Stokes equations) with vorticity slip boundary condition (which is close to boundary condition (1.1.4)) has been studied for non-negative smooth function α and existence of weak solution, global in time is obtained.

Further, in [10], Benes has established a unique weak solution, local in time for the Navier–Stokes system with mixed boundary condition: on some part of the boundary Navier condition with $\alpha = 0$ is considered and on other part, Neumann type boundary condition. Similar result for Navier–Stokes problem with Navier-type boundary condition (which corresponds to $\alpha = 0$) has been studied in, for example [26]. Also for $\alpha = 0$ the semigroup associated with Eq. (1.1.6) with (1.1.2)–(1.1.5) has been studied in [4].

In this article, we wish to study the general semigroup theory for any $p \in (1,\infty)$ for the Stokes operator with Navier boundary condition (NBC) with (possibly) minimal regularity on α which gives us existence, uniqueness, and regularity of both strong and weak solutions of (1.1.6), (1.1.2)–(1.1.5). We start with introducing the strong and weak Stokes operators for each fixed α and show that they generate analytic semigroups on the respective spaces for all $p \in (1,\infty)$. The proof of this is not very complicated and mostly use the existence and estimate results studied in [6] for the steady problem. Further we study the imaginary and fractional powers of the Stokes operators. To show that these operators are of bounded imaginary power (BIP) is not straightforward; Here we do not use pseudo-differential operator theory or Fourier multiplier theory as done by Giga [20], but chose a rather different approach: we show that the Stokes operator with (full) slip boundary condition (in its weak form) can be written as a lower-order perturbation of the Navier-type boundary condition (1.3.30) for which the result (that the operator is BIP) is known (cf. [3, Theorem 6.1]); And then with the help of Amann's interpolation-extrapolation theory [5], we recover the boundedness of imaginary power of $A_{p,\alpha}$. This method has been used in [34] to establish that the Stokes operator with NBC for $\alpha > 0$ constant, possesses a bounded $\mathcal{H}^\infty$-calculus on $\boldsymbol{L}^p_{\sigma,\boldsymbol{\tau}}(\Omega)$. As a consequence, we prove that the Stokes operator has maximal L^q-regularity and establish various types of $L^p - L^q$ estimates which in turn helps to develop an L^p-theory for the Navier–Stokes equations. We have used the abstract theory by Giga [18] for semi-linear parabolic equations in L^p to achieve the similar existence and regularity of a local strong solution and global weak solution.

The interesting part is to show the resolvent estimate

$$|\lambda| \|(\lambda I - A_{p,\alpha})^{-1} \boldsymbol{f}\|_{\boldsymbol{L}^p_{\sigma,\boldsymbol{\tau}}(\Omega)} \leq C \|\boldsymbol{f}\|_{\boldsymbol{L}^p_{\sigma,\boldsymbol{\tau}}(\Omega)} \qquad \forall\, \lambda \in \mathbb{C}\backslash\{0\} \text{ with } \operatorname{Re}\lambda \geq 0.$$

Adopting the standard method, multiplying the equation by $|\boldsymbol{u}|^{p-2}\bar{\boldsymbol{u}}$, one can easily obtain the above estimate, provided some suitable pressure estimate is available. For example, with the Navier-type boundary condition (1.3.30), the pressure term disappears from the Stokes operator (cf. [2, Proposition 3.1]) and for the perfect slip condition ($\alpha = 0$), the Stokes operator has the pressure term whose estimate is straightforward, as it contains the lower-order term (cf. (1.3.21)). In our case, taking into account the low regularity of α, it is not obvious at all to deduce the pressure estimate suitably. In the Hilbert case, this follows (cf. Theorem 1.3.1) from the variational formulation as expected.

Finally, it seems very natural to let $\alpha \to \infty$ in some sense in order to obtain the Dirichlet boundary condition on Γ from the condition (1.1.4). One of the goals of the present article is to study the behavior of the solutions of the unsteady Stokes and Navier–Stokes equation with NBC with respect to α, in particular what happens when α goes to ∞. This problem is considered in [24] in 2D where the author shows in Theorem 9.2 that, for $\boldsymbol{u}_0 \in \boldsymbol{H}^3(\Omega)$, when $\|1/\alpha\|_{L^\infty(\Gamma)} \to 0$, the solution of problem (1.1.1)–(1.1.5) converges to the solution of the Navier–Stokes problem with Dirichlet boundary condition in suitable spaces (cf. Sect. 1.9

for details). This encourages us to consider the case where α is a constant function and $\alpha \to \infty$ in both the linear problem (1.1.6), (1.1.2)–(1.1.5) and the nonlinear problem (1.1.1)–(1.1.5). We show that the solutions of the problems with NBC converge strongly in the energy space to the solutions of corresponding problem with Dirichlet boundary condition as α goes to ∞.

We state now our main results, for which the following notations are needed:

$$\boldsymbol{L}^p_{\sigma,\boldsymbol{\tau}}(\Omega) = \{\boldsymbol{v} \in \boldsymbol{L}^p(\Omega); \operatorname{div} \boldsymbol{v} = 0 \text{ in } \Omega, \boldsymbol{v} \cdot \boldsymbol{n} = 0 \text{ on } \Gamma\}$$

equipped with the norm of $\boldsymbol{L}^p(\Omega)$ and

$$\mathbf{D}(A_{p,\alpha}) = \left\{\boldsymbol{u} \in \boldsymbol{W}^{2,p}(\Omega) \cap \boldsymbol{L}^p_{\sigma,\boldsymbol{\tau}}(\Omega);\ 2[(\mathbb{D}\boldsymbol{u})\boldsymbol{n}]_{\boldsymbol{\tau}} + \alpha \boldsymbol{u}_{\boldsymbol{\tau}} = \mathbf{0} \text{ on } \Gamma\right\}.$$

The space $\mathbf{D}(A_{p,\alpha})$ is nothing but the domain of $A_{p,\alpha}$, the Stokes operator on $\boldsymbol{L}^p_{\sigma,\boldsymbol{\tau}}(\Omega)$ with the boundary conditions (1.1.3)–(1.1.4).

Theorem 1.1.1. Suppose that $\alpha \in W^{1-\frac{1}{r_1},r_1}(\Gamma)$ for some $r_1 \geq 3$ and $\alpha \geq 0$. Then, for every $\boldsymbol{u}_0 \in \boldsymbol{L}^{r_2}_{\sigma,\boldsymbol{\tau}}(\Omega)$ with $r_2 \geq 3$, there exists a unique solution $\boldsymbol{u}$ of (1.1.1)–(1.1.5) defined on a maximal time interval $[0, T_\star)$ such that

$$\boldsymbol{u} \in C([0,T_\star); \boldsymbol{L}^r_{\sigma,\boldsymbol{\tau}}(\Omega)) \cap L^q(0,T_\star; \boldsymbol{L}^p_{\sigma,\boldsymbol{\tau}}(\Omega))$$

$$t^{1/q}\boldsymbol{u} \in C([0,T_\star); \boldsymbol{L}^p_{\sigma,\boldsymbol{\tau}}(\Omega)) \quad \text{and} \quad t^{1/q}\|\boldsymbol{u}\|_{\boldsymbol{L}^p(\Omega)} \to 0 \text{ as } t \to 0$$

with $r = \min(r_1, r_2)$, $p > r$, $q > r$ and $\frac{2}{q} = \frac{3}{r} - \frac{3}{p}$. Moreover,

$$\boldsymbol{u} \in C((0,T_\star), \mathbf{D}(A_{r,\alpha})) \cap C^1((0,T_\star);\ \boldsymbol{L}^r_{\sigma,\boldsymbol{\tau}}(\Omega)).$$

If $r > 3$ and $T_\star < \infty$,

$$\|u(t)\|_{\boldsymbol{L}^r(\Omega)} \geq C(T_\star - t)^{(3-r)/2r},$$

where C is independent of $T_\star$ and t.

Also, there exists a constant $\delta > 0$ such that if $\|\boldsymbol{u}_0\|_{\boldsymbol{L}^3(\Omega)} < \delta$, then $T_\star = \infty$.

Under weaker conditions on Ω and α, a similar theorem holds for initial data in the space of distributions $\boldsymbol{u}_0 = \boldsymbol{\psi} + \nabla\chi$ where $\boldsymbol{\psi} \in L^r(\Omega)$ and $\chi \in L^r(\Omega)$ (denoted by $[\mathbf{H}^{r'}_0(\operatorname{div},\Omega)]'$, cf. Proposition 1.2.1), with $r \geq 3$.

The proof of these results is based on a careful study of the semigroup associated with the linear equation

$$\frac{\partial \boldsymbol{u}}{\partial t} - \Delta \boldsymbol{u} + \nabla \pi = \boldsymbol{f} \tag{1.1.6}$$

with conditions (1.1.2)–(1.1.5). For that we first study the strong and weak Stokes operators $A_{p,\alpha}$ and $B_{p,\alpha}$ and deduce that both of them have bounded inverse on $\boldsymbol{L}^p_{\sigma,\boldsymbol{\tau}}(\Omega)$ and $[\mathbf{H}^{p'}_0(\operatorname{div},\Omega)]'$ respectively for all $p \in (1,\infty)$. Also $-A_{p,\alpha}$ and $-B_{p,\alpha}$ generate bounded analytic semigroups on their respective spaces (cf. Theorems 1.3.6 and 1.4.3) and their pure imaginary powers are uniformly bounded as well (cf. Theorem 1.5.1). We obtain the following theorems, if $\boldsymbol{f} = \mathbf{0}$:

Theorem 1.1.2. Let $1 < p < \infty$ and $\alpha \geq 0$ be as in (1.3.5). Then for $\boldsymbol{u}_0 \in \boldsymbol{L}^p_{\sigma,\boldsymbol{\tau}}(\Omega)$, the problem (1.6.1) has a unique solution $\boldsymbol{u}(t)$ satisfying

$$\boldsymbol{u} \in C([0,\infty), \boldsymbol{L}^p_{\sigma,\boldsymbol{\tau}}(\Omega)) \cap C((0,\infty), \mathbf{D}(A_{p,\alpha})) \cap C^1((0,\infty), \boldsymbol{L}^p_{\sigma,\boldsymbol{\tau}}(\Omega))$$

and

$$\boldsymbol{u} \in C^k((0,\infty), \mathbf{D}(A^\ell_{p,\alpha})) \quad \forall\, k \in \mathbb{N},\ \forall\, \ell \in \mathbb{N}\backslash\{0\}.$$

Also, for all $t > 0$ and $q \geq p$, $\boldsymbol{u}(t) \in L^q(\Omega)$ and there exists $\delta > 0$ independent of t and q such that:

$$\|\boldsymbol{u}(t)\|_{\boldsymbol{L}^q(\Omega)} \leq C(\Omega,\alpha,p)\; e^{-\delta t} t^{-3/2(1/p-1/q)} \|\boldsymbol{u}_0\|_{\boldsymbol{L}^p(\Omega)}.$$

Moreover, the following estimates also hold:

$$\|\mathbb{D}\boldsymbol{u}(t)\|_{\boldsymbol{L}^q(\Omega)} \leq C(\Omega,\alpha,p)\; e^{-\delta t} t^{-3/2(1/p-1/q)-1/2} \|\boldsymbol{u}_0\|_{\boldsymbol{L}^p(\Omega)},$$

$$\forall\, m,n \in \mathbb{N},\ \ \|\frac{\partial^m}{\partial t^m} A^n_{p,\alpha}\boldsymbol{u}(t)\|_{\boldsymbol{L}^q(\Omega)} \leq C(\Omega,\alpha,p)\; e^{-\delta t} t^{-(m+n)-3/2(1/p-1/q)} \|\boldsymbol{u}_0\|_{\boldsymbol{L}^p(\Omega)}.$$

For $\boldsymbol{f} \neq \mathbf{0}$, and if we denote E_q the following real interpolation space:

$$E_q \equiv (D(A_{p,\alpha}), \boldsymbol{L}^p_{\sigma,\boldsymbol{\tau}}(\Omega))_{\frac{1}{q},q}$$

we have the result:

Theorem 1.1.3. Let $1 < p,q < \infty$. Also assume that $0 < T \leq \infty$ and $\alpha \geq 0$ be as in (1.3.5). Then for $\boldsymbol{f} \in L^q(0,T;\boldsymbol{L}^p_{\sigma,\boldsymbol{\tau}}(\Omega))$ and $\boldsymbol{u}_0 \in E_q$, there exists a unique solution $(\boldsymbol{u},\pi)$ of (1.7.1) satisfying:

$$\boldsymbol{u} \in L^q(0,T_0;\boldsymbol{W}^{2,p}(\Omega)) \ \text{ for all } \ T_0 \leq T \ \text{ if } \ T < \infty \ \text{ and } \ T_0 < \infty \ \text{ if } \ T = \infty,$$

$$\pi \in L^q(0,T;W^{1,p}(\Omega)/\mathbb{R}), \qquad \frac{\partial \boldsymbol{u}}{\partial t} \in L^q(0,T,\boldsymbol{L}^p_{\sigma,\boldsymbol{\tau}}(\Omega)),$$

$$\int_0^T \left\|\frac{\partial \boldsymbol{u}}{\partial t}\right\|^q_{\boldsymbol{L}^p(\Omega)} \mathrm{d}t + \int_0^T \|\boldsymbol{u}\|^q_{\boldsymbol{W}^{2,p}(\Omega)}\mathrm{d}t + \int_0^T \|\pi\|^q_{W^{1,p}(\Omega)}\mathrm{d}t \leq C\left(\int_0^T \|\boldsymbol{f}\|^q_{\boldsymbol{L}^p(\Omega)}\mathrm{d}t + \|\boldsymbol{u}_0\|^q_{E_q}\right), \tag{1.1.7}$$

where $C > 0$ is independent of $\boldsymbol{f}, \boldsymbol{u}_0$, and T, but may depend on α.

Similar results hold for less regular data (cf. Theorems 1.6.3 and 1.7.1). And the last interesting result of our work is the following limit problem which improves the result in [24, Theorem 9.2]:

Theorem 1.1.4. Let α be a constant and $(\boldsymbol{u}_\alpha, \pi_\alpha)$ be the solution of the problem (1.8.1) with $\boldsymbol{u}_0 \in \boldsymbol{L}^2_{\sigma,\boldsymbol{\tau}}(\Omega)$ and $(\boldsymbol{u}_\infty, \pi_\infty) \in L^2(0,T;\boldsymbol{H}^1_0(\Omega)) \times L^2(0,T;L^2(\Omega))$ a solution of the following Navier–Stokes problem with Dirichlet boundary condition:

$$\begin{cases} \dfrac{\partial \boldsymbol{u}_\infty}{\partial t} - \Delta \boldsymbol{u}_\infty + (\boldsymbol{u}_\infty \cdot \nabla)\boldsymbol{u}_\infty + \nabla \pi_\infty = \boldsymbol{0}, \quad \text{div } \boldsymbol{u}_\infty = 0 & \text{in } \Omega \times (0,T); \\ \boldsymbol{u}_\infty = \boldsymbol{0} & \text{on } \Gamma \times (0,T); \\ \boldsymbol{u}_\infty(0) = \boldsymbol{u}_0 & \text{in } \Omega \end{cases} \tag{1.1.8}$$

(whose existence has been proved in [38, Theorem 3.1, Chapter III]). Then for any $T < T_*$,

$$\boldsymbol{u}_\alpha \to \boldsymbol{u}_\infty \quad \text{in} \quad L^2(0,T;\boldsymbol{H}^1(\Omega)) \quad \text{as} \quad \alpha \to \infty$$

and

$$\int_0^T \int_\Gamma |\boldsymbol{u}_\alpha - \boldsymbol{u}_\infty|^2 \leq \frac{C}{\alpha}. \tag{1.1.9}$$

Moreover, if Γ is $\mathcal{C}^{2,1}$ and $\boldsymbol{u}_0 \in \boldsymbol{H}^2(\Omega) \cap \boldsymbol{H}^1_{0,\sigma}(\Omega)$, we also have

$$\sup_{t\in[0,T]} \int_\Omega |\boldsymbol{u}_\alpha - \boldsymbol{u}_\infty|^2 + \int_0^T \int_\Omega |\mathbb{D}(\boldsymbol{u}_\alpha - \boldsymbol{u}_\infty)|^2 + \int_0^T \int_\Gamma |\boldsymbol{u}_\alpha - \boldsymbol{u}_\infty|^2 \leq \frac{C}{\alpha}. \tag{1.1.10}$$

Similar result holds for the linear problem as well (cf. Sect. 1.9).

1.2 Preliminaries

First, we review some basic notations and functional framework we will use in the study. Throughout the work, if we do not specify otherwise, Ω is an open bounded set in $\mathbb{R}^3$ with boundary Γ of class $\mathcal{C}^{1,1}$ possibly multiply connected. Also if not otherwise stated, we assume

$$\alpha \geq 0 \quad \text{on } \Gamma \quad \text{and } \alpha > 0 \quad \text{on some } \Gamma_0 \subset \Gamma \quad \text{with } |\Gamma_0| > 0.$$

We follow the convention that C is an unspecified positive constant that may vary from expression to expression, even across an inequality (but not across an equality); Also C depends on Ω and p generally and the dependence on other parameters will be specified in the parenthesis when necessary.

The vector-valued Laplace operator of a vector field $\boldsymbol{v}$ can be equivalently defined by

$$\Delta \boldsymbol{v} = 2 \text{ div } \mathbb{D}\boldsymbol{v} - \textbf{grad} \text{ (div } \boldsymbol{v}).$$

We will denote by $\mathfrak{D}(\Omega)$ the set of smooth functions (infinitely differentiable) with compact support in Ω. Define

$$\mathfrak{D}_\sigma(\Omega) := \{\boldsymbol{v} \in \mathfrak{D}(\Omega) : \text{div } \boldsymbol{v} = 0 \text{ in } \Omega\}$$

and

$$L_0^p(\Omega) := \left\{ v \in L^p(\Omega) : \int_\Omega v = 0 \right\}.$$

For $p \in [1, \infty)$, p' denotes the conjugate exponent of p i.e. $\frac{1}{p} + \frac{1}{p'} = 1$. We also introduce the following space:

$$\boldsymbol{H}^p(\mathrm{div}, \Omega) := \{\boldsymbol{v} \in \boldsymbol{L}^p(\Omega) : \mathrm{div}\ \boldsymbol{v} \in L^p(\Omega)\}$$

equipped with the norm

$$\|\boldsymbol{v}\|_{\boldsymbol{H}^p(\mathrm{div},\Omega)} = \|\boldsymbol{v}\|_{\boldsymbol{L}^p(\Omega)} + \|\mathrm{div}\ \boldsymbol{v}\|_{L^p(\Omega)}.$$

It can be shown that $\mathfrak{D}(\overline{\Omega})$ is dense in $\boldsymbol{H}^p(\mathrm{div}, \Omega)$ for all $p \in [1, \infty)$. The closure of $\mathfrak{D}(\Omega)$ in $\boldsymbol{H}^p(\mathrm{div}, \Omega)$ is denoted by $\boldsymbol{H}_0^p(\mathrm{div}, \Omega)$ and it can be characterized by

$$\boldsymbol{H}_0^p(\mathrm{div}, \Omega) = \{\boldsymbol{v} \in \boldsymbol{H}^p(\mathrm{div}, \Omega) : \boldsymbol{v} \cdot \boldsymbol{n} = 0 \quad \text{on} \quad \Gamma\}. \tag{1.2.1}$$

For $p \in (1, \infty)$, we denote by $[\boldsymbol{H}_0^p(\mathrm{div}, \Omega)]'$, the dual space of $\boldsymbol{H}_0^p(\mathrm{div}, \Omega)$, which can be characterized as (for details, see [36, Proposition 1.0.4]):

Proposition 1.2.1. A distribution $\boldsymbol{f}$ belongs to $[\boldsymbol{H}_0^p(\mathrm{div}, \Omega)]'$ iff there exists $\boldsymbol{\psi} \in \boldsymbol{L}^{p'}(\Omega)$ and $\chi \in L^{p'}(\Omega)$ such that $\boldsymbol{f} = \boldsymbol{\psi} + \nabla\chi$. Moreover, we have the estimate:

$$\|\boldsymbol{f}\|_{[\boldsymbol{H}_0^p(\mathrm{div},\Omega)]'} = \inf_{\boldsymbol{f}=\boldsymbol{\psi}+\nabla\chi} \max\{\|\boldsymbol{\psi}\|_{\boldsymbol{L}^{p'}(\Omega)}, \|\chi\|_{L^{p'}(\Omega)}\}.$$

Next we introduce the spaces:

$$\boldsymbol{L}^p_{\sigma,\boldsymbol{\tau}}(\Omega) := \{\boldsymbol{v} \in \boldsymbol{L}^p(\Omega) : \mathrm{div}\ \boldsymbol{v} = 0 \text{ in } \Omega, \boldsymbol{v} \cdot \boldsymbol{n} = 0 \text{ on } \Gamma\}$$

equipped with the norm of $\boldsymbol{L}^p(\Omega)$;

$$\mathbf{W}^{1,p}_{\sigma,\boldsymbol{\tau}}(\Omega) := \{\boldsymbol{v} \in \boldsymbol{W}^{1,p}(\Omega) : \mathrm{div}\ \boldsymbol{v} = 0 \text{ in } \Omega, \boldsymbol{v} \cdot \boldsymbol{n} = 0 \text{ on } \Gamma\}$$

equipped with the norm of $\boldsymbol{W}^{1,p}(\Omega)$ and $\mathbf{H}^1_{\sigma,\boldsymbol{\tau}}(\Omega) := \mathbf{W}^{1,2}_{\sigma,\boldsymbol{\tau}}(\Omega)$. Also let us define

$$\mathbf{E}^p(\Omega) := \left\{(\boldsymbol{u}, \pi) \in \mathbf{W}^{1,p}(\Omega) \times L^p(\Omega); -\Delta\boldsymbol{u} + \nabla\pi \in \boldsymbol{L}^{r(p)}(\Omega)\right\},$$

where

$$\begin{cases} r(p) = \max\left\{1, \frac{3p}{p+3}\right\} & \text{if } p \neq \frac{3}{2} \\ r(p) > 1 & \text{if } p = \frac{3}{2}. \end{cases}$$

Let us now introduce some notations to describe the boundary. Consider any point P on Γ and choose an open neighborhood W of P in Γ, small enough to allow the existence of 2 families of $\mathcal{C}^2$ curves on W with the following properties: a curve of each family passes through every point of W and the unit tangent vectors to these curves form an orthogonal system (which we assume to have the direct orientation) at every point of W. The lengths s_1, s_2 along each family of curves, respectively, are a possible system of coordinates in W. We denote by $\boldsymbol{\tau}_1, \boldsymbol{\tau}_2$ the unit tangent vectors to each family of curves.

With these notations, we have $\boldsymbol{v} = \sum_{k=1}^{2} v_k \boldsymbol{\tau}_k + (\boldsymbol{v} \cdot \boldsymbol{n})\boldsymbol{n}$ where $\boldsymbol{\tau}_k = (\tau_{k1}, \tau_{k2}, \tau_{k3})$ and $v_k = \boldsymbol{v} \cdot \boldsymbol{\tau}_k$. For simplicity of notations, we will denote,

$$\Lambda \boldsymbol{v} = \sum_{k=1}^{2} \left(\boldsymbol{v}_{\boldsymbol{\tau}} \cdot \frac{\partial \boldsymbol{n}}{\partial s_k} \right) \boldsymbol{\tau}_k \, . \tag{1.2.2}$$

Here we state a relation between the Navier boundary condition and another type of boundary condition involving **curl** (often called as "Navier-type boundary condition") which will be used in later work. For proof, see [7, Appendix A].

Lemma 1.2.2. For any $\boldsymbol{v} \in \boldsymbol{W}^{2,p}(\Omega)$, we have the following equalities:

$$2\left[(\mathbb{D}\boldsymbol{v})\boldsymbol{n}\right]_{\boldsymbol{\tau}} = \nabla_{\boldsymbol{\tau}}(\boldsymbol{v} \cdot \boldsymbol{n}) + \left(\frac{\partial \boldsymbol{v}}{\partial \boldsymbol{n}} \right)_{\boldsymbol{\tau}} - \boldsymbol{\Lambda} \boldsymbol{v}$$

and

$$\mathbf{curl}\, \boldsymbol{v} \times \boldsymbol{n} = -\nabla_{\boldsymbol{\tau}}(\boldsymbol{v} \cdot \boldsymbol{n}) + \left(\frac{\partial \boldsymbol{v}}{\partial \boldsymbol{n}} \right)_{\boldsymbol{\tau}} + \boldsymbol{\Lambda} \boldsymbol{v}.$$

Remark 1.2.3. In the particular case $\boldsymbol{v} \cdot \boldsymbol{n} = 0$ on Γ, we have the following equalities for all $\boldsymbol{v} \in \boldsymbol{W}^{2,p}(\Omega)$,

$$2\left[(\mathbb{D}\boldsymbol{v})\boldsymbol{n}\right]_{\boldsymbol{\tau}} = \left(\frac{\partial \boldsymbol{v}}{\partial \boldsymbol{n}} \right)_{\boldsymbol{\tau}} - \boldsymbol{\Lambda} \boldsymbol{v} \quad \text{and} \quad \mathbf{curl}\, \boldsymbol{v} \times \boldsymbol{n} = \left(\frac{\partial \boldsymbol{v}}{\partial \boldsymbol{n}} \right)_{\boldsymbol{\tau}} + \boldsymbol{\Lambda} \boldsymbol{v} \tag{1.2.3}$$

which implies that

$$2\left[(\mathbb{D}\boldsymbol{v})\boldsymbol{n}\right]_{\boldsymbol{\tau}} = \mathbf{curl}\, \boldsymbol{v} \times \boldsymbol{n} - 2\boldsymbol{\Lambda} \boldsymbol{v}. \tag{1.2.4}$$

Note that on a flat boundary, $\boldsymbol{\Lambda} = \boldsymbol{0}$ and $2\left[(\mathbb{D}\boldsymbol{v})\boldsymbol{n}\right]_{\boldsymbol{\tau}}$ is actually equal to $\mathbf{curl}\, \boldsymbol{v} \times \boldsymbol{n}$.

Let us recall the Green's formula that plays an important role in this work, which is proved in [6, Lemma 3.5].

Theorem 1.2.1. Let $\Omega \subset \mathbb{R}^3$ be a $C^{0,1}$ bounded domain. Then,

(i) $\boldsymbol{\mathcal{D}}(\overline{\Omega}) \times \mathcal{D}(\overline{\Omega})$ is dense in $\mathbf{E}^p(\Omega)$.

(ii) The linear mapping $(\boldsymbol{v}, \pi) \mapsto \left[(\mathbb{D}\boldsymbol{v})\boldsymbol{n}\right]_{\boldsymbol{\tau}}$, defined on $\boldsymbol{\mathcal{D}}(\overline{\Omega}) \times \mathcal{D}(\overline{\Omega})$ can be extended to a linear, continuous map from $\boldsymbol{E}^p(\Omega)$ to $\boldsymbol{W}^{-\frac{1}{p},p}(\Gamma)$. Moreover, we have the following relation: for all $(\boldsymbol{v}, \pi) \in \boldsymbol{E}^p(\Omega)$ and $\boldsymbol{\varphi} \in \mathbf{W}^{1,p'}_{\sigma,\boldsymbol{\tau}}(\Omega)$,

$$\int_{\Omega} (-\Delta \boldsymbol{v} + \nabla \pi) \cdot \boldsymbol{\varphi} = 2 \int_{\Omega} \mathbb{D}\boldsymbol{v} : \mathbb{D}\boldsymbol{\varphi} - 2 \left\langle \left[(\mathbb{D}\boldsymbol{v})\boldsymbol{n}\right]_{\boldsymbol{\tau}}, \boldsymbol{\varphi} \right\rangle_{\boldsymbol{W}^{-\frac{1}{p},p}(\Gamma) \times \boldsymbol{W}^{\frac{1}{p},p'}(\Gamma)} . \tag{1.2.5}$$

The next classical identity (cf. [2, Lemma 2.5]) plays a key role in deriving the resolvent estimate of the Stokes operator.

Lemma 1.2.4. For $\boldsymbol{u} \in \boldsymbol{W}^{2,p}(\Omega)$ and Ω of class $\mathcal{C}^{1,1}$, we have,

$$-\int_\Omega |\boldsymbol{u}|^{p-2}\Delta\boldsymbol{u}\cdot\bar{\boldsymbol{u}} = \int_\Omega |\boldsymbol{u}|^{p-2}|\nabla\boldsymbol{u}|^2 + \frac{4(p-2)}{p^2}\int_\Omega |\nabla|\boldsymbol{u}|^{p/2}|^2$$

$$+ (p-2)i\sum_{k=1}^{3}\int_\Omega |\boldsymbol{u}|^{p-4}\,\mathrm{Re}\left(\frac{\partial\boldsymbol{u}}{\partial x_k}\cdot\bar{\boldsymbol{u}}\right)\mathrm{Im}\left(\frac{\partial\boldsymbol{u}}{\partial x_k}\cdot\bar{\boldsymbol{u}}\right) - \int_\Gamma \frac{\partial\boldsymbol{u}}{\partial\boldsymbol{n}}\cdot|\boldsymbol{u}|^{p-2}\bar{\boldsymbol{u}}.$$

Finally, we recall that the infinitesimal generator of an analytic semigroup can be characterized by the following theorem [8, Theorem 3.2, Chapter I]:

Theorem 1.2.2. Let A be a densely defined linear operator in a Banach space $\mathcal{E}$. Then A generates an analytic semigroup on $\mathcal{E}$ if and only if there exists $M > 0$ such that

$$\|\,(\lambda I - A)^{-1}\,\|_{\mathcal{L}(\mathcal{E})} \leq \frac{M}{|\lambda|}$$

for all $\lambda \in \mathbb{C}$ with $\mathrm{Re}\,\lambda > w$ for some $w \geq 0$.

1.3 Stokes Operator on $L^p_{\sigma,\tau}(\Omega)$: Strong Solutions

It is known that the closure of $\boldsymbol{\mathfrak{D}}_\sigma(\Omega)$ in $\boldsymbol{L}^p(\Omega)$ is the Banach space $\boldsymbol{L}^p_{\sigma,\boldsymbol{\tau}}(\Omega)$ [38, Theorem 1.4]. We introduce now the unbounded operator $(A_{p,\alpha}, \mathbf{D}(A_{p,\alpha}))$ on $\boldsymbol{L}^p_{\sigma,\boldsymbol{\tau}}(\Omega)$ whose definition depends on the regularity of the function α.

1. If $\alpha \in L^{t(p)}(\Gamma)$ with

$$t(p) = \begin{cases} \frac{2}{3}p' + \rho & \text{if } \ 1 < p < \frac{3}{2} \\ 2 + \rho & \text{if } \ \frac{3}{2} \leq p \leq 3, p \neq 2 \\ 2 & \text{if } \ p = 2 \\ \frac{2}{3}p + \rho & \text{if } \ p > 3 \end{cases} \tag{1.3.1}$$

with $\rho > 0$ arbitrarily small, we define the Stokes operator $A_{p,\alpha}$ on $\boldsymbol{L}^p_{\sigma,\boldsymbol{\tau}}(\Omega)$ as

$$\begin{cases} \mathbf{D}(A_{p,\alpha}) = \left\{\boldsymbol{u} \in \boldsymbol{W}^{1,p}_{\sigma,\boldsymbol{\tau}}(\Omega) : \Delta\boldsymbol{u} \in \boldsymbol{L}^p(\Omega), 2[(\mathbb{D}\boldsymbol{u})\boldsymbol{n}]_{\boldsymbol{\tau}} + \alpha\boldsymbol{u}_{\boldsymbol{\tau}} = \boldsymbol{0} \text{ on } \Gamma\right\} & (1.3.2) \\ A_{p,\alpha}(\boldsymbol{u}) = -P(\Delta\boldsymbol{u}) \quad \text{for } \boldsymbol{u} \in \mathbf{D}(A_{p,\alpha}), & (1.3.3) \end{cases}$$

where $P : \boldsymbol{L}^p(\Omega) \to \boldsymbol{L}^p_{\sigma,\boldsymbol{\tau}}(\Omega)$ is the orthogonal projection on $\boldsymbol{L}^p_{\sigma,\boldsymbol{\tau}}(\Omega)$. More precisely, for all $\boldsymbol{\psi} \in \boldsymbol{L}^p(\Omega)$, $P(\boldsymbol{\psi}) = \boldsymbol{\psi} - \nabla\pi$ where $\pi \in W^{1,p}(\Omega)$ is a solution of

$$\begin{cases} \mathrm{div}(\nabla\pi - \boldsymbol{\psi}) = 0 \ \text{ in } \ \Omega \\ (\nabla\pi - \boldsymbol{\psi})\cdot\boldsymbol{n} = 0 \ \text{ on } \Gamma. \end{cases} \tag{1.3.4}$$

Notice that, when $\boldsymbol{u} \in \mathbf{D}(A_{p,\alpha})$ but $\boldsymbol{u} \notin \boldsymbol{W}^{2,p}(\Omega)$, the boundary term $[(\mathbb{D}\boldsymbol{u})\boldsymbol{n}]_{\boldsymbol{\tau}}$ is still well defined as shown in [7, Lemma 2.4].

2. If α is such that

$$\alpha \in \begin{cases} W^{1-\frac{1}{\frac{3}{2}+\rho},\frac{3}{2}+\rho}(\Gamma) & \text{if } \ 1 < p \leq \frac{3}{2} \\ W^{1-\frac{1}{p},p}(\Gamma) & \text{if } \ p > \frac{3}{2} \end{cases} \tag{1.3.5}$$

with $\rho > 0$ arbitrarily small, then we define the Stokes operator $A_{p,\alpha}$ on $\boldsymbol{L}^p_{\sigma,\boldsymbol{\tau}}(\Omega)$ as

$$\begin{cases} \mathbf{D}(A_{p,\alpha}) = \{\boldsymbol{u} \in \boldsymbol{W}^{2,p}(\Omega) \cap \boldsymbol{L}^p_{\sigma,\boldsymbol{\tau}}(\Omega), 2[(\mathbb{D}\boldsymbol{u})\boldsymbol{n}]_{\boldsymbol{\tau}} + \alpha \boldsymbol{u}_{\boldsymbol{\tau}} = \mathbf{0} \text{ on } \Gamma\} & (1.3.6) \\ A_{p,\alpha}(\boldsymbol{u}) = -P(\Delta \boldsymbol{u}) \quad \text{for } \boldsymbol{u} \in \mathbf{D}(A_{p,\alpha}). & (1.3.7) \end{cases}$$

Remark 1.3.1. If α satisfies (1.3.5), then $\alpha \in L^{t(p)}(\Gamma)$ as well.

Remark 1.3.2. When α satisfies (1.3.5) and $\boldsymbol{u} \in \mathbf{D}(A_{p,\alpha})$, $A_{p,\alpha}\boldsymbol{u} = A\boldsymbol{u}$ where A is the following operator defined in [6, just before Proposition 5.6]:

$$A \in \mathcal{L}\left(\mathbf{W}^{1,p}_{\sigma,\boldsymbol{\tau}}(\Omega), \left(\mathbf{W}^{1,p'}_{\sigma,\boldsymbol{\tau}}(\Omega)\right)'\right)$$

$$\text{defined by} \qquad \langle A\boldsymbol{u}, \boldsymbol{v}\rangle = a(\boldsymbol{u}, \boldsymbol{v})$$

$$\text{where} \qquad a(\boldsymbol{u}, \boldsymbol{v}) = 2\int_\Omega \mathbb{D}\boldsymbol{u} : \mathbb{D}\bar{\boldsymbol{v}} + \int_\Gamma \alpha \boldsymbol{u}_{\boldsymbol{\tau}} \cdot \bar{\boldsymbol{v}}_{\boldsymbol{\tau}}.$$

More precisely, if $\boldsymbol{u} \in \mathbf{D}(A_{p,\alpha})$, then using Green's formula (1.2.5) and the relation $2[(\mathbb{D}\boldsymbol{u})\boldsymbol{n}]_{\boldsymbol{\tau}} = -\alpha \boldsymbol{u}_{\boldsymbol{\tau}}$ on Γ, we can show $\langle A\boldsymbol{u}, \boldsymbol{v}\rangle = \langle A_{p,\alpha}\boldsymbol{u}, \boldsymbol{v}\rangle$ for any $\boldsymbol{v} \in \mathbf{W}^{1,p'}_{\sigma,\boldsymbol{\tau}}(\Omega)$.

In order to show that $(\mathbf{D}(A_{p,\alpha}), A_{p,\alpha})$ generates an analytic semigroup on $\boldsymbol{L}^p_{\sigma,\boldsymbol{\tau}}(\Omega)$, $\lambda \in \mathbb{C}$, some estimate on the resolvent $(\lambda I + A_p)^{-1}$ is needed.

Suppose that $\alpha \in L^{t(p)}(\Gamma)$ and $\boldsymbol{f} \in \boldsymbol{L}^p_{\sigma,\boldsymbol{\tau}}(\Omega)$. Then by (1.3.2)–(1.3.3), $\boldsymbol{u} \in \mathbf{D}(A_{p,\alpha})$ and $(\lambda I + A_{p,\alpha})\boldsymbol{u} = \boldsymbol{f}$ is equivalent to $\boldsymbol{u} \in \boldsymbol{W}^{1,p}(\Omega)$ satisfying

$$\begin{cases} \lambda \boldsymbol{u} - \Delta \boldsymbol{u} + \nabla \pi = \boldsymbol{f} & \text{in } \Omega & (1.3.8) \\ \operatorname{div} \boldsymbol{u} = 0 & \text{in } \Omega & (1.3.9) \\ \boldsymbol{u} \cdot \boldsymbol{n} = 0 & \text{on } \Gamma & (1.3.10) \\ 2[(\mathbb{D}\boldsymbol{u})\boldsymbol{n}]_{\boldsymbol{\tau}} + \alpha \boldsymbol{u}_{\boldsymbol{\tau}} = \mathbf{0} & \text{on } \Gamma & (1.3.11) \end{cases}$$

for some $\pi \in L^p(\Omega)$.

If on the other hand, α satisfies (1.3.5), $\boldsymbol{u} \in \mathbf{D}(A_{p,\alpha})$ and $(\lambda I + A_{p,\alpha})\boldsymbol{u} = \boldsymbol{f}$ for $\boldsymbol{f} \in \boldsymbol{L}^p_{\sigma,\boldsymbol{\tau}}(\Omega)$, is equivalent to $(\boldsymbol{u}, \pi) \in \boldsymbol{W}^{2,p}(\Omega) \times W^{1,p}(\Omega)$ satisfying (1.3.8)–(1.3.11).

Proposition 1.3.3. Suppose that $\alpha \in L^{t(p)}(\Gamma)$, $\boldsymbol{f} \in [\boldsymbol{H}_0^{p'}(\mathrm{div},\Omega)]'$ with $p \in (1,\infty)$ and $\lambda \in \mathbb{C}$. Then, $\boldsymbol{u} \in \boldsymbol{W}^{1,p}(\Omega)$ solves (1.3.8)–(1.3.11) for some $\pi \in L^p(\Omega)$ is equivalent to $\boldsymbol{u} \in \mathbf{W}^{1,p}_{\sigma,\boldsymbol{T}}(\Omega)$ satisfies

$$\begin{cases} a_\lambda(\boldsymbol{u},\boldsymbol{\varphi}) = \langle \boldsymbol{f}, \bar{\boldsymbol{\varphi}}\rangle \quad \forall\ \boldsymbol{\varphi} \in \mathbf{W}^{1,p'}_{\sigma,\boldsymbol{T}}(\Omega) \\ \text{where:} \\ a_\lambda(\boldsymbol{u},\boldsymbol{\varphi}) = \lambda \displaystyle\int_\Omega \boldsymbol{u}\cdot\bar{\boldsymbol{\varphi}} + 2\int_\Omega \mathbb{D}\boldsymbol{u} : \mathbb{D}\bar{\boldsymbol{\varphi}} + \langle \alpha \boldsymbol{u}_{\boldsymbol{T}}, \bar{\boldsymbol{\varphi}}_{\boldsymbol{T}}\rangle_\Gamma \end{cases} \tag{1.3.12}$$

and $\langle,\rangle_\Gamma$ denotes the duality product between $\boldsymbol{W}^{-\frac{1}{p},p}(\Gamma)$ and $\boldsymbol{W}^{\frac{1}{p},p'}(\Gamma)$.

Proof. By Amrouche et al. [6, Lemma 5.1], for all $\boldsymbol{u} \in \mathbf{W}^{1,p}_{\sigma,\boldsymbol{T}}\Omega)$ and $\boldsymbol{\varphi} \in \mathbf{W}^{1,p'}_{\sigma,\boldsymbol{T}}(\Omega)$ we have $\alpha \boldsymbol{u}_{\boldsymbol{T}} \cdot \bar{\boldsymbol{\varphi}}_{\boldsymbol{T}} \in L^1(\Gamma)$ and

$$\int_\Gamma \alpha \boldsymbol{u}_{\boldsymbol{T}} \cdot \bar{\boldsymbol{\varphi}}_{\boldsymbol{T}} \leq C\|\alpha\|_{L^{t(p)}(\Gamma)} \|\boldsymbol{u}\|_{\mathbf{W}^{1,p}_{\sigma,\boldsymbol{T}}(\Omega)} \|\boldsymbol{\varphi}\|_{\mathbf{W}^{1,p'}_{\sigma,\boldsymbol{T}}(\Omega)}.$$

It easily then follows that $a_\lambda(\cdot,\cdot)$ is a continuous sesqui-linear form on $\mathbf{W}^{1,p}_{\sigma,\boldsymbol{T}}(\Omega) \times \mathbf{W}^{1,p'}_{\sigma,\boldsymbol{T}}(\Omega)$. When $\lambda = 0$, it is the sesqui-linear form $a(\cdot,\cdot)$ introduced in [6]. The proof of this proposition then follows exactly the same steps and uses the same arguments as in [6, Proposition 5.2]. ∎

The following theorem gives the existence of a unique solution of the resolvent problem and also the resolvent estimate.

Theorem 1.3.1. For any $\varepsilon \in (0,\pi)$, let $\lambda \in \Sigma_\varepsilon := \{\lambda \in \mathbb{C} : |\arg\lambda| \leq \pi - \varepsilon\}$, $\boldsymbol{f} \in \boldsymbol{L}^2(\Omega)$ and $\alpha \in L^2(\Gamma)$. Then,
1. the problem (1.3.8)–(1.3.11) has a unique solution $(\boldsymbol{u},\pi) \in \boldsymbol{H}^1(\Omega) \times L^2_0(\Omega)$.
2. there exists a constant $C_\varepsilon > 0$, independent of $\boldsymbol{f}$, α and λ, such that the solution $\boldsymbol{u}$ satisfies the following estimates, for $\lambda \neq 0$:

$$\|\boldsymbol{u}\|_{\boldsymbol{L}^2(\Omega)} \leq \frac{1}{C_\varepsilon|\lambda|}\, \|\boldsymbol{f}\|_{\boldsymbol{L}^2(\Omega)} \tag{1.3.13}$$

$$\|\mathbb{D}\boldsymbol{u}\|_{\boldsymbol{L}^2(\Omega)} \leq \frac{1}{C_\varepsilon\sqrt{|\lambda|}}\, \|\boldsymbol{f}\|_{\boldsymbol{L}^2(\Omega)} \tag{1.3.14}$$

and

$$\|\pi\|_{L^2(\Omega)} \leq C(\Omega)\left[1 + \frac{1}{C_\varepsilon}\left(1 + \frac{1}{\sqrt{|\lambda|}}\right)\right] \|\boldsymbol{f}\|_{\boldsymbol{L}^2(\Omega)}. \tag{1.3.15}$$

3. moreover, if either (i) Ω is not axisymmetric or (ii) Ω is axisymmetric and $\alpha \geq \alpha_* > 0$, then

$$\|\boldsymbol{u}\|_{\boldsymbol{H}^1(\Omega)} + \|\pi\|_{L^2(\Omega)} \leq C(\Omega)\left(1 + \frac{1}{C_\varepsilon}\right) \|\boldsymbol{f}\|_{\boldsymbol{L}^2(\Omega)} \tag{1.3.16}$$

and if α is a constant, then

$$\|\boldsymbol{u}\|_{\boldsymbol{H}^2(\Omega)} + \|\pi\|_{H^1(\Omega)} \le C(\Omega)\left(1 + \frac{1}{C_\varepsilon}\right)\|\boldsymbol{f}\|_{\boldsymbol{L}^2(\Omega)} \tag{1.3.17}$$

where $C(\Omega)$ does not depend on α.

Remark 1.3.4. Note that the estimates (1.3.13) and (1.3.14) give better decay for λ large and enable us having a good semigroup theory, in general. On the other hand, estimate (1.3.16) gives uniform bound on the solution, especially when λ is small.

Proof. **1.** In view of Proposition 1.3.3, it is enough to prove the existence and uniqueness of a solution to (1.3.12). Also $\lambda = 0$ case corresponds to the existence result for the stationary Stokes problem [6, Theorem 4.1]. So we may consider $\lambda \neq 0$.

By Korn inequality, $\|\boldsymbol{u}\|_{\boldsymbol{L}^2(\Omega)} + \|\mathbb{D}\boldsymbol{u}\|_{\boldsymbol{L}^2(\Omega)}$ is an equivalent norm on $\mathbf{H}^1(\Omega)$. Then using the following inequality (cf. [3, Lemma 4.1]):

$$\forall\ \lambda \in \Sigma_\varepsilon, \forall\ a, b > 0, \qquad |\lambda a + b| \ge C_\varepsilon\left(|\lambda| a + b\right) \text{ for some constant } C_\varepsilon > 0$$

and $\alpha \ge 0$, we get,

$$\begin{aligned} |a_\lambda(\boldsymbol{u}, \boldsymbol{u})| &\ge C_\varepsilon\left(|\lambda| \|\boldsymbol{u}\|^2_{\boldsymbol{L}^2(\Omega)} + 2\|\mathbb{D}\boldsymbol{u}\|^2_{\boldsymbol{L}^2(\Omega)} + \int_\Gamma \alpha |\boldsymbol{u}_{\boldsymbol{\tau}}|^2\right) \\ &\ge C_\varepsilon\ \min\left(|\lambda|, 2\right)\left(\|\boldsymbol{u}\|^2_{\boldsymbol{L}^2(\Omega)} + \|\mathbb{D}\boldsymbol{u}\|^2_{\boldsymbol{L}^2(\Omega)}\right) \\ &\ge C_\varepsilon\ \min\left(|\lambda|, 2\right)\|\boldsymbol{u}\|^2_{\boldsymbol{H}^1(\Omega)}. \end{aligned}$$

Hence, for all $\lambda \in \Sigma_\varepsilon$, a_λ is coercive on $\mathbf{H}^1_{\sigma,\boldsymbol{\tau}}(\Omega)$ and therefore, by Lax–Milgram lemma, we get a unique solution in $\mathbf{H}^1_{\sigma,\boldsymbol{\tau}}(\Omega)$ of the problem (1.3.12) which proves **1.**

2. From the variational formulation, we have,

$$a_\lambda(\boldsymbol{u}, \boldsymbol{u}) = \int_\Omega \boldsymbol{f} \cdot \bar{\boldsymbol{u}}$$

which gives

$$|a_\lambda(\boldsymbol{u}, \boldsymbol{u})| \le \|\boldsymbol{f}\|_{\boldsymbol{L}^2(\Omega)} \|\boldsymbol{u}\|_{\boldsymbol{L}^2(\Omega)}.$$

But we also have,

$$|a_\lambda(\boldsymbol{u}, \boldsymbol{u})| \ge C_\varepsilon\left(|\lambda| \|\boldsymbol{u}\|^2_{\boldsymbol{L}^2(\Omega)} + 2\|\mathbb{D}\boldsymbol{u}\|^2_{\boldsymbol{L}^2(\Omega)}\right).$$

Note that the above constant depends only on ε. Thus

$$\|\boldsymbol{u}\|_{\boldsymbol{L}^2(\Omega)} \le \frac{1}{C_\varepsilon |\lambda|}\|\boldsymbol{f}\|_{\boldsymbol{L}^2(\Omega)}$$

and then

$$\|\mathbb{D}\boldsymbol{u}\|^2_{\boldsymbol{L}^2(\Omega)} \leq \frac{1}{2\,C_\varepsilon}\|\boldsymbol{f}\|_{\boldsymbol{L}^2(\Omega)}\|\boldsymbol{u}\|_{\boldsymbol{L}^2(\Omega)} \leq \frac{1}{2\,|\lambda|\,C_\varepsilon^2}\|\boldsymbol{f}\|^2_{\boldsymbol{L}^2(\Omega)}$$

prove the inequalities (1.3.13) and (1.3.14).

From Eq. (1.3.8), we may write, using (1.3.13) and (1.3.14),

$$\begin{aligned}\|\pi\|_{L^2(\Omega)} \leq \|\nabla\pi\|_{\boldsymbol{H}^{-1}(\Omega)} &\leq C(\Omega)\left(\|\boldsymbol{f}\|_{\boldsymbol{L}^2(\Omega)} + \|\Delta\boldsymbol{u}\|_{\boldsymbol{H}^{-1}(\Omega)} + |\lambda|\|\boldsymbol{u}\|_{\boldsymbol{L}^2(\Omega)}\right)\\ &\leq C(\Omega)\left(\|\boldsymbol{f}\|_{\boldsymbol{L}^2(\Omega)} + \|\mathbb{D}\boldsymbol{u}\|_{\boldsymbol{L}^2(\Omega)} + |\lambda|\|\boldsymbol{u}\|_{\boldsymbol{L}^2(\Omega)}\right)\\ &\leq C(\Omega)\left[1+\frac{1}{C_\varepsilon}\left(1+\frac{1}{\sqrt{2|\lambda|}}\right)\right]\|\boldsymbol{f}\|_{\boldsymbol{L}^2(\Omega)}\end{aligned}$$

which gives (1.3.15).

3. Moreover, writing (1.3.8) as $-\Delta\boldsymbol{u}+\nabla\pi = \boldsymbol{f}-\lambda\boldsymbol{u}$, we deduce from the stationary Stokes estimate [6, Proposition 4.3] in the case either (i) Ω is not axisymmetric or (ii) Ω is axisymmetric and $\alpha \geq \alpha_* > 0$, the existence of a constant $C > 0$ which depends only on Ω such that

$$\|\boldsymbol{u}\|_{\boldsymbol{H}^1(\Omega)}+\|\pi\|_{L^2(\Omega)} \leq C(\Omega)\left(\|\boldsymbol{f}\|_{\boldsymbol{L}^2(\Omega)}+|\lambda|\|\boldsymbol{u}\|_{\boldsymbol{L}^2(\Omega)}\right) \leq C(\Omega)\left(1+\frac{1}{C_\varepsilon}\right)\|\boldsymbol{f}\|_{\boldsymbol{L}^2(\Omega)}.$$

This provides the better bound (1.3.16) on $\boldsymbol{u}$ and π when λ is small.

Similarly, for constant α, the H^2 estimate for the stationary Stokes problem [6, Theorem 4.5] yields

$$\|\boldsymbol{u}\|_{\boldsymbol{H}^2(\Omega)} + \|\pi\|_{H^1(\Omega)} \leq C(\Omega)\left(1+\frac{1}{C_\varepsilon}\right)\|\boldsymbol{f}\|_{\boldsymbol{L}^2(\Omega)}.$$

This completes the proof. ■

In the next theorem we prove the analyticity of the semigroup generated by the Stokes operator with Navier boundary condition on $\boldsymbol{L}^2_{\sigma,\boldsymbol{\tau}}(\Omega)$.

Theorem 1.3.2. For any $\alpha \in L^2(\Gamma)$, the operator $-A_{2,\alpha}$, defined in (1.3.2)–(1.3.3) with $p = 2$, generates a bounded analytic semigroup on $\boldsymbol{L}^2_{\sigma,\boldsymbol{\tau}}(\Omega)$.

Proof. Obviously $\mathbf{D}(A_{2,\alpha})$ is dense in $\boldsymbol{L}^2_{\sigma,\boldsymbol{\tau}}(\Omega)$. Therefore, according to Theorem 1.2.2, it is enough to prove the resolvent estimate. Now, by definition and from the previous theorem, we have,

$$\|(\lambda I + A_{2,\alpha})^{-1}\| = \sup_{\substack{\boldsymbol{f}\in\boldsymbol{L}^2_{\sigma,\boldsymbol{\tau}}(\Omega)\\ \boldsymbol{f}\neq\boldsymbol{0}}} \frac{\|(\lambda I+A_{2,\alpha})^{-1}\boldsymbol{f}\|_{\boldsymbol{L}^2(\Omega)}}{\|\boldsymbol{f}\|_{\boldsymbol{L}^2(\Omega)}} = \sup_{\boldsymbol{f}\in\boldsymbol{L}^2_{\sigma,\boldsymbol{\tau}}(\Omega)} \frac{\|\boldsymbol{u}\|_{\boldsymbol{L}^2(\Omega)}}{\|\boldsymbol{f}\|_{\boldsymbol{L}^2(\Omega)}} \leq \frac{1}{C_\varepsilon|\lambda|}.$$

Hence, the result. ■

Next we extend the results of Theorem 1.3.1 for all $p \in (1,\infty)$.

Theorem 1.3.3. Suppose that $p \in (1, \infty)$, α satisfies (1.3.5) and $\lambda \in \Sigma_\varepsilon$ where Σ_ε is defined as in Theorem 1.3.1. Then for every $\boldsymbol{f} \in \boldsymbol{L}^p(\Omega)$, the problem (1.3.8)–(1.3.11) has a unique solution $(\boldsymbol{u}, \pi) \in \boldsymbol{W}^{2,p}(\Omega) \times \big(W^{1,p}(\Omega) \cap L^p_0(\Omega)\big)$.

Proof. **Case (i):** $p > 2$. Since from the assumption, $\boldsymbol{f} \in \boldsymbol{L}^2(\Omega)$ and $\alpha \in L^2(\Gamma)$, there exists a unique solution $(\boldsymbol{u}, \pi) \in \boldsymbol{H}^1(\Omega) \times L^2_0(\Omega)$ of (1.3.8)–(1.3.11) by Theorem 1.3.1. Now writing Eq. (1.3.8) as $-\Delta \boldsymbol{u} + \nabla \pi = \boldsymbol{f} - \lambda \boldsymbol{u}$ and since $\boldsymbol{u} \in \boldsymbol{H}^1(\Omega) \hookrightarrow \boldsymbol{L}^6(\Omega)$, we have $\boldsymbol{f} - \lambda \boldsymbol{u} \in \boldsymbol{L}^p(\Omega)$ for all $p \leq 6$. Thus, using the regularity result in [6, Theorem 5.11], we obtain $\boldsymbol{u} \in \boldsymbol{W}^{2,p}(\Omega)$ for all $p \leq 6$.

Now for $p > 6$, we have $\boldsymbol{u} \in \boldsymbol{W}^{2,6}(\Omega) \hookrightarrow \boldsymbol{L}^\infty(\Omega)$. Hence $\boldsymbol{f} - \lambda \boldsymbol{u} \in \boldsymbol{L}^p(\Omega)$ and by the same regularity result, we get $\boldsymbol{u} \in \boldsymbol{W}^{2,p}(\Omega)$ for all $p > 6$.

Case (ii): $1 < p < 2$. We first claim that $(\lambda I + A)$ is an isomorphism from $\boldsymbol{W}^{1,q}_{\sigma,\boldsymbol{\tau}}(\Omega)$ to $(\boldsymbol{W}^{1,q'}_{\sigma,\boldsymbol{\tau}}(\Omega))'$ for all $q \geq 2$. Then the adjoint operator, $\lambda I + A^*$ is also an isomorphism from $\boldsymbol{W}^{1,q'}_{\sigma,\boldsymbol{\tau}}(\Omega)$ to $(\boldsymbol{W}^{1,q}_{\sigma,\boldsymbol{\tau}}(\Omega))'$ with $q' \leq 2$. Then, for any $\boldsymbol{f} \in \mathbf{L}^p_{\sigma,\boldsymbol{\tau}}(\Omega) \subset (\boldsymbol{W}^{1,p'}_{\sigma,\boldsymbol{\tau}}(\Omega))'$, there exists a unique $\boldsymbol{u} \in \boldsymbol{W}^{1,p}_{\sigma,\boldsymbol{\tau}}(\Omega)$ such that $(\lambda I + A^*)\boldsymbol{u} = \boldsymbol{f}$. Our second claim is that, since $\boldsymbol{f} \in \boldsymbol{L}^p_{\sigma,\boldsymbol{\tau}}(\Omega)$ it follows that $\boldsymbol{u} \in \mathbf{D}(A_{p,\alpha})$ and $A^*\boldsymbol{u} = A_{p,\alpha}\boldsymbol{u}$. This finally implies that $(\lambda I + A_{p,\alpha})\boldsymbol{u} = \boldsymbol{f}$.

First claim: For $q > 2$ and $\boldsymbol{\ell} \in (\boldsymbol{W}^{1,q'}_{\sigma,\boldsymbol{\tau}}(\Omega))' \subset (\boldsymbol{H}^1_{\sigma,\boldsymbol{\tau}}(\Omega))'$, from Lax–Milgram lemma there is a unique $\boldsymbol{u} \in \boldsymbol{H}^1_{\sigma,\boldsymbol{\tau}}(\Omega)$ such that $a_\lambda(\boldsymbol{u}, \boldsymbol{\varphi}) = \langle \boldsymbol{\ell}, \boldsymbol{\varphi} \rangle$ for all $\boldsymbol{\varphi} \in \boldsymbol{H}^1_{\sigma,\boldsymbol{\tau}}(\Omega)$. Then, $a(\boldsymbol{u}, \boldsymbol{\varphi}) = \langle \boldsymbol{\ell} - \lambda \boldsymbol{u}, \boldsymbol{\varphi} \rangle$ with $\boldsymbol{\ell} - \lambda \boldsymbol{u} \in (\boldsymbol{W}^{1,q'}_{\sigma,\boldsymbol{\tau}}(\Omega))'$. On the other hand, by Amrouche et al. [6, Theorem 5.5] there exists a unique $\boldsymbol{w} \in \boldsymbol{W}^{1,q}_{\sigma,\boldsymbol{\tau}}(\Omega) \subset \boldsymbol{W}^{1,2}_{\sigma,\boldsymbol{\tau}}(\Omega)$ such that

$$a(\boldsymbol{w}, \boldsymbol{\varphi}) = \langle \boldsymbol{\ell} - \lambda \boldsymbol{u}, \boldsymbol{\varphi} \rangle \ \ \forall \boldsymbol{\varphi} \in \boldsymbol{W}^{1,q'}_{\sigma,\boldsymbol{\tau}}(\Omega).$$

It then follows that

$$a(\boldsymbol{w} - \boldsymbol{u}, \boldsymbol{\varphi}) = 0 \ \ \forall \boldsymbol{\varphi} \in \boldsymbol{H}^1_{\sigma,\boldsymbol{\tau}}(\Omega)$$

and by the uniqueness result [6, Theorem 5.5], $\boldsymbol{u} = \boldsymbol{w}$ in $\boldsymbol{H}^1_{\sigma,\boldsymbol{\tau}}(\Omega)$ and thus $\boldsymbol{u} \in \boldsymbol{W}^{1,q}_{\sigma,\boldsymbol{\tau}}(\Omega)$.

Second claim: If $\boldsymbol{f} \in \boldsymbol{L}^p_{\sigma,\boldsymbol{\tau}}(\Omega)$ with $1 < p < 2$ and $(\lambda I + A^*)\boldsymbol{u} = \boldsymbol{f}$ with $\boldsymbol{u} \in \boldsymbol{W}^{1,p}_{\sigma,\boldsymbol{\tau}}(\Omega)$, then $A^*\boldsymbol{u} = \boldsymbol{f} - \lambda \boldsymbol{u} =: \boldsymbol{g} \in \boldsymbol{L}^p_{\sigma,\boldsymbol{\tau}}(\Omega)$ which means $a(\boldsymbol{u}, \boldsymbol{\varphi}) = (\boldsymbol{g}, \boldsymbol{\varphi})$ for all $\boldsymbol{\varphi} \in \boldsymbol{W}^{1,p'}_{\sigma,\boldsymbol{\tau}}(\Omega)$. It then follows by Amrouche et al. [6, Theorem 5.11] that $(\boldsymbol{u}, \pi) \in \boldsymbol{W}^{2,p}(\Omega) \times W^{1,p}(\Omega)$ with π defined by (1.3.4) for $\boldsymbol{\psi} = \Delta \boldsymbol{u}$ and $\boldsymbol{u}$ satisfies the boundary condition. In particular $\boldsymbol{u} \in \mathbf{D}(A_{p,\alpha})$. Then, using Remark 1.3.2, $A_p \boldsymbol{u} = A^* \boldsymbol{u}$. ■

In the following lemma, we obtain an estimate of the pressure in L^p-Space.

Lemma 1.3.5. Let $2 \leq p \leq 6$, Ω be of class $\mathcal{C}^{2,1}$, α satisfying (1.3.5), $\boldsymbol{f} \in \boldsymbol{L}^p(\Omega)$ and $(\boldsymbol{u}, \pi) \in D(\mathbf{A}_p), W^{1,p}(\Omega)$ the solution to the problem (1.3.8)–(1.3.11) given by Theorem 1.3.1 or Theorem 1.3.3. Then there exists a positive constant $C = C(\Omega, \alpha, p)$ independent of λ, such that the pressure satisfies the estimate:

$$\|\nabla \pi\|_{L^p(\Omega)} \leq C \, \|\boldsymbol{f}\|_{\boldsymbol{L}^p(\Omega)}. \tag{1.3.18}$$

Remark 1.3.6. In the cases where it is possible to apply [6, Proposition 4.21] to the solutions of (1.3.8)–(1.3.11), an explicit dependence of the constant C in (1.3.18) with respect to α in terms $\|\alpha\|_{W^{1-\frac{1}{p},p}(\Gamma)}$ can be obtained.

Proof. From the resolvent problem (1.3.8)–(1.3.11), it follows that the pressure term satisfies the following problem:

$$\begin{cases} \operatorname{div}(\nabla\pi - \boldsymbol{f}) = 0 & \text{in } \Omega \qquad (1.3.19)\\ (\nabla\pi - \boldsymbol{f})\cdot\boldsymbol{n} = \Delta\boldsymbol{u}\cdot\boldsymbol{n} & \text{on } \Gamma. \qquad (1.3.20)\end{cases}$$

We re-write the term in right hand side of (1.3.20) using the boundary condition (1.9.4). To this end, we notice that $-\Delta\boldsymbol{u}\cdot\boldsymbol{n} = \mathbf{curl}\,\mathbf{curl}\,\boldsymbol{u}\cdot\boldsymbol{n}$ in $W^{-\frac{1}{p},p}(\Gamma)$. Then, for all $\chi \in W^{2,p'}(\Omega)$,

$$\begin{aligned}\langle \mathbf{curl}\,\mathbf{curl}\,\boldsymbol{u}\cdot\boldsymbol{n}, \chi\rangle_\Gamma = \langle \mathbf{curl}\,\mathbf{curl}\,\boldsymbol{u}, \nabla\chi\rangle_\Omega &= \langle \mathbf{curl}\,\boldsymbol{u}\times\boldsymbol{n}, \nabla\chi\rangle_\Gamma\\ &= -\langle \operatorname{div}_\Gamma(\mathbf{curl}\,\boldsymbol{u}\times\boldsymbol{n}), \chi\rangle_\Gamma .\end{aligned}$$

Hence, from the relation (1.9.4), we get $-\Delta\boldsymbol{u}\cdot\boldsymbol{n} = \operatorname{div}_\Gamma(\mathbf{curl}\,\boldsymbol{u}\times\boldsymbol{n}) = -\operatorname{div}_\Gamma(\alpha\boldsymbol{u}_{\boldsymbol{\tau}} - 2\Lambda\boldsymbol{u})$ on Γ and π satisfy the system,

$$\begin{cases} \operatorname{div}(\nabla\pi - \boldsymbol{f}) = 0 & \text{in } \Omega\\ (\nabla\pi - \boldsymbol{f})\cdot\boldsymbol{n} = \operatorname{div}_\Gamma(\alpha\boldsymbol{u}_{\boldsymbol{\tau}} - 2\Lambda\boldsymbol{u}) & \text{on } \Gamma.\end{cases} \qquad (1.3.21)$$

But, since $\boldsymbol{u} \in \boldsymbol{W}^{2,p}(\Omega)$, we have $\alpha\boldsymbol{u}_{\boldsymbol{\tau}} - 2\Lambda\boldsymbol{u} \in \boldsymbol{W}^{1-\frac{1}{p},p}(\Gamma)$. Hence (cf. [37]),

$$\begin{aligned}\|\nabla\pi\|_{\boldsymbol{L}^p(\Omega)} &\le C\left(\|\boldsymbol{f}\|_{\boldsymbol{L}^p(\Omega)} + \|\operatorname{div}_\Gamma(\alpha\boldsymbol{u}_{\boldsymbol{\tau}} - 2\Lambda\boldsymbol{u})\|_{\boldsymbol{W}^{-\frac{1}{p},p}(\Gamma)}\right)\\ &\le C\left(\|\boldsymbol{f}\|_{\boldsymbol{L}^p(\Omega)} + \|\alpha\boldsymbol{u}_{\boldsymbol{\tau}} - 2\Lambda\boldsymbol{u}\|_{\boldsymbol{W}^{1-\frac{1}{p},p}(\Gamma)}\right)\\ &\le C\left(\|\boldsymbol{f}\|_{\boldsymbol{L}^p(\Omega)} + \|\alpha\boldsymbol{u}_{\boldsymbol{\tau}}\|_{\boldsymbol{W}^{1-\frac{1}{p},p}(\Gamma)} + 2\|\boldsymbol{u}\|_{\boldsymbol{W}^{1,p}(\Omega)}\right). \qquad (1.3.22)\end{aligned}$$

First we estimate the term $\|\alpha\boldsymbol{u}_{\boldsymbol{\tau}}\|_{\boldsymbol{W}^{1-\frac{1}{p},p}(\Gamma)}$. Since using the lift operator, we can always assume $\alpha \in W^{1,p}(\Omega)$, we can write,

$$\begin{aligned}\|\alpha\boldsymbol{u}_{\boldsymbol{\tau}}\|_{\boldsymbol{W}^{1-\frac{1}{p},p}(\Gamma)} &\le C\|\alpha\boldsymbol{u}\|_{\boldsymbol{W}^{1,p}(\Omega)}\\ &\le C\left(\|\alpha\boldsymbol{u}\|_{\boldsymbol{L}^p(\Omega)} + \|\sum_{i=1}^{3}\frac{\partial\alpha}{\partial x_i}\boldsymbol{u}\|_{\boldsymbol{L}^p(\Omega)} + \|\sum_{i=1}^{3}\alpha\frac{\partial\boldsymbol{u}}{\partial x_i}\|_{\boldsymbol{L}^p(\Omega)}\right).\end{aligned} \qquad (1.3.23)$$

Consider, $2 \le p \le 3$. Since $\boldsymbol{u} \in \boldsymbol{W}^{2,p}(\Omega)$, it follows that $\boldsymbol{u} \in \boldsymbol{L}^\infty(\Omega)$ and $\nabla\boldsymbol{u} \in \boldsymbol{L}^p(\Omega)\cap\boldsymbol{L}^{p^*}(\Omega)$. Then, $\nabla\boldsymbol{u} \in \boldsymbol{L}^q(\Omega)$ for some $q \in (p, 6)$. Therefore, for all $1 \le i \le 3$,

$$\alpha\boldsymbol{u},\ \frac{\partial\alpha}{\partial x_i}\boldsymbol{u},\ \alpha\frac{\partial\boldsymbol{u}}{\partial x_i} \in \boldsymbol{L}^p(\Omega)$$

and using the result in [6, Corollary 5.8], we have

$$\begin{aligned}\|\alpha \boldsymbol{u}\|_{\boldsymbol{L}^p(\Omega)} + \|\sum_{i=1}^{3} \frac{\partial \alpha}{\partial x_i}\boldsymbol{u}\|_{\boldsymbol{L}^p(\Omega)} + \|\sum_{i=1}^{3} \alpha \frac{\partial \boldsymbol{u}}{\partial x_i}\|_{\boldsymbol{L}^p(\Omega)} &\leq 3C\|\alpha\|_{W^{1,p}(\Omega)}\|\boldsymbol{u}\|_{\boldsymbol{W}^{1,q}(\Omega)} \\ &\leq C(\alpha)\|\alpha\|_{W^{1-\frac{1}{p},p}(\Gamma)}\|\boldsymbol{f}-\lambda\boldsymbol{u}\|_{\boldsymbol{L}^{r(q)}(\Omega)}.\end{aligned} \tag{1.3.24}$$

If we plug the estimate (1.3.24) in (1.3.23) and use the embedding $L^2(\Omega) \hookrightarrow L^{r(q)}(\Omega)$ and $W^{1-\frac{1}{p},p}(\Gamma) \hookrightarrow L^{t(q)}(\Gamma)$, we get

$$\|\alpha \boldsymbol{u}_{\boldsymbol{\tau}}\|_{\boldsymbol{W}^{1-\frac{1}{p},p}(\Gamma)} \leq C(\alpha)\|\boldsymbol{f} - \lambda\boldsymbol{u}\|_{\boldsymbol{L}^2(\Omega)}.$$

Now using (1.3.13) and $p > 2$, we obtain

$$\|\alpha \boldsymbol{u}_{\boldsymbol{\tau}}\|_{\boldsymbol{W}^{1-\frac{1}{p},p}(\Gamma)} \leq C(\alpha)\left(1 + \frac{1}{C_\varepsilon}\right)\|\boldsymbol{f}\|_{\boldsymbol{L}^p(\Omega)}. \tag{1.3.25}$$

We consider now $p > 3$. Then, by the lift operator $\alpha \in W^{1,p}(\Omega) \hookrightarrow L^\infty(\Omega)$ and $\boldsymbol{u} \in \boldsymbol{W}^{2,p}(\Omega) \hookrightarrow \boldsymbol{W}^{1,p}(\Omega) \hookrightarrow \boldsymbol{L}^\infty(\Omega)$ and we get easily

$$\alpha\boldsymbol{u},\ \frac{\partial \alpha}{\partial x_i}\boldsymbol{u},\ \alpha\frac{\partial \boldsymbol{u}}{\partial x_i} \in \boldsymbol{L}^p(\Omega) \quad \forall\, 1 \leq i \leq 3$$

and by Amrouche et al. [6, Corollary 5.8] again

$$\begin{aligned}\|\alpha \boldsymbol{u}\|_{\boldsymbol{L}^p(\Omega)} + \|\sum_{i=1}^{3} \frac{\partial \alpha}{\partial x_i}\boldsymbol{u}\|_{\boldsymbol{L}^p(\Omega)} + \|\sum_{i=1}^{3} \alpha \frac{\partial \boldsymbol{u}}{\partial x_i}\|_{\boldsymbol{L}^p(\Omega)} &\leq 3C\|\alpha\|_{W^{1,p}(\Omega)}\|\boldsymbol{u}\|_{\boldsymbol{W}^{1,p}(\Omega)} \\ &\leq C(\alpha)\|\alpha\|_{W^{1-\frac{1}{p},p}(\Gamma)}\|\boldsymbol{f}-\lambda\boldsymbol{u}\|_{\boldsymbol{L}^{r(p)}(\Omega)}.\end{aligned} \tag{1.3.26}$$

Since $p \leq 6$, it follows that $\boldsymbol{L}^2(\Omega) \hookrightarrow \boldsymbol{L}^{r(p)}(\Omega)$. We may then estimate the term $\|\lambda\boldsymbol{u}\|_{\boldsymbol{L}^{r(p)}(\Omega)}$ in (1.3.26) using (1.3.13). Plugging this in (1.3.23) and using $W^{1-\frac{1}{p},p}(\Gamma) \hookrightarrow L^{t(p)}(\Gamma)$ we obtain,

$$\|\alpha \boldsymbol{u}_{\boldsymbol{\tau}}\|_{\boldsymbol{W}^{1-\frac{1}{p},p}(\Gamma)} \leq C(\alpha)\|\boldsymbol{f} - \lambda\boldsymbol{u}\|_{\boldsymbol{L}^2(\Omega)} \leq C(\alpha)\left(1 + \frac{1}{C_\varepsilon}\right)\|\boldsymbol{f}\|_{\boldsymbol{L}^p(\Omega)}. \tag{1.3.27}$$

Therefore, by (1.3.25) and (1.3.27), we get: for all $2 \leq p \leq 6$,

$$\|\alpha \boldsymbol{u}_{\boldsymbol{\tau}}\|_{\boldsymbol{W}^{1-\frac{1}{p},p}(\Gamma)} \leq\ C(\alpha, p, \Omega)\ \|\boldsymbol{f}\|_{\boldsymbol{L}^p(\Omega)}. \tag{1.3.28}$$

Now to estimate the term $\|\boldsymbol{u}\|_{\boldsymbol{W}^{1,p}(\Omega)}$, writing (1.3.8) as $-\Delta\boldsymbol{u} + \nabla\pi = \boldsymbol{f} - \lambda\boldsymbol{u}$ and using [6, Corollary 5.8], we get for $2 \leq p \leq 6$,

$$\|\boldsymbol{u}\|_{\boldsymbol{W}^{1,p}(\Omega)} \leq C\|\boldsymbol{f} - \lambda\boldsymbol{u}\|_{\boldsymbol{L}^{r(p)}(\Omega)} \leq C\|\boldsymbol{f}\|_{\boldsymbol{L}^p(\Omega)}. \tag{1.3.29}$$

We deduce (1.3.18) from (1.3.22), (1.3.28), and (1.3.29). ■

Remark 1.3.7. Notice that though the two boundary conditions

$$\mathbf{curl}\,\boldsymbol{u}\times\boldsymbol{n}=0 \quad \text{on } \Gamma \tag{1.3.30}$$

and

$$2[\mathbb{D}(\boldsymbol{u})\boldsymbol{n}]_{\boldsymbol{\tau}}+\alpha\boldsymbol{u}_{\boldsymbol{\tau}}=\mathbf{0} \quad \text{on } \Gamma \tag{1.3.31}$$

are very much similar, as described in (1.2.4), but in the case of the Stokes problem with Navier-type boundary condition (1.3.30) on $\boldsymbol{L}^p_{\sigma,\tau}(\Omega)$ the pressure is constant and hence does not appear in the operator (see [2, Proposition 3.1]). On the contrary, the pressure term does appear in the Stokes operator with Navier boundary condition (1.3.31).

Remark 1.3.8. The restriction $2\leq p\leq 6$ comes from the fact that we do not know yet the estimate (1.3.13) for $p\neq 2$.

Next we deduce resolvent estimate for all $1<p<\infty$, similar to (1.3.13) which has been done in several steps in the following Propositions.

Proposition 1.3.9. Suppose $2\leq p\leq 6$, Ω is of class $\mathcal{C}^{2,1}$ and α satisfies (1.3.5). Then, there exists $\lambda_0\in\mathbb{R}$ satisfying the following: for all $\lambda\in\mathbb{C}^*$ with $\operatorname{Re}\lambda\geq 0$ and $|\lambda|\geq\lambda_0$ and all $\boldsymbol{f}\in\boldsymbol{L}^p(\Omega)$, if $(\boldsymbol{u},\pi)$ is the unique solution of (1.3.8)–(1.3.11) in $\mathbf{D}(A_{p,\alpha})\times W^{1,p}(\Omega)$, then $\boldsymbol{u}$ satisfies the estimate:

$$\|\boldsymbol{u}\|_{\boldsymbol{L}^p(\Omega)}\leq\frac{C}{|\lambda|}\|\boldsymbol{f}\|_{\boldsymbol{L}^p(\Omega)},$$

where $C=C(\Omega,\alpha,p)$ is independent of λ.

Remark 1.3.10. In the proof of Proposition 1.3.9, the hypothesis $p\leq 6$ is needed only in order to apply the estimate (1.3.18), where that condition appears. As soon as the estimate (1.3.18) is extended for all $p\in(6,\infty)$ (done in Theorem 1.3.4 below), the above proof of Proposition 1.3.9 for $3<p\leq 6$ will be also valid for all $p>6$.

Proof. Multiplying (1.3.8) by $|\boldsymbol{u}|^{p-2}\bar{\boldsymbol{u}}$ and using Lemma 1.2.4, we get

$$\begin{aligned}\lambda\int_\Omega|\boldsymbol{u}|^p&+\int_\Omega|\boldsymbol{u}|^{p-2}|\nabla\boldsymbol{u}|^2+\frac{4(p-2)}{p^2}\int_\Omega|\nabla|\boldsymbol{u}|^{p/2}|^2\\&+(p-2)i\sum_{k=1}^{3}\int_\Omega|\boldsymbol{u}|^{p-4}\operatorname{Re}\left(\frac{\partial\boldsymbol{u}}{\partial x_k}\cdot\bar{\boldsymbol{u}}\right)\operatorname{Im}\left(\frac{\partial\boldsymbol{u}}{\partial x_k}\cdot\bar{\boldsymbol{u}}\right)\\&=\int_\Gamma\frac{\partial\boldsymbol{u}}{\partial\boldsymbol{n}}\cdot|\boldsymbol{u}|^{p-2}\bar{\boldsymbol{u}}-\int_\Omega|\boldsymbol{u}|^{p-2}\nabla\pi\cdot\bar{\boldsymbol{u}}+\int_\Omega|\boldsymbol{u}|^{p-2}\boldsymbol{f}\cdot\bar{\boldsymbol{u}}.\end{aligned}\tag{1.3.32}$$

Since it follows from (1.2.3) and the boundary condition (1.3.11) that

$$\left(\frac{\partial\boldsymbol{u}}{\partial\boldsymbol{n}}\right)_{\boldsymbol{\tau}}\cdot\bar{\boldsymbol{u}}_{\boldsymbol{\tau}}=(2[\mathbb{D}(\boldsymbol{u})\boldsymbol{n}]_{\boldsymbol{\tau}}+\Lambda\boldsymbol{u})\cdot\bar{\boldsymbol{u}}_{\boldsymbol{\tau}}=(-\alpha\boldsymbol{u}_{\boldsymbol{\tau}}+\Lambda\boldsymbol{u})\cdot\bar{\boldsymbol{u}}_{\boldsymbol{\tau}}=-\alpha|\boldsymbol{u}_{\boldsymbol{\tau}}|^2+\Lambda\boldsymbol{u}\cdot\bar{\boldsymbol{u}}_{\boldsymbol{\tau}},$$

and

$$\Lambda \boldsymbol{u} \cdot \bar{\boldsymbol{u}}_{\boldsymbol{\tau}} = \sum_{k=1}^{2} \left(\boldsymbol{u}_{\boldsymbol{\tau}} \cdot \frac{\partial \boldsymbol{n}}{\partial s_k} \right) \boldsymbol{\tau}_k \cdot \bar{\boldsymbol{u}}_{\boldsymbol{\tau}} = - \sum_{k=1}^{2} \left(\frac{\partial \boldsymbol{u}_{\boldsymbol{\tau}}}{\partial s_k} \cdot \boldsymbol{n} \right) \bar{\boldsymbol{u}}_k = - \sum_{j,k=1}^{2} \boldsymbol{u}_j \bar{\boldsymbol{u}}_k \frac{\partial \boldsymbol{\tau}_j}{\partial s_k} \cdot \boldsymbol{n},$$

we have,

$$\int_\Omega |\boldsymbol{u}|^{p-2} \left(\frac{\partial \boldsymbol{u}}{\partial \boldsymbol{n}} \right)_{\boldsymbol{\tau}} \cdot \bar{\boldsymbol{u}}_{\boldsymbol{\tau}} = - \int_\Gamma \alpha |\boldsymbol{u}|^p - \sum_{j,k=1}^{2} \int_\Gamma |\boldsymbol{u}|^{p-2} \left(\boldsymbol{u}_j \bar{\boldsymbol{u}}_k \frac{\partial \boldsymbol{\tau}_j}{\partial s_k} \cdot \boldsymbol{n} \right). \tag{1.3.33}$$

Now, plugging (1.3.33) in (1.3.32) and taking real and imaginary parts, we get,

$$\begin{aligned} &\operatorname{Re} \lambda \int_\Omega |\boldsymbol{u}|^p + \int_\Omega |\boldsymbol{u}|^{p-2} |\nabla \boldsymbol{u}|^2 + \frac{4(p-2)}{p^2} \int_\Omega |\nabla |\boldsymbol{u}|^{p/2}|^2 + \int_\Gamma \alpha |\boldsymbol{u}|^p = \\ &- \operatorname{Re} \sum_{j,k=1}^{2} \int_\Gamma |\boldsymbol{u}|^{p-2} \, \boldsymbol{u}_j \bar{\boldsymbol{u}}_k \frac{\partial \boldsymbol{\tau}_j}{\partial s_k} \cdot \boldsymbol{n} - \operatorname{Re} \int_\Omega |\boldsymbol{u}|^{p-2} \nabla \pi \cdot \bar{\boldsymbol{u}} + \operatorname{Re} \int_\Omega |\boldsymbol{u}|^{p-2} \boldsymbol{f} \cdot \bar{\boldsymbol{u}} \end{aligned} \tag{1.3.34}$$

and

$$\begin{aligned} &\operatorname{Im} \lambda \int_\Omega |\boldsymbol{u}|^p \\ &= -(p-2) \sum_{k=1}^{3} \int_\Omega |\boldsymbol{u}|^{p-4} \operatorname{Re} \left(\frac{\partial \boldsymbol{u}}{\partial x_k} \cdot \bar{\boldsymbol{u}} \right) \operatorname{Im} \left(\frac{\partial \boldsymbol{u}}{\partial x_k} \cdot \bar{\boldsymbol{u}} \right) - \operatorname{Im} \sum_{j,k=1}^{2} \int_\Gamma |\boldsymbol{u}|^{p-2} \, \boldsymbol{u}_j \bar{\boldsymbol{u}}_k \frac{\partial \boldsymbol{\tau}_j}{\partial s_k} \cdot \boldsymbol{n} \\ &\quad - \operatorname{Im} \int_\Omega |\boldsymbol{u}|^{p-2} \nabla \pi \cdot \bar{\boldsymbol{u}} + \operatorname{Im} \int_\Omega |\boldsymbol{u}|^{p-2} \boldsymbol{f} \cdot \bar{\boldsymbol{u}}. \end{aligned} \tag{1.3.35}$$

Since $\alpha \geq 0$, we deduce from (1.3.34),

$$\begin{aligned} &\operatorname{Re} \lambda \, \|\boldsymbol{u}\|^p_{\boldsymbol{L}^p(\Omega)} + \int_\Omega |\boldsymbol{u}|^{p-2} |\nabla \boldsymbol{u}|^2 + \frac{4(p-2)}{p^2} \int_\Omega |\nabla |\boldsymbol{u}|^{p/2}|^2 \\ &\leq \left(\|\boldsymbol{f}\|_{\boldsymbol{L}^p(\Omega)} + \|\nabla \pi\|_{\boldsymbol{L}^p(\Omega)} \right) \|\boldsymbol{u}\|^{p-1}_{\boldsymbol{L}^p(\Omega)} + C(\Omega) \int_\Gamma |\boldsymbol{u}|^p \end{aligned} \tag{1.3.36}$$

for some constant $C(\Omega) > 0$. Also (1.3.35) yields,

$$\begin{aligned}
&|\operatorname{Im}\lambda|\,\|\boldsymbol{u}\|^p_{\boldsymbol{L}^p(\Omega)}\\
&\le \sum_{j,k=1}^{2}\int_\Gamma |\boldsymbol{u}|^{p-2}\,|\boldsymbol{u}_j\bar{\boldsymbol{u}}_k|\left|\frac{\partial \boldsymbol{\tau}_j}{\partial s_k}\cdot\boldsymbol{n}\right| ds + \int_\Omega |\boldsymbol{u}|^{p-2}\,|\nabla\pi\cdot\bar{\boldsymbol{u}}|\,dx\\
&\qquad + \int_\Omega |\boldsymbol{u}|^{p-2}|\boldsymbol{f}\cdot\bar{\boldsymbol{u}}|\,dx + (p-2)\sum_{k=1}^{3}\int_\Omega |\boldsymbol{u}|^{p-2}\left|\frac{\partial\boldsymbol{u}}{\partial x_k}\right|^2\\
&\le\; C(\Omega)\int_\Gamma |\boldsymbol{u}|^p + \left(\|\boldsymbol{f}\|_{\boldsymbol{L}^p(\Omega)} + \|\nabla\pi\|_{\boldsymbol{L}^p(\Omega)}\right)\|\boldsymbol{u}\|^{p-1}_{\boldsymbol{L}^p(\Omega)} + (p-2)\int_\Omega |\boldsymbol{u}|^{p-2}|\nabla\boldsymbol{u}|^2.
\end{aligned} \tag{1.3.37}$$

Now, adding (1.3.36) and (1.3.37) we get,

$$\begin{aligned}
&|\lambda|\,\|\boldsymbol{u}\|^p_{\boldsymbol{L}^p(\Omega)} + \int_\Omega |\boldsymbol{u}|^{p-2}|\nabla\boldsymbol{u}|^2 + \frac{4(p-2)}{p^2}\int_\Omega |\nabla|\boldsymbol{u}|^{p/2}|^2\\
&\le 2\left(\|\boldsymbol{f}\|_{\boldsymbol{L}^p(\Omega)} + \|\nabla\pi\|_{\boldsymbol{L}^p(\Omega)}\right)\|\boldsymbol{u}\|^{p-1}_{\boldsymbol{L}^p(\Omega)} + C(\Omega)\int_\Gamma |\boldsymbol{u}|^p + (p-2)\int_\Omega |\boldsymbol{u}|^{p-2}|\nabla\boldsymbol{u}|^2.
\end{aligned} \tag{1.3.38}$$

Moreover, thanks to [21, Chapter 1, Theorem 1.5.1.10], we know that for all $\varepsilon > 0$ there exists $C_\varepsilon > 0$ such that

$$\int_\Gamma |w|^2 \le \varepsilon\int_\Omega |\nabla w|^2 + C_\varepsilon\int_\Omega |w|^2 \tag{1.3.39}$$

for all $w \in H^1(\Omega)$. Applying formula (1.3.39) with $w = |\boldsymbol{u}|^{p/2}$ and substituting in (1.3.38), we get,

$$\begin{aligned}
&|\lambda|\,\|\boldsymbol{u}\|^p_{\boldsymbol{L}^p(\Omega)} + \int_\Omega |\boldsymbol{u}|^{p-2}|\nabla\boldsymbol{u}|^2 + \frac{4(p-2)}{p^2}\int_\Omega |\nabla|\boldsymbol{u}|^{p/2}|^2\\
&\le 2\left(\|\boldsymbol{f}\|_{\boldsymbol{L}^p(\Omega)}+\|\nabla\pi\|_{\boldsymbol{L}^p(\Omega)}\right)\|\boldsymbol{u}\|^{p-1}_{\boldsymbol{L}^p(\Omega)}+(p-2)\int_\Omega |\boldsymbol{u}|^{p-2}|\nabla\boldsymbol{u}|^2+\varepsilon\int_\Omega |\nabla|\boldsymbol{u}|^{p/2}|^2+C_\varepsilon\int_\Omega |\boldsymbol{u}|^p.
\end{aligned}$$

Choosing $\varepsilon = \frac{2(p-2)}{p^2}$, we get,

$$\begin{aligned}
&|\lambda|\,\|\boldsymbol{u}\|^p_{\boldsymbol{L}^p(\Omega)} + \int_\Omega |\boldsymbol{u}|^{p-2}|\nabla\boldsymbol{u}|^2 + \frac{2(p-2)}{p^2}\int_\Omega |\nabla|\boldsymbol{u}|^{p/2}|^2\\
&\le 2\left(\|\boldsymbol{f}\|_{\boldsymbol{L}^p(\Omega)}+\|\nabla\pi\|_{\boldsymbol{L}^p(\Omega)}\right)\|\boldsymbol{u}\|^{p-1}_{\boldsymbol{L}^p(\Omega)}+(p-2)\int_\Omega |\boldsymbol{u}|^{p-2}|\nabla\boldsymbol{u}|^2+C(\Omega,p)\|\boldsymbol{u}\|^p_{\boldsymbol{L}^p(\Omega)}.
\end{aligned}$$

This implies, for some constant $C_* > 0$,

$$(|\lambda| - C_*)\ \|\boldsymbol{u}\|^p_{\boldsymbol{L}^p(\Omega)} + \int_\Omega |\boldsymbol{u}|^{p-2}|\nabla \boldsymbol{u}|^2 + \frac{2(p-2)}{p^2}\int_\Omega |\nabla |\boldsymbol{u}|^{p/2}|^2$$

$$\leq 2\left(\|\boldsymbol{f}\|_{\boldsymbol{L}^p(\Omega)} + \|\nabla\pi\|_{\boldsymbol{L}^p(\Omega)}\right)\|\boldsymbol{u}\|^{p-1}_{\boldsymbol{L}^p(\Omega)} + (p-2)\int_\Omega |\boldsymbol{u}|^{p-2}|\nabla \boldsymbol{u}|^2.$$

Let us define

$$\lambda_0 = 2C_*.$$

Then, for $|\lambda| > \lambda_0$, we have,

$$\frac{|\lambda|}{2}\ \|\boldsymbol{u}\|^p_{\boldsymbol{L}^p(\Omega)} + \int_\Omega |\boldsymbol{u}|^{p-2}|\nabla \boldsymbol{u}|^2 + \frac{2(p-2)}{p^2}\int_\Omega |\nabla |\boldsymbol{u}|^{p/2}|^2$$

$$\leq 2\left(\|\boldsymbol{f}\|_{\boldsymbol{L}^p(\Omega)} + \|\nabla\pi\|_{\boldsymbol{L}^p(\Omega)}\right)\|\boldsymbol{u}\|^{p-1}_{\boldsymbol{L}^p(\Omega)} + (p-2)\int_\Omega |\boldsymbol{u}|^{p-2}|\nabla \boldsymbol{u}|^2$$

which in turn gives,

$$\frac{|\lambda|}{2}\ \|\boldsymbol{u}\|^p_{\boldsymbol{L}^p(\Omega)} + (3-p)\int_\Omega |\boldsymbol{u}|^{p-2}|\nabla \boldsymbol{u}|^2 + \frac{2(p-2)}{p^2}\int_\Omega |\nabla |\boldsymbol{u}|^{p/2}|^2$$

$$\leq 2\left(\|\boldsymbol{f}\|_{\boldsymbol{L}^p(\Omega)} + \|\nabla\pi\|_{\boldsymbol{L}^p(\Omega)}\right)\|\boldsymbol{u}\|^{p-1}_{\boldsymbol{L}^p(\Omega)}\ .$$

We now consider two cases $2 \leq p \leq 3$ and $3 < p \leq 6$.

Case (i): $2 \leq p \leq 3$. We have,

$$\frac{|\lambda|}{2}\ \|\boldsymbol{u}\|^p_{\boldsymbol{L}^p(\Omega)} \leq 2\left(\|\boldsymbol{f}\|_{\boldsymbol{L}^p(\Omega)} + \|\nabla\pi\|_{\boldsymbol{L}^p(\Omega)}\right)\|\boldsymbol{u}\|^{p-1}_{\boldsymbol{L}^p(\Omega)}\ .$$

and using the estimate (1.3.18), we obtain,

$$\|\boldsymbol{u}\|_{\boldsymbol{L}^p(\Omega)} \leq \frac{C(\Omega,\alpha,p)}{|\lambda|}\|\boldsymbol{f}\|_{\boldsymbol{L}^p(\Omega)}\ . \tag{1.3.40}$$

Case (ii): $3 < p \leq 6$. We deduce from (1.3.36), using again (1.3.39)

$$\operatorname{Re}\lambda\ \|\boldsymbol{u}\|^p_{\boldsymbol{L}^p(\Omega)} + \int_\Omega |\boldsymbol{u}|^{p-2}|\nabla \boldsymbol{u}|^2 + \frac{4(p-2)}{p^2}\int_\Omega |\nabla |\boldsymbol{u}|^{p/2}|^2$$

$$\leq \left(\|\boldsymbol{f}\|_{\boldsymbol{L}^p(\Omega)} + \|\nabla\pi\|_{\boldsymbol{L}^p(\Omega)}\right)\|\boldsymbol{u}\|^{p-1}_{\boldsymbol{L}^p(\Omega)} + C(\Omega)\int_\Gamma |\boldsymbol{u}|^p$$

$$\leq \left(\|\boldsymbol{f}\|_{\boldsymbol{L}^p(\Omega)} + \|\nabla\pi\|_{\boldsymbol{L}^p(\Omega)}\right)\|\boldsymbol{u}\|^{p-1}_{\boldsymbol{L}^p(\Omega)} + \varepsilon\int_\Omega |\nabla |\boldsymbol{u}|^{p/2}|^2 + C_\varepsilon\int_\Omega |\boldsymbol{u}|^p$$

$$\leq \left(\|\boldsymbol{f}\|_{\boldsymbol{L}^p(\Omega)} + \|\nabla\pi\|_{\boldsymbol{L}^p(\Omega)}\right)\|\boldsymbol{u}\|^{p-1}_{\boldsymbol{L}^p(\Omega)} + \frac{p-2}{p^2}\int_\Omega |\nabla |\boldsymbol{u}|^{p/2}|^2 + C\int_\Omega |\boldsymbol{u}|^p$$

which gives

$$\begin{aligned}&\operatorname{Re}\lambda\ \|\boldsymbol{u}\|^p_{\boldsymbol{L}^p(\Omega)}+\int\limits_{\Omega}|\boldsymbol{u}|^{p-2}|\nabla\boldsymbol{u}|^2+\frac{3(p-2)}{p^2}\int\limits_{\Omega}|\nabla|\boldsymbol{u}|^{p/2}|^2\\&\leq\left(\|\boldsymbol{f}\|_{\boldsymbol{L}^p(\Omega)}+\|\nabla\pi\|_{\boldsymbol{L}^p(\Omega)}\right)\|\boldsymbol{u}\|^{p-1}_{\boldsymbol{L}^p(\Omega)}+C\|\boldsymbol{u}\|^p_{\boldsymbol{L}^p(\Omega)}.\end{aligned}\tag{1.3.41}$$

Since $(\boldsymbol{u},\pi)$ satisfies (1.3.8), it follows that

$$-\Delta\boldsymbol{u}+\nabla\pi=\boldsymbol{f}-\lambda\boldsymbol{u}$$

and by Amrouche at al. [6, Theorem 5.11], $\|\boldsymbol{u}\|_{\boldsymbol{W}^{2,3}(\Omega)}\leq C\|\boldsymbol{f}-\lambda\boldsymbol{u}\|_{\boldsymbol{L}^3(\Omega)}$. We use now for $3\leq p\leq 6$, $\boldsymbol{W}^{2,p}(\Omega)\hookrightarrow\boldsymbol{W}^{2,3}(\Omega)\hookrightarrow\boldsymbol{L}^p(\Omega)$ as well as the estimate (1.3.40) for $p=3$ to obtain,

$$\|\boldsymbol{u}\|_{\boldsymbol{L}^p(\Omega)}\leq C\|\boldsymbol{u}\|_{\boldsymbol{W}^{2,3}(\Omega)}\leq C\|\boldsymbol{f}-\lambda\boldsymbol{u}\|_{\boldsymbol{L}^3(\Omega)}\leq C\|\boldsymbol{f}\|_{\boldsymbol{L}^p(\Omega)}.$$

Therefore,

$$\|\boldsymbol{u}\|^p_{\boldsymbol{L}^p(\Omega)}=\|\boldsymbol{u}\|_{\boldsymbol{L}^p(\Omega)}\|\boldsymbol{u}\|^{p-1}_{\boldsymbol{L}^p(\Omega)}\leq C\|\boldsymbol{f}\|_{\boldsymbol{L}^p(\Omega)}\|\boldsymbol{u}\|^{p-1}_{\boldsymbol{L}^p(\Omega)}.$$

Hence, from (1.3.41) and using the pressure estimate (1.3.18), we deduce,

$$\begin{aligned}&\operatorname{Re}\lambda\ \|\boldsymbol{u}\|^p_{\boldsymbol{L}^p(\Omega)}+\int\limits_{\Omega}|\boldsymbol{u}|^{p-2}|\nabla\boldsymbol{u}|^2+\frac{3(p-2)}{p^2}\int\limits_{\Omega}|\nabla|\boldsymbol{u}|^{p/2}|^2\\&\leq\left(\|\boldsymbol{f}\|_{\boldsymbol{L}^p(\Omega)}+\|\nabla\pi\|_{\boldsymbol{L}^p(\Omega)}+C\|\boldsymbol{f}\|_{\boldsymbol{L}^p(\Omega)}\right)\|\boldsymbol{u}\|^{p-1}_{\boldsymbol{L}^p(\Omega)}\\&\leq C\|\boldsymbol{f}\|_{\boldsymbol{L}^p(\Omega)}\|\boldsymbol{u}\|^{p-1}_{\boldsymbol{L}^p(\Omega)}.\end{aligned}\tag{1.3.42}$$

Now from (1.3.37), using (1.3.39) with suitably chosen $\varepsilon>0$ and using (1.3.42), we deduce

$$\begin{aligned}&|\operatorname{Im}\lambda|\ \|\boldsymbol{u}\|^p_{\boldsymbol{L}^p(\Omega)}\\&\leq(1+C(\Omega,\alpha,p))\,\|\boldsymbol{f}\|_{\boldsymbol{L}^p(\Omega)}\|\boldsymbol{u}\|^{p-1}_{\boldsymbol{L}^p(\Omega)}+(p-2)\int\limits_{\Omega}|\boldsymbol{u}|^{p-2}|\nabla\boldsymbol{u}|^2+\varepsilon\int\limits_{\Omega}|\nabla|\boldsymbol{u}|^{p/2}|^2+C_\varepsilon\int\limits_{\Omega}|\boldsymbol{u}|^p\\&\leq C(\Omega,\alpha,p)\|\boldsymbol{f}\|_{\boldsymbol{L}^p(\Omega)}\|\boldsymbol{u}\|^{p-1}_{\boldsymbol{L}^p(\Omega)}+(p-2)\int\limits_{\Omega}|\boldsymbol{u}|^{p-2}|\nabla\boldsymbol{u}|^2+\frac{p-2}{p^2}\int\limits_{\Omega}|\nabla|\boldsymbol{u}|^{p/2}|^2+C\int\limits_{\Omega}|\boldsymbol{u}|^p\\&\leq C\|\boldsymbol{f}\|_{\boldsymbol{L}^p(\Omega)}\|\boldsymbol{u}\|^{p-1}_{\boldsymbol{L}^p(\Omega)}.\end{aligned}\tag{1.3.43}$$

Hence, combining (1.3.42) and (1.3.43), we obtain,

$$|\lambda|\|\boldsymbol{u}\|^p_{\boldsymbol{L}^p(\Omega)}\leq C\|\boldsymbol{f}\|_{\boldsymbol{L}^p(\Omega)}\|\boldsymbol{u}\|^{p-1}_{\boldsymbol{L}^p(\Omega)}$$

which is the required estimate. ∎

The next proposition gives the resolvent estimate for $|\lambda| < \lambda_0$ to complete the case $p \in [2, 6]$, together with the proposition above.

Proposition 1.3.11. Let $2 \leq p \leq 6$, Ω be of class $\mathcal{C}^{2,1}$ and α as in (1.3.5). Also, let $\lambda \in \mathbb{C}^*$ with $\operatorname{Re}\lambda \geq 0$ and $|\lambda| < \lambda_0$ where λ_0 is given in Proposition 1.3.9 and $\boldsymbol{f} \in \boldsymbol{L}^p(\Omega)$. If $(\boldsymbol{u}, \pi)$ is the unique solution of (1.3.8)–(1.3.11) in $\mathbf{D}(A_{p,\alpha}) \times W^{1,p}(\Omega)$, then $\boldsymbol{u}$ satisfies the estimate:

$$\|\boldsymbol{u}\|_{\boldsymbol{L}^p(\Omega)} \leq \frac{C}{|\lambda|}\|\boldsymbol{f}\|_{\boldsymbol{L}^p(\Omega)}$$

with $C = C(\Omega, p)$ independent of λ and α.

Proof. We know from Korn inequality and using Theorem 1.3.1,

$$\|\boldsymbol{u}\|^2_{\boldsymbol{H}^1(\Omega)} \leq C\left(\|\boldsymbol{u}\|^2_{\boldsymbol{L}^2(\Omega)} + \|\mathbb{D}(\boldsymbol{u})\|^2_{\boldsymbol{L}^2(\Omega)}\right) \leq C\left(\frac{1}{|\lambda|^2} + \frac{1}{2|\lambda|}\right)\|\boldsymbol{f}\|^2_{\boldsymbol{L}^2(\Omega)}.$$

Note that the above constant does not depend on α. But, as $|\lambda| < \lambda_0$, we obtain,

$$\|\boldsymbol{u}\|_{\boldsymbol{H}^1(\Omega)} \leq \frac{C(\lambda_0)}{|\lambda|}\|\boldsymbol{f}\|_{\boldsymbol{L}^2(\Omega)}.$$

Now $\boldsymbol{u} \in \boldsymbol{W}^{2,p}(\Omega) \hookrightarrow \boldsymbol{H}^1(\Omega) \hookrightarrow \boldsymbol{L}^p(\Omega)$ as $2 \leq p \leq 6$. Therefore

$$\|\boldsymbol{u}\|_{\boldsymbol{L}^p(\Omega)} \leq C\|\boldsymbol{u}\|_{\boldsymbol{H}^1(\Omega)} \leq \frac{C(\lambda_0)}{|\lambda|}\|\boldsymbol{f}\|_{\boldsymbol{L}^2(\Omega)} \leq \frac{C(\lambda_0)}{|\lambda|}\|\boldsymbol{f}\|_{\boldsymbol{L}^p(\Omega)}.$$

This completes the proof. ∎

The next theorem extends the result of Lemma 1.3.5 to all $p \in (6, \infty)$. The arguments in its proof are then largely borrowed from the proof of that lemma.

Theorem 1.3.4. Let $2 \leq p < \infty$, Ω be of class $\mathcal{C}^{2,1}$ and α satisfying (1.3.5). Then, for $\boldsymbol{f} \in \boldsymbol{L}^p(\Omega)$, the unique solution $(\boldsymbol{u}, \pi)$ of (1.3.8)–(1.3.11) in $\mathbf{D}(A_{p,\alpha}) \times W^{1,p}(\Omega)$ is such that the pressure $\pi \in W^{1,p}(\Omega)$ satisfies the following estimate:

$$\|\nabla\pi\|_{\boldsymbol{L}^p(\Omega)} \leq C\ \|\boldsymbol{f}\|_{\boldsymbol{L}^p(\Omega)} \tag{1.3.44}$$

for some constant $C = C(\Omega, \alpha, p)$ independent of λ.

Proof. The result for $2 \leq p \leq 6$ is proved in Lemma 1.3.5 and we may then assume $p > 6$. As in the case $p > 3$ in Lemma 1.3.5, we use $\alpha \in W^{1,p}(\Omega)$ with the help of the lift operator since by hypothesis $\alpha \in W^{1-\frac{1}{p},p}(\Gamma)$ and $\boldsymbol{u} \in \boldsymbol{W}^{2,p}(\Omega) \hookrightarrow \boldsymbol{L}^\infty(\Omega)$, to obtain

$$\alpha\boldsymbol{u}, \frac{\partial\alpha}{\partial x_i}\boldsymbol{u}, \alpha\frac{\partial\boldsymbol{u}}{\partial x_i} \in \boldsymbol{L}^p(\Omega) \quad \forall\, 1 \leq i \leq 3$$

and

$$\begin{aligned}\|\alpha \boldsymbol{u}_{\boldsymbol{\tau}}\|_{\boldsymbol{W}^{1-\frac{1}{p},p}(\Gamma)} &\leq C\left(\|\alpha \boldsymbol{u}\|_{\boldsymbol{L}^p(\Omega)} + \|\sum_{i=1}^{3}\frac{\partial \alpha}{\partial x_i}\boldsymbol{u}\|_{\boldsymbol{L}^p(\Omega)} + \|\sum_{i=1}^{3}\alpha\frac{\partial \boldsymbol{u}}{\partial x_i}\|_{\boldsymbol{L}^p(\Omega)}\right)\\ &\leq 3C\ \|\alpha\|_{W^{1,p}(\Omega)}\|\boldsymbol{u}\|_{\boldsymbol{W}^{1,p}(\Omega)} \leq C(\alpha)\|\boldsymbol{f}-\lambda\boldsymbol{u}\|_{\boldsymbol{L}^{r(p)}(\Omega)},\end{aligned}$$

where to get the last inequality, we use [6, Corollary 5.8]. Since $r(p) < 3$ for all $1 < p < \infty$, we deduce, using Propositions 1.3.9 and 1.3.11,

$$\|\alpha \boldsymbol{u}_{\boldsymbol{\tau}}\|_{\boldsymbol{W}^{1-\frac{1}{p},p}(\Gamma)} \leq C\|\alpha\|_{W^{1-\frac{1}{p},p}(\Gamma)}\left(1+\|\alpha\|^2_{\boldsymbol{W}^{1-\frac{1}{p},p}(\Omega)}\right)\|\boldsymbol{f}\|_{\boldsymbol{L}^p(\Omega)}.$$

The estimate (1.3.44) follows now from

$$\|\nabla\pi\|_{\boldsymbol{L}^p(\Omega)} \leq C\left(\|\boldsymbol{f}\|_{\boldsymbol{L}^p(\Omega)} + \|\alpha \boldsymbol{u}_{\boldsymbol{\tau}}\|_{\boldsymbol{W}^{1-\frac{1}{p},p}(\Gamma)} + 2\|\boldsymbol{u}\|_{\boldsymbol{W}^{1,p}(\Omega)}\right).$$

■

As a conclusion, we have the following theorem:

Theorem 1.3.5. Let $1 < p < \infty$, Ω be of class $\mathcal{C}^{2,1}$ and α as in (1.3.5). Then for all $\lambda \in \mathbb{C}^*$ with $\operatorname{Re}\lambda \geq 0$ and $\boldsymbol{f} \in \boldsymbol{L}^p(\Omega)$, the unique solution $(\boldsymbol{u},\pi) \in \mathbf{D}(A_{p,\alpha}) \times W^{1,p}(\Omega)$ of (1.3.8)–(1.3.11) satisfies

$$\|\boldsymbol{u}\|_{\boldsymbol{L}^p(\Omega)} \leq \frac{C}{|\lambda|}\|\boldsymbol{f}\|_{\boldsymbol{L}^p(\Omega)} \tag{1.3.45}$$

$$\|\mathbb{D}(\boldsymbol{u})\|_{\boldsymbol{L}^p(\Omega)} \leq \frac{C}{\sqrt{|\lambda|}}\|\boldsymbol{f}\|_{\boldsymbol{L}^p(\Omega)} \tag{1.3.46}$$

and

$$\|\boldsymbol{u}\|_{\boldsymbol{W}^{2,p}(\Omega)} \leq C\ \|\boldsymbol{f}\|_{\boldsymbol{L}^p(\Omega)}. \tag{1.3.47}$$

The above constants $C = C(\Omega,\alpha,p)$ do not depend on λ.

Proof. For $2 \leq p \leq 6$, estimate (1.3.45) follows from Propositions 1.3.9 and 1.3.11.

Let us consider then $p > 6$. For $|\lambda| > \lambda_0$, as indicated in Remark 1.3.10, the proof of (1.3.45) for $p \in (3,6]$ in Proposition 1.3.9 extends straightforwardly to all $p > 3$ using the pressure estimate (1.3.44). We suppose now that $|\lambda| < \lambda_0$. Then, we use [6, Corollary 5.8] to write

$$\|\boldsymbol{u}\|_{\mathbf{W}^{1,\mathbf{p}}(\Omega)} \leq C\|\boldsymbol{f}-\lambda\boldsymbol{u}\|_{\mathbf{L}^{r(p)}(\Omega)} \leq C\left(\|\boldsymbol{f}\|_{\mathbf{L}^{r(p)}(\Omega)} + \lambda_0\|\boldsymbol{u}\|_{\mathbf{L}^{r(p)}(\Omega)}\right). \tag{1.3.48}$$

By estimate (1.3.13) in Theorem (1.3.1), there exists a positive constant C such that for all $\lambda \in \mathbb{C}^*$, $\operatorname{Re}\lambda \geq 0$,

$$\|\boldsymbol{u}\|_{\boldsymbol{L}^2(\Omega)} \leq \frac{C}{|\lambda|}\ \|\boldsymbol{f}\|_{\boldsymbol{L}^2(\Omega)} \leq \frac{C'}{|\lambda|}\ \|\boldsymbol{f}\|_{\boldsymbol{L}^p(\Omega)}. \tag{1.3.49}$$

Estimate (1.3.45) for $p > 6$ and $|\lambda| < \lambda_0$ follows now from (1.3.48) and (1.3.49).

For $p \leq 2$ the estimate (1.3.45) follows from a duality argument.

Let us now prove the estimate (1.3.46). From Gagliardo–Nirenberg inequality [1, Chapter IV, Theorem 4.14, Theorem 4.17], and [6, Theorem 5.11] we have,

$$\begin{aligned}\|\mathbb{D}(\boldsymbol{u})\|_{\boldsymbol{L}^p(\Omega)} &\leq C\,\|\boldsymbol{u}\|_{\boldsymbol{W}^{1,p}(\Omega)} \leq C\,\|\boldsymbol{u}\|^{1/2}_{\boldsymbol{W}^{2,p}(\Omega)}\|\boldsymbol{u}\|^{1/2}_{\boldsymbol{L}^p(\Omega)}\\ &\leq C(\Omega,\alpha,p)\|\boldsymbol{f}-\lambda\boldsymbol{u}\|^{1/2}_{\boldsymbol{L}^p(\Omega)}\|\boldsymbol{u}\|^{1/2}_{\boldsymbol{L}^p(\Omega)}\\ &\leq C(\Omega,\alpha,p)\left(\|\boldsymbol{f}\|_{\boldsymbol{L}^p(\Omega)}+|\lambda|\|\boldsymbol{u}\|_{\boldsymbol{L}^p(\Omega)}\right)^{1/2}\|\boldsymbol{u}\|^{1/2}_{\boldsymbol{L}^p(\Omega)}.\end{aligned}$$

and estimate (1.3.46) follows using (1.3.45).

To prove (1.3.47) we use that, for $\boldsymbol{u} \in \mathbf{D}(A_{p,\alpha})$, we have the norm equivalence $\|\boldsymbol{u}\|_{\boldsymbol{W}^{2,p}(\Omega)} \simeq \|A_{p,\alpha}\boldsymbol{u}\|_{\boldsymbol{L}^p(\Omega)}$. Hence, by Amrouche et al. [6, Theorem 5.11] and (1.3.45),

$$\|\boldsymbol{u}\|_{\boldsymbol{W}^{2,p}(\Omega)} \leq C\,\|\boldsymbol{f}-\lambda\boldsymbol{u}\|_{\boldsymbol{L}^p(\Omega)} \leq C\,\|\boldsymbol{f}\|_{\boldsymbol{L}^p(\Omega)}.$$

This completes the proof. ■

Remark 1.3.12. We emphasize here that the regularity (1.3.5) of α is needed to obtain the resolvent estimate (1.3.45); mainly to estimate suitably the pressure term. Without that regularity of α, we do not know for the moment how to proceed to get the resolvent estimate which is in turn necessary to obtain the solution of the time-dependent Stokes problem.

By Theorems 1.3.5 and 1.2.2 the following result follows:

Theorem 1.3.6. Let Ω be of class $\mathcal{C}^{2,1}$ and α be as in (1.3.5). The operator $-A_p$ generates a bounded analytic semigroup on $\boldsymbol{L}^p_{\sigma,\boldsymbol{\tau}}(\Omega)$ for all $1 < p < \infty$.

Proof. In view of Theorem 1.3.5, to apply Theorem 1.2.2 it remains to check that $\mathbf{D}(A_p)$ is dense in $\boldsymbol{L}^p_{\sigma,\boldsymbol{\tau}}(\Omega)$. But this is immediate since $\boldsymbol{\mathcal{D}}_\sigma(\Omega) \hookrightarrow \mathbf{D}(A_p) \hookrightarrow \boldsymbol{L}^p_{\sigma,\boldsymbol{\tau}}(\Omega)$ and by definition $\boldsymbol{\mathcal{D}}_\sigma(\Omega)$ is dense in $\boldsymbol{L}^p_{\sigma,\boldsymbol{\tau}}(\Omega)$. ■

1.4 Stokes Operator on $[H_0^{p'}(\mathrm{div},\Omega)]'_{\sigma,\tau}$

We first recall that if $\boldsymbol{f} \in [\boldsymbol{H}_0^{p'}(\mathrm{div},\Omega)]'$ (defined in (1.2.1)) and is such that $\mathrm{div}\boldsymbol{f} \in \boldsymbol{L}^p(\Omega)$ for some $p \in (1,\infty)$, then its normal trace $(\boldsymbol{f}\cdot\boldsymbol{n})_{|\Gamma}$ is well defined and belongs to $W^{-1-\frac{1}{p},p}(\Gamma)$ [3, Corollary 3.7].

Let $\mathcal{B}$ be the closure of $\boldsymbol{\mathcal{D}}_\sigma(\Omega)$ in $[\boldsymbol{H}_0^{p'}(\mathrm{div},\Omega)]'$. Then, it can be shown that

$$\mathcal{B} = [\boldsymbol{H}_0^{p'}(\mathrm{div},\Omega)]'_{\sigma,\boldsymbol{\tau}} := \left\{\boldsymbol{f} \in [\boldsymbol{H}_0^{p'}(\mathrm{div},\Omega)]' : \mathrm{div}\,\boldsymbol{f} = 0 \text{ in } \Omega, \boldsymbol{f}\cdot\boldsymbol{n} = 0 \text{ on } \Gamma\right\}$$

which is a Banach space with the norm of $[\boldsymbol{H}_0^{p'}(\mathrm{div},\Omega)]'$ [3, Proposition 3.9]. Let $Q : [\boldsymbol{H}_0^{p'}(\mathrm{div},\Omega)]' \to \mathcal{B}$ be the orthogonal projection on $\mathcal{B}$. We define the Stokes operator $B_{p,\alpha}$ on $\mathcal{B}$, as

$$\begin{cases} \mathbf{D}(B_{p,\alpha}) = \Big\{ \boldsymbol{u} \in \boldsymbol{W}^{1,p}(\Omega) \cap \mathcal{B} : \Delta \boldsymbol{u} \in [\boldsymbol{H}_0^{p'}(\mathrm{div}, \Omega)]', 2[(\mathbb{D}\boldsymbol{u})\boldsymbol{n}]_{\boldsymbol{\tau}} + \alpha \boldsymbol{u}_{\boldsymbol{\tau}} = \boldsymbol{0} \text{ on } \Gamma \Big\}; \\ B_{p,\alpha}(\boldsymbol{u}) = -Q(\Delta \boldsymbol{u}) \quad \text{ for } \boldsymbol{u} \in \mathbf{D}(B_{p,\alpha}) \end{cases}$$

with $\alpha \in L^{t(p)}(\Gamma)$ where $t(p)$ is defined in (1.3.1).

As in the previous section, we will now discuss the analyticity of the semigroup generated by the Stokes operator $B_{p,\alpha}$ on $[\boldsymbol{H}_0^{p'}(\mathrm{div}, \Omega)]'_{\sigma,\boldsymbol{\tau}}$.

Theorem 1.4.1. Let $p \in (1, \infty)$ and $\alpha \in L^{t(p)}(\Gamma)$. Then, for all $\lambda \in \mathbb{C}^*$ such that $\mathrm{Re}\,\lambda \geq 0$, and all $\boldsymbol{f} \in [\boldsymbol{H}_0^{p'}(\mathrm{div}, \Omega)]'$, the problem (1.3.8)–(1.3.11) has a unique solution $(\boldsymbol{u}, \pi) \in \boldsymbol{W}^{1,p}(\Omega) \times L_0^p(\Omega)$ satisfying

$$\|\boldsymbol{u}\|_{[\boldsymbol{H}_0^{p'}(\mathrm{div},\Omega)]'} \leq \frac{C}{|\lambda|} \|\boldsymbol{f}\|_{[\boldsymbol{H}_0^{p'}(\mathrm{div},\Omega)]'} \tag{1.4.1}$$

for some constant $C = C(\Omega, \alpha, p)$ independent of λ.

Proof. **1. Existence:** The proof of the existence and uniqueness of the solution follows similar arguments as in the proof of Theorems 1.3.1 and 1.3.3. For $p = 2$ the existence and uniqueness of solution comes from Lax–Milgram lemma and De Rham theorem for the pressure.

When $p > 2$, since $\boldsymbol{f} \in [\boldsymbol{H}_0^{p'}(\mathrm{div}, \Omega)]' \subset [\boldsymbol{H}_0^2(\mathrm{div}, \Omega)]'$ and $\alpha \in L^{t(p)}(\Gamma) \subset L^2(\Gamma)$, we have the existence of the unique solution $(\boldsymbol{u}, \pi) \in \boldsymbol{H}^1(\Omega) \times L_0^2(\Omega)$. We can now apply [6, Corollary 5.8], since $\boldsymbol{f} - \lambda \boldsymbol{u} \in [\boldsymbol{H}_0^{p'}(\mathrm{div}, \Omega)]'$ to obtain $\boldsymbol{u} \in \boldsymbol{W}^{1,p}(\Omega)$.

For $p < 2$, the proof follows in the same way as in [6, Corollary 5.8].

2. Estimate: Now to prove the estimate (1.4.1), consider the problem,

$$\begin{cases} \lambda \boldsymbol{v} - \Delta \boldsymbol{v} + \nabla \theta = \boldsymbol{F}, \quad \mathrm{div}\, \boldsymbol{v} = 0 & \text{in } \Omega \\ \boldsymbol{v} \cdot \boldsymbol{n} = 0, \quad 2[(\mathbb{D}\boldsymbol{v})\boldsymbol{n}]_{\boldsymbol{\tau}} + \tilde{\alpha} \boldsymbol{v}_{\boldsymbol{\tau}} = \boldsymbol{0} & \text{on } \Gamma \end{cases}$$

where $\boldsymbol{F} \in \boldsymbol{H}_0^{p'}(\mathrm{div}, \Omega)$ and $\tilde{\alpha}$ as in (1.3.5). Thanks to Theorem 1.3.3 and the estimate (1.3.45), there exists unique $(\boldsymbol{v}, \theta) \in \boldsymbol{W}^{2,p'}(\Omega) \times (W^{1,p'}(\Omega) \cap L_0^{p'}(\Omega))$ with the estimate

$$\|\boldsymbol{v}\|_{\boldsymbol{L}^{p'}(\Omega)} \leq \frac{C}{|\lambda|} \|\boldsymbol{F}\|_{\boldsymbol{L}^{p'}(\Omega)}\,.$$

As a result, we get,

$$\|\boldsymbol{v}\|_{\boldsymbol{H}_0^{p'}(\mathrm{div},\Omega)} \leq \frac{C}{|\lambda|} \|\boldsymbol{F}\|_{\boldsymbol{H}_0^{p'}(\mathrm{div},\Omega)}.$$

Now, for the solution $(\boldsymbol{u}, \pi) \in \boldsymbol{W}^{1,p}(\Omega) \times L^p(\Omega)$ of the problem (1.3.8)–(1.3.11), we obtain,

$$
\begin{aligned}
\|\boldsymbol{u}\|_{[\boldsymbol{H}_0^{p'}(\mathrm{div},\Omega)]'} &= \sup_{\substack{\boldsymbol{F}\in \boldsymbol{H}_0^{p'}(\mathrm{div},\Omega)\\ \boldsymbol{F}\neq \boldsymbol{0}}} \frac{|\langle \boldsymbol{u}, \boldsymbol{F}\rangle|}{\|\boldsymbol{F}\|_{\boldsymbol{H}_0^{p'}(\mathrm{div},\Omega)}} = \sup_{\substack{\boldsymbol{F}\in \boldsymbol{H}_0^{p'}(\mathrm{div},\Omega)\\ \boldsymbol{F}\neq \boldsymbol{0}}} \frac{|\langle \boldsymbol{u}, \lambda\boldsymbol{v} - \Delta\boldsymbol{v} + \nabla\theta\rangle|}{\|\boldsymbol{F}\|_{\boldsymbol{H}_0^{p'}(\mathrm{div},\Omega)}} \\
&= \sup_{\substack{\boldsymbol{F}\in \boldsymbol{H}_0^{p'}(\mathrm{div},\Omega)\\ \boldsymbol{F}\neq \boldsymbol{0}}} \frac{|\langle \lambda\boldsymbol{u} - \Delta\boldsymbol{u} + \nabla\pi, \boldsymbol{v}\rangle|}{\|\boldsymbol{F}\|_{\boldsymbol{H}_0^{p'}(\mathrm{div},\Omega)}} = \sup_{\substack{\boldsymbol{F}\in \boldsymbol{H}_0^{p'}(\mathrm{div},\Omega)\\ \boldsymbol{F}\neq \boldsymbol{0}}} \frac{|\langle \boldsymbol{f}, \boldsymbol{v}\rangle|}{\|\boldsymbol{F}\|_{\boldsymbol{H}_0^{p'}(\mathrm{div},\Omega)}} \\
&\leq \frac{C}{|\lambda|}\,\|\boldsymbol{f}\|_{[\boldsymbol{H}_0^{p'}(\mathrm{div},\Omega)]'}
\end{aligned}
$$

which is the required estimate. ∎

Theorem 1.4.2. Let $p \in (1, \infty)$ and $\alpha \in L^{t(p)}(\Gamma)$ with $t(p)$ defined in (1.3.1). Then, for $\boldsymbol{f} \in [\boldsymbol{H}_0^{p'}(\mathrm{div}, \Omega)]'$, the unique solution $(\boldsymbol{u}, \pi)$ of (1.3.8)–(1.3.11) in $\mathbf{D}(B_{p,\alpha}) \times L_0^p(\Omega)$ is such that the pressure π satisfies the following estimate:

$$\|\pi\|_{L^p(\Omega)} \leq C\|\boldsymbol{f}\|_{[\boldsymbol{H}_0^{p'}(\mathrm{div},\Omega)]'}$$

for some constant C independent of λ.

Proof. By the regularity result of the stationary Stokes problem [6, Corollary 5.8] and [6, Remark 5.10, i)], we can write,

$$\|\pi\|_{L^p(\Omega)} \leq C(\Omega, \alpha)\|\boldsymbol{f} - \lambda\boldsymbol{u}\|_{[\boldsymbol{H}_0^{p'}(\mathrm{div},\Omega)]'}$$

and the result follows using estimate (1.4.1). ∎

The analyticity of the semigroup generated by the Stokes operator $B_{p,\alpha}$ on $[\boldsymbol{H}_0^{p'}(\mathrm{div}, \Omega)]'_{\sigma,\boldsymbol{\tau}}$ is now easily deduced from Theorem 1.4.1.

Theorem 1.4.3. Let $\alpha \in L^{t(p)}(\Gamma)$ with $t(p)$ defined in (1.3.1). The operator $-B_{p,\alpha}$ generates a bounded analytic semigroup on $[\boldsymbol{H}_0^{p'}(\mathrm{div}, \Omega)]'_{\sigma,\boldsymbol{\tau}}$ for all $1 < p < \infty$.

Proof. In view of Theorem 1.4.1, to apply Theorem 1.2.2 it remains to check that $\mathbf{D}(B_{p,\alpha})$ is dense in $[\boldsymbol{H}_0^{p'}(\mathrm{div}, \Omega)]'_{\sigma,\boldsymbol{\tau}}$. But this is immediate since $\mathfrak{D}_\sigma(\Omega) \hookrightarrow \mathbf{D}(B_{p,\alpha}) \hookrightarrow [\boldsymbol{H}_0^{p'}(\mathrm{div}, \Omega)]'_{\sigma,\boldsymbol{\tau}}$ and by definition $\mathfrak{D}_\sigma(\Omega)$ is dense in $[\boldsymbol{H}_0^{p'}(\mathrm{div}, \Omega)]'_{\sigma,\boldsymbol{\tau}}$. ∎

1.5 Imaginary and Fractional Powers

1.5.1 Imaginary Powers

Our main purpose in this section is to prove local bounds on pure imaginary powers $A_{p,\alpha}^{is}$ and $B_{p,\alpha}^{is}$ of the Stokes operators defined in Sects. 1.3 and 1.4

respectively. A complete theory on fractional powers of an operator (bounded or unbounded) can be found in Komatsu [25].

Since these operators are non-negative operators, it then follows from the results in [25] and in [39] that their powers are well, densely defined, and closed linear operators on $\boldsymbol{L}^p_{\sigma,\boldsymbol{\tau}}(\Omega)$ and $[\boldsymbol{H}_0^{p'}(\mathrm{div},\Omega)]'$ with domain $\mathbf{D}(A^{is}_{p,\alpha})$ and $\mathbf{D}(B^{is}_{p,\alpha})$ respectively.

Notice that in [2], it was comparatively straightforward to obtain the bounds on pure imaginary powers, since with Navier-type boundary condition, the Stokes operator actually reduces to Laplace operator and thus they could borrow the well-established theory for elliptic operators, which is not our case. Therefore we use the theory of interpolation-extrapolation to make use of the established theory for Stokes operator with Navier-type boundary condition and implement a perturbation argument.

Theorem 1.5.1. Let α be as in (1.3.5) and if $p \in (1,3]$ suppose also that $\alpha \in L^\infty(\Gamma)$. Then there exists an angle $0 < \theta < \pi/2$ and a constant $C > 0$, which may be dependent on α, such that for any $s \in \mathbb{R}$,

$$\|A^{is}_{p,\alpha}\| \leq C\ e^{|s|\theta}. \tag{1.5.1}$$

Similarly, for $\alpha \in L^\infty(\Gamma)$, there exists an angle $0 < \theta' < \pi/2$ and a constant $C' > 0$ such that for any $s \in \mathbb{R}$,

$$\|B^{is}_{p,\alpha}\| \leq C'\ e^{|s|\theta'}. \tag{1.5.2}$$

Proof. Since the proof of (1.5.2) is exactly similar to that of (1.5.1), we only show (1.5.1). The proof of (1.5.1) is based on the theory of interpolation-extrapolation scales from [5]. A similar approach has been followed in [34], considering the perturbation of a different operator than ours and for α constant.

1. Let us define $X_0 := \boldsymbol{L}^p_{\sigma,\boldsymbol{\tau}}(\Omega)$ and $A_0 := \lambda I + A_{NT}$, for all $\lambda > 0$ and where A_{NT} is the Stokes operator with Navier-type boundary condition

$$\boldsymbol{u}\cdot\boldsymbol{n} = 0, \quad \mathbf{curl}\ \boldsymbol{u}\times\boldsymbol{n} = \mathbf{0} \quad \text{on } \Gamma$$

i.e.

$$\begin{cases} \mathbf{D}(A_{NT}) = \left\{\boldsymbol{u}\in \boldsymbol{W}^{2,p}(\Omega)\cap \boldsymbol{L}^p_{\sigma,\boldsymbol{\tau}}(\Omega), \mathbf{curl}\ \boldsymbol{u}\times\boldsymbol{n} = \mathbf{0} \text{ on } \Gamma\right\} \\ A_{NT}(\boldsymbol{u}) = -P(\Delta \boldsymbol{u}) \quad \text{for } \boldsymbol{u}\in\mathbf{D}(A_{NT}). \end{cases}$$

Note that from [25, Theorem 6.4], it follows that for $\lambda \geq 0$, the domain $\mathbf{D}(\lambda I + A_p)$ does not depend on λ and

$$\mathbf{D}(\lambda I + A_p) = \mathbf{D}(A_p) \quad \text{for } \lambda > 0.$$

As indicated in the introduction to this section, the powers A_0^a of the operator A_0 are well, densely defined, and closed linear operators on $\boldsymbol{L}^p_{\sigma,\boldsymbol{\tau}}(\Omega)$ with domain $\mathbf{D}(A_0^a)$.

Now by Amann [5, Theorems V.1.5.1 and V.1.5.4], (X_0, A_0) generates an interpolation-extrapolation scale $(X_a, A_a), a \in \mathbb{R}$ with respect to the complex interpolation functor since A_0 is a closed operator on X_0 with bounded inverse (cf. [2, Theorem 4.8]). More precisely, for every $a \in \mathbb{R}$, X_a is a Banach space, $X_a \hookrightarrow X_{a-1}$ and A_a is an unbounded linear operator on X_a with domain X_{a+1} and for $a > 0$:

$$(i)\ X_a = (\mathbf{D}(A_0^a), \|A_0^a \cdot \|)$$
$$(ii)\ A_a \text{ is the restriction of } A_0 \text{ on } X_a.$$

Moreover, for any $b \in (a, a+1)$,

$$X_b = [X_a, X_{a+1}]_\theta \quad \text{where} \quad \frac{1}{b} = \frac{1-\theta}{a} + \frac{\theta}{a+1}.$$

Similarly, let $X_0^\sharp := (X_0)' = \boldsymbol{L}^{p'}_{\sigma,\boldsymbol{\tau}}(\Omega)$, $A_0^\sharp := (A_0)'$. Then $(X_0^\sharp, A_0^\sharp)$ generates another interpolation-extrapolation scale $(X_a^\sharp, A_a^\sharp)$, the dual scale by Amann [5, Theorem V.1.5.12] and

$$(X_a)' = X^\sharp_{-a} \quad \text{and} \quad (A_a)' = A^\sharp_{-a} \quad \text{for } a \in \mathbb{R}$$

where A' denotes the dual of A. In the particular case $a = -1/2$, we obtain by definition, an operator $A_{-1/2} : X_{-1/2} \to X_{-1/2}$ with

$$\mathbf{D}(A_{-1/2}) = X_{1/2} = [X_0, X_1]_{1/2}. \tag{1.5.3}$$

We now claim that

$$[X_0, X_1]_{1/2} = \boldsymbol{W}^{1,p}_{\sigma,\boldsymbol{\tau}}(\Omega) \tag{1.5.4}$$

and then,

$$X_{-1/2} = \left[\boldsymbol{W}^{1,p'}_{\sigma,\boldsymbol{\tau}}(\Omega)\right]' \tag{1.5.5}$$

will follow.

To prove (1.5.4), one inclusion is obvious. Indeed,

$$[X_0, X_1]_{1/2} \subset [\boldsymbol{L}^p_{\sigma,\boldsymbol{\tau}}(\Omega), \boldsymbol{W}^{2,p}_{\sigma,\boldsymbol{\tau}}(\Omega)]_{1/2} = \boldsymbol{W}^{1,p}_{\sigma,\boldsymbol{\tau}}(\Omega).$$

And for the other inclusion, by (1.5.3) it is enough to prove that $\boldsymbol{W}^{1,p}_{\sigma,\boldsymbol{\tau}}(\Omega) \subset \mathbf{D}(A_0^{1/2})$. To this end, first consider the operator $A_0^{1/2}$ on $\boldsymbol{L}^{p'}_{\sigma,\boldsymbol{\tau}}(\Omega)$. Since A_0 has a bounded inverse, $A_0^{1/2}$ is an isomorphism from $\mathbf{D}(A_0^{1/2})$ to $\boldsymbol{L}^{p'}_{\sigma,\boldsymbol{\tau}}(\Omega)$ [39, Theorem 1.15.2, part(e)] and thus, for any $\mathbf{F} \in \boldsymbol{L}^{p'}_{\sigma,\boldsymbol{\tau}}(\Omega)$, there exists a unique $\boldsymbol{v} \in \mathbf{D}(A_0^{1/2})$ such that $A_0^{1/2}\boldsymbol{v} = \mathbf{F}$. So, for all $\boldsymbol{u} \in \mathbf{D}(A_0)$,

$$
\begin{aligned}
\|A_0^{1/2}\boldsymbol{u}\|_{\boldsymbol{L}^p_{\sigma,\boldsymbol{\tau}}(\Omega)} &= \sup_{\substack{\mathbf{F}\in \boldsymbol{L}^{p'}_{\sigma,\boldsymbol{\tau}}(\Omega)\\ \mathbf{F}\neq \mathbf{0}}} \frac{\left|\left\langle A_0^{1/2}\boldsymbol{u},\mathbf{F}\right\rangle\right|}{\|\mathbf{F}\|_{\boldsymbol{L}^{p'}(\Omega)}} = \sup_{\substack{\boldsymbol{v}\in \mathbf{D}(A_0^{1/2})\\ \boldsymbol{v}\neq \mathbf{0}}} \frac{\left|\left\langle A_0^{1/2}\boldsymbol{u}, A_0^{\frac{1}{2}}\boldsymbol{v}\right\rangle\right|}{\|A_0^{\frac{1}{2}}\boldsymbol{v}\|_{\boldsymbol{L}^{p'}(\Omega)}}\\
&= \sup_{\substack{\boldsymbol{v}\in \mathbf{D}(A_0^{1/2})\\ \boldsymbol{v}\neq \mathbf{0}}} \frac{|\langle A_0\boldsymbol{u},\boldsymbol{v}\rangle|}{\|A_0^{1/2}\boldsymbol{v}\|_{\boldsymbol{L}^{p'}(\Omega)}}\\
&= \sup_{\substack{\boldsymbol{v}\in \mathbf{D}(A_0^{1/2})\\ \boldsymbol{v}\neq \mathbf{0}}} \frac{\left|\int_\Omega \lambda\boldsymbol{u}\cdot\boldsymbol{v} + \mathbf{curl}\,\boldsymbol{u}\cdot\mathbf{curl}\,\boldsymbol{v}\right|}{\|A_0^{1/2}\boldsymbol{v}\|_{\boldsymbol{L}^{p'}(\Omega)}}\\
&\le C\|\boldsymbol{u}\|_{\boldsymbol{W}^{1,p}(\Omega)}. \qquad (1.5.6)
\end{aligned}
$$

Now as $\mathbf{D}(A_0)$ is dense in $\boldsymbol{W}^{1,p}_{\sigma,\boldsymbol{\tau}}(\Omega)$, we get the inequality (1.5.6) for all $\boldsymbol{u}\in \boldsymbol{W}^{1,p}_{\sigma,\boldsymbol{\tau}}(\Omega)$ which gives the required embedding.

Now from [3, Theorem 6.1], we know that there exist constants $M>0$ and $\theta\in(0,\frac{\pi}{2})$ such that

$$\forall s\in\mathbb{R},\ \ \|A_0^{is}\|_{\mathcal{L}(X_0)}\le Me^{|s|\theta}.$$

It then follows from [5, Theorem V.1.5.5 (ii)] that

$$\forall s\in\mathbb{R},\ \ \|\left(A_{-1/2}\right)^{is}\|_{\mathcal{L}(X_{-1/2})}\le Me^{|s|\theta}.$$

We call the operator $A_{-1/2}$ the weak Stokes operator subject to Navier-type boundary condition. Since $A_{-1/2}$ is the closure of A_0 in $X_{-1/2}$ and $X_1\hookrightarrow X_{1/2}$, it follows that $A_{-1/2}\boldsymbol{u}=A_0\boldsymbol{u}$ for $\boldsymbol{u}\in X_1$ and thus, for all $\boldsymbol{v}\in \boldsymbol{W}^{1,p'}_{\sigma,\boldsymbol{\tau}}(\Omega)$,

$$\left\langle \boldsymbol{v}, A_{-1/2}\boldsymbol{u}\right\rangle_{(X_{-1/2})'\times X_{-1/2}} = \langle \boldsymbol{v}, A_0\boldsymbol{u}\rangle = \lambda\int_\Omega \boldsymbol{u}\cdot\boldsymbol{v} + \int_\Omega \mathbf{curl}\,\boldsymbol{u}\cdot\mathbf{curl}\,\boldsymbol{v},$$

where we only used integration by parts. Now using the density of X_1 in $X_{1/2}$, we obtain the relation, for all $(\boldsymbol{u},\boldsymbol{v})\in \boldsymbol{W}^{1,p}_{\sigma,\boldsymbol{\tau}}(\Omega)\times \boldsymbol{W}^{1,p'}_{\sigma,\boldsymbol{\tau}}(\Omega)$,

$$\left\langle A_{-1/2}\boldsymbol{u},\boldsymbol{v}\right\rangle = \lambda\int_\Omega \boldsymbol{u}\cdot\boldsymbol{v} + \int_\Omega \mathbf{curl}\,\boldsymbol{u}\cdot\mathbf{curl}\,\boldsymbol{v}. \qquad (1.5.7)$$

2. Next let us define an unbounded operator $A_{N,w}$ on $X_{-1/2}$, with domain $X_{1/2}$, as, for all $(\boldsymbol{u},\boldsymbol{v})\in \boldsymbol{W}^{1,p}_{\sigma,\boldsymbol{\tau}}(\Omega)\times \boldsymbol{W}^{1,p'}_{\sigma,\boldsymbol{\tau}}(\Omega)$,

$$\langle A_{N,w}\boldsymbol{u},\boldsymbol{v}\rangle = \int_\Omega \mathbf{curl}\,\boldsymbol{u}\cdot\mathbf{curl}\,\boldsymbol{v} + \langle \Lambda\boldsymbol{u},\boldsymbol{v}\rangle_\Gamma + \int_\Gamma \alpha\boldsymbol{u}\cdot\boldsymbol{v} \qquad (1.5.8)$$

where Λ is defined in (1.2.2). We call the operator $A_{N,w}$ the weak Stokes operator subject to Navier boundary conditions. Comparing (1.5.8) with (1.5.7) implies

$$\langle(\lambda I + A_{N,w})\,\boldsymbol{u},\boldsymbol{v}\rangle = \left\langle A_{-1/2}\boldsymbol{u},\boldsymbol{v}\right\rangle + \langle \Lambda_\alpha\boldsymbol{u},\boldsymbol{v}\rangle_\Gamma$$

where the linear operator $\Lambda_\alpha : X_{-1/2} \to X_{-1/2}$, given by,

$$\langle \Lambda_\alpha \boldsymbol{u}, \boldsymbol{v} \rangle_\Gamma = \langle \Lambda \boldsymbol{u}, \boldsymbol{v} \rangle_\Gamma + \int_\Gamma \alpha \boldsymbol{u} \cdot \boldsymbol{v}$$

is a lower-order perturbation of $A_{-1/2}$. Therefore, as $\alpha \in L^\infty(\Gamma)$ (due to the assumption of the theorem), it follows from [34, Proposition 3.3.9],

$$\forall s \in \mathbb{R}, \quad \| \left[(\lambda I + A_{N,w})\right]^{is} \|_{\mathcal{L}(X_{-1/2})} \le M e^{|s|\theta_A}$$

for some constant $\theta_A \in (0, \pi/2)$. Here, the above constant M may depend on α. Since, from [6, Theorem 5.8], $A_{N,w}$ has a bounded inverse, it follows from [34, Proposition 3.3.9] again that

$$\|A_{N,w}^{is}\|_{\mathcal{L}(X_{-1/2})} \le M e^{|s|\theta_A}.$$

3. Now we want to transfer this "bounded imaginary power" property to the strong Stokes operator A_p with Navier boundary condition, defined in (1.3.6)–(1.3.7) on $\boldsymbol{L}^p_{\sigma,\boldsymbol{\tau}}(\Omega)$. For that we will apply again Amann's theory of interpolation-extrapolation scales. Let $X_0^w := \left[\boldsymbol{W}^{1,p'}_{\sigma,\boldsymbol{\tau}}(\Omega)\right]'$, $A_0^w := A_{N,w}$, and $X_1^w := \boldsymbol{W}^{1,p}_{\sigma,\boldsymbol{\tau}}(\Omega)$. By Amann [5, Theorems V.1.5.1 and V.1.5.4], the pair (X_0^w, A_0^w) generates an interpolation-extrapolation scale $(X_a^w, A_a^w), a \in \mathbb{R}$ with respect to the complex interpolation functor and by Amann [5, Theorem V.1.5.5 (ii)], for any $a \in \mathbb{R}$,

$$\forall s \in \mathbb{R}, \quad \|(A_a^w)^{is}\|_{\mathcal{L}(X_a^w)} \le M e^{|s|\theta_A}.$$

We will show in the remaining part of this proof that the operator $A_{1/2}^w : X_{3/2}^w \subset X_{1/2}^w \to X_{1/2}^w$ coincides with A_p where the strong Stokes operator $A_p : \mathbf{D}(A_p) \subset \boldsymbol{L}^p_{\sigma,\boldsymbol{\tau}}(\Omega) \to \boldsymbol{L}^p_{\sigma,\boldsymbol{\tau}}(\Omega)$ is defined in (1.3.6)–(1.3.7). Observe that, by (1.5.3), (1.5.5),

$$X_0^w = X_{-1/2} \quad \text{and} \quad X_1^w = X_{1/2}.$$

Therefore,

$$X_{1/2}^w = [X_0^w, X_1^w]_{1/2} = \left[X_{-1/2}, X_{1/2}\right]_{1/2} = X_0 = \boldsymbol{L}^p_{\sigma,\boldsymbol{\tau}}(\Omega)$$

and the operator $A_{1/2}^w$ is the restriction of A_0^w on $X_{1/2}^w$. Hence, $A_{1/2}^w \boldsymbol{u} = A_0^w \boldsymbol{u} = A_{N,w} \boldsymbol{u}$ for any $\boldsymbol{u} \in \mathbf{D}(A_{1/2}^w) = X_{3/2}^w$ and then, for any $\boldsymbol{\varphi} \in \boldsymbol{W}^{1,p'}_{\sigma,\boldsymbol{\tau}}(\Omega)$,

$$\left\langle \boldsymbol{\varphi}, A_{1/2}^w \boldsymbol{u} \right\rangle_{\left(X_{1/2}^w\right)' \times X_{1/2}^w} = \langle \boldsymbol{\varphi}, A_{N,w} \boldsymbol{u} \rangle_{\left(X_{1/2}^w\right)' \times X_{1/2}^w} \tag{1.5.9}$$

$$= \int_\Omega \mathbf{curl}\, \boldsymbol{u} \cdot \mathbf{curl}\, \boldsymbol{\varphi} + \langle \Lambda \boldsymbol{u}, \boldsymbol{\varphi} \rangle_\Gamma + \int_\Gamma \alpha \boldsymbol{u} \cdot \boldsymbol{\varphi}. \tag{1.5.10}$$

On the other hand, for any $(\boldsymbol{v}, \boldsymbol{\varphi}) \in \mathbf{D}(A_p) \times \boldsymbol{W}^{1,p'}_{\sigma,\boldsymbol{\tau}}(\Omega)$, it follows from integration by parts that

$$\langle \boldsymbol{\varphi}, A_p \boldsymbol{v} \rangle_{\left(X^w_{1/2}\right)' \times X^w_{1/2}} = \int_\Omega \mathbf{curl}\, \boldsymbol{v} \cdot \mathbf{curl}\, \boldsymbol{\varphi} + \langle \Lambda \boldsymbol{v}, \boldsymbol{\varphi} \rangle_\Gamma + \int_\Gamma \alpha \boldsymbol{v} \cdot \boldsymbol{\varphi}. \quad (1.5.11)$$

Now for any given $\boldsymbol{u} \in \mathbf{D}(A^w_{1/2})$, $A^w_{1/2} \boldsymbol{u} \in \boldsymbol{L}^p_{\sigma,\boldsymbol{\tau}}(\Omega)$ and then there exists a unique $\boldsymbol{v} \in \mathbf{D}(A_p)$ such that

$$A_p \boldsymbol{v} = A^w_{1/2} \boldsymbol{u}$$

since A_p is onto. Thus it follows from (1.5.9) that for any $\boldsymbol{\varphi} \in \boldsymbol{W}^{1,p'}_{\sigma,\boldsymbol{\tau}}(\Omega)$,

$$\langle \boldsymbol{\varphi}, A_p \boldsymbol{v} \rangle_{\left(X^w_{1/2}\right)' \times X^w_{1/2}} = \langle \boldsymbol{\varphi}, A_{N,w} \boldsymbol{u} \rangle_{\left(X^w_{1/2}\right)' \times X^w_{1/2}}.$$

This in turn implies by (1.5.10) and (1.5.11) that

$$\langle \boldsymbol{\varphi}, A_{N,w} \boldsymbol{u} \rangle_{\left(X^w_{1/2}\right)' \times X^w_{1/2}} = \langle \boldsymbol{\varphi}, A_{N,w} \boldsymbol{v} \rangle_{\left(X^w_{1/2}\right)' \times X^w_{1/2}}.$$

Hence, $\boldsymbol{v} = \boldsymbol{u}$ by injectivity of $A_{N,w}$. Similarly, if $\boldsymbol{v} \in \mathbf{D}(A_p)$ is given, then there exists a unique $\boldsymbol{u} \in \mathbf{D}(A^w_{1/2})$ such that $A^w_{1/2} \boldsymbol{u} = A_p \boldsymbol{v}$ since $A_p \boldsymbol{v} \in \boldsymbol{L}^p_{\sigma,\boldsymbol{\tau}}(\Omega)$ and $A^w_{1/2}$ is onto. By the same argument as above, we obtain $\boldsymbol{u} = \boldsymbol{v}$ showing that $\mathbf{D}(A_p) = \mathbf{D}(A^w_{1/2})$ and $A_p = A^w_{1/2}$. Thus finally we get that,

$$\forall s \in \mathbb{R}, \quad \|A_p^{is}\|_{\mathcal{L}(\boldsymbol{L}^p_{\sigma,\boldsymbol{\tau}}(\Omega))} \leq M e^{|s| \theta_A}.$$

∎

1.5.2 Fractional Powers

The above result allows us to study the domains of A_p^β, $\beta \in \mathbb{R}$. It can be shown that $\boldsymbol{D}(A^\beta_{p,\alpha})$ is a Banach space with the graph norm which is equivalent to the norm $\|A^\beta_{p,\alpha} \cdot \|_{\boldsymbol{L}^p(\Omega)}$, since $A_{p,\alpha}$ has bounded inverse. Note that for any $\beta \in \mathbb{R}$, the map $\boldsymbol{u} \to \|A^\beta_{p,\alpha} \boldsymbol{u}\|_{\boldsymbol{L}^p(\Omega)}$ defines a norm on $\boldsymbol{D}(A^\beta_{p,\alpha})$ due to the injectivity of $A^\beta_{p,\alpha}$.

Theorem 1.5.2. For all $p \in (1, \infty)$, $\mathbf{D}(A^{1/2}_{p,\alpha}) = \boldsymbol{W}^{1,p}_{\sigma,\boldsymbol{\tau}}(\Omega)$ with equivalent norms.

Proof. Since the pure imaginary power of $A_{p,\alpha}$ is bounded and satisfies estimate (1.5.1), using the result [39, Theorem 1.15.3], we get that

$$\mathbf{D}(A^{1/2}_{p,\alpha}) = [\boldsymbol{L}^p_{\sigma,\boldsymbol{\tau}}(\Omega), \mathbf{D}(A_{p,\alpha})]_{\frac{1}{2}}. \quad (1.5.12)$$

Then it is enough to show that

$$[\boldsymbol{L}^p_{\sigma,\boldsymbol{\tau}}(\Omega), \mathbf{D}(A_{p,\alpha})]_{\frac{1}{2}} = \mathbf{W}^{1,p}_{\sigma,\boldsymbol{\tau}}(\Omega)$$

with equivalent norms, which is already proved in (1.5.4). ∎

Remark 1.5.1. If Ω is not obtained by rotation around an axis i.e. if Ω is not axisymmetric, the norms $\|\boldsymbol{u}\|_{\boldsymbol{W}^{1,p}(\Omega)}$ and $\|\mathbb{D}\boldsymbol{u}\|_{\boldsymbol{L}^p(\Omega)}$ are equivalent for $\boldsymbol{u} \in \boldsymbol{W}^{1,p}(\Omega)$ with $\boldsymbol{u} \cdot \boldsymbol{n} = 0$ on Γ, as shown in [6, Proposition 3.7]. As a result we have the following equivalence for all $\boldsymbol{u} \in \mathbf{D}(A_{p,\alpha}^{1/2})$:

$$\|\mathbb{D}\boldsymbol{u}\|_{\boldsymbol{L}^p(\Omega)} \simeq \|A_{p,\alpha}^{1/2}\boldsymbol{u}\|_{\boldsymbol{L}^p(\Omega)}.$$

Our next result is an embedding theorem of Sobolev type for domains of fractional powers which will be applied to deduce the so-called $L^p - L^q$ estimates for the solution of the evolutionary Stokes equation.

Theorem 1.5.3. For all $1 < p < \infty$ and for all $\beta \in \mathbb{R}$ such that $0 < \beta < \frac{3}{2p}$, the following embedding holds:

$$\mathbf{D}(A_{p,\alpha}^{\beta}) \hookrightarrow \boldsymbol{L}^q(\Omega) \quad \text{where} \quad \frac{1}{q} = \frac{1}{p} - \frac{2\beta}{3}.$$

Proof. First observe that for $0 \leq \theta \leq 1$, by the result [39, Theorem 1.15.3] and the estimate (1.5.1), we can write

$$\mathbf{D}(A_{p,\alpha}^{\theta}) = [\boldsymbol{L}_{\sigma,\boldsymbol{\tau}}^{p}(\Omega), \mathbf{D}(\lambda I + A_{p,\alpha})]_\theta \hookrightarrow [\boldsymbol{L}^p(\Omega), \boldsymbol{W}^{2,p}(\Omega)]_\theta \hookrightarrow \boldsymbol{W}^{2\theta,p}(\Omega) \hookrightarrow \boldsymbol{L}^q(\Omega), \tag{1.5.13}$$

where

$$\frac{1}{q} = \frac{1}{p} - \frac{2\theta}{3} \quad \text{when} \quad p < \frac{3}{2\theta}.$$

Now let $\beta = \theta + k$ where $0 \leq \theta < 1$ and $k \in \mathbb{N} \cup \{0\}$. Consider m large so that $\mathbf{D}(A_{p,\alpha}^{m}) \subset \mathbf{D}(A_{q,\alpha}^{\beta})$ where $\frac{1}{q} = \frac{1}{p} - \frac{2\beta}{3}$. Also, by the definition of q, it is obvious that $\mathbf{D}(A_{q,\alpha}^{\beta}) \subset \mathbf{D}(A_{p,\alpha}^{\beta})$. If we set

$$\frac{1}{q_0} = \frac{1}{p} - \frac{2\theta}{3} \quad \text{and} \quad \frac{1}{q_j} = \frac{1}{q_0} - \frac{2j}{\theta} \quad \text{for} \ \ 0 \leq j \leq k,$$

then we have $\frac{1}{q_j} = \frac{1}{q_{j-1}} - \frac{2}{3}$ for $1 \leq j \leq k$ and $q_k = q$. Moreover, $q_{j-1} < \frac{3}{2}$ for $1 \leq j \leq k$ by assumptions on p and β. Hence as the consequence of the embedding (1.5.13), we get that

$$\mathbf{D}(A_{p,\alpha}^{\theta}) \hookrightarrow \boldsymbol{L}^{q_0}(\Omega)$$

and

$$\mathbf{D}(A_{q_{j-1},\alpha}) \hookrightarrow \boldsymbol{L}^{q_j}(\Omega) \ \text{ for } \ 1 \leq j \leq k.$$

Thus it follows that for all $\boldsymbol{u} \in \mathbf{D}(A_p^m)$,

$$\|\boldsymbol{u}\|_{\boldsymbol{L}^q(\Omega)} \leq C\|A_{q_{k-1},\alpha}\boldsymbol{u}\|_{\boldsymbol{L}^{q_{k-1}}(\Omega)} \leq \ldots \leq C\|A_{q_0,\alpha}^{k}\boldsymbol{u}\|_{\boldsymbol{L}^{q_0}(\Omega)} \leq C\|A_{p,\alpha}^{\beta}\boldsymbol{u}\|_{\boldsymbol{L}^p(\Omega)}.$$

By density of $\mathbf{D}(A_{p,\alpha}^{m})$ in $\mathbf{D}(A_{p,\alpha}^{\beta})$, we get the final result. ■

1.6 The Homogeneous Stokes Problem

In this section, with the help of the semigroup theory, we solve the homogeneous time-dependent Stokes problem:

$$\begin{cases} \dfrac{\partial \boldsymbol{u}}{\partial t} - \Delta \boldsymbol{u} + \nabla \pi = \mathbf{0}, \quad \operatorname{div} \boldsymbol{u} = 0 & \text{in } \Omega \times (0,T), \\ \boldsymbol{u} \cdot \boldsymbol{n} = 0, \quad 2[(\mathbb{D}\boldsymbol{u})\boldsymbol{n}]_{\boldsymbol{\tau}} + \alpha \boldsymbol{u}_{\boldsymbol{\tau}} = \mathbf{0} & \text{on } \Gamma \times (0,T), \\ \boldsymbol{u}(0) = \boldsymbol{u}_0 & \text{in } \Omega \end{cases} \tag{1.6.1}$$

for which the analyticity of the semigroups considered before gives a unique solution satisfying the usual regularity.

1.6.1 Strong Solution

We start with the strong solution of the problem (1.6.1).

Theorem 1.6.1. Let $p \in (1,\infty)$ and α be as in (1.3.5). Then for $\boldsymbol{u}_0 \in \boldsymbol{L}^p_{\sigma,\boldsymbol{\tau}}(\Omega)$, the problem (1.6.1) has a unique solution $\boldsymbol{u}(t)$ satisfying

$$\boldsymbol{u} \in C([0,\infty), \boldsymbol{L}^p_{\sigma,\boldsymbol{\tau}}(\Omega)) \cap C((0,\infty), \mathbf{D}(A_{p,\alpha})) \cap C^1((0,\infty), \boldsymbol{L}^p_{\sigma,\boldsymbol{\tau}}(\Omega)) \tag{1.6.2}$$

and

$$\boldsymbol{u} \in C^k((0,\infty), \mathbf{D}(A^{\ell}_{p,\alpha})) \quad \forall\, k \in \mathbb{N},\ \forall\, \ell \in \mathbb{N}\backslash\{0\}. \tag{1.6.3}$$

Also we have the estimates, for some constant $C = C(\Omega, \alpha, p) > 0$ which is independent of t,

$$\|\boldsymbol{u}(t)\|_{\boldsymbol{L}^p(\Omega)} \leq C \|\boldsymbol{u}_0\|_{\boldsymbol{L}^p(\Omega)}, \tag{1.6.4}$$

$$\left\| \frac{\partial \boldsymbol{u}(t)}{\partial t} \right\|_{\boldsymbol{L}^p(\Omega)} + \|\nabla \pi\|_{L^p(\Omega)} + \|\boldsymbol{u}(t)\|_{\boldsymbol{W}^{2,p}(\Omega)} \leq \frac{C}{t} \|\boldsymbol{u}_0\|_{\boldsymbol{L}^p(\Omega)}, \tag{1.6.5}$$

and

$$\|\mathbb{D}\boldsymbol{u}(t)\|_{\boldsymbol{L}^p(\Omega)} \leq \frac{C}{\sqrt{t}} \|\boldsymbol{u}_0\|_{\boldsymbol{L}^p(\Omega)}. \tag{1.6.6}$$

Proof. Since $-A_{p,\alpha}$ generates an analytic semigroup $\{T(t)\}_{t\geq 0}$ for every $\boldsymbol{u}_0 \in \boldsymbol{L}^p_{\sigma,\boldsymbol{\tau}}(\Omega)$, the initial value problem (1.6.1) has a unique solution $\boldsymbol{u}(t) = T(t)\boldsymbol{u}_0$, by Pazy [33, Corollary 1.5, Chapter 4, page 104]. Also, from [33, Theorem 7.7, Chapter 1, Page 30], we get that

$$\|T(t)\| \leq C \quad \text{for some constant } C > 0.$$

As a result, we obtain the estimate (1.6.4). Also, with the help of [33, point (d), Theorem 5.2, Chapter 2], we get the estimate of the first term in (1.6.5). To prove the estimate (1.6.6), we need to proceed as in the proof of (1.3.46), hence we skip it. The estimate for the third term of (1.6.5) follows from the estimate on the first term in (1.6.5) and using the fact that $\|\boldsymbol{v}\|_{\boldsymbol{W}^{2,p}(\Omega)} \simeq \|A_{p,\alpha}\boldsymbol{v}\|_{\boldsymbol{L}^p(\Omega)}$ for $\boldsymbol{v} \in \mathbf{D}(A_{p,\alpha})$.

Further, using the usual regularity properties of semigroup and by Pazy [33, Lemma 4.2, chapter 2], we can deduce the regularity (1.6.2) and (1.6.3). The estimate on the pressure term in (1.6.5) can be deduced from the equation using the estimates for the first and third term in (1.6.5). ■

The estimates (1.6.4)–(1.6.6) allow us to deduce the following regularity result.

Corollary 1.6.1. Let $p \in (1, \infty)$ and α be as in (1.3.5). Moreover, $\boldsymbol{u}_0 \in \boldsymbol{L}^p_{\sigma,\boldsymbol{\tau}}(\Omega), 0 < T < \infty$ and $(\boldsymbol{u}, \pi)$ be the unique solution of problem (1.6.1) given by theorem 1.6.1. Then, for all $1 \leq q < 2$, we have,

$$\boldsymbol{u} \in L^q(0,T;\boldsymbol{W}^{1,p}(\Omega)), \pi \in L^q(0,T;L^p_0(\Omega)) \text{ and } \frac{\partial \boldsymbol{u}}{\partial t} \in L^q(0,T;[\boldsymbol{H}^{p'}_0(\text{div},\Omega)]').$$

Proof. Since we have the Korn inequality

$$\|\boldsymbol{u}(t)\|_{\boldsymbol{W}^{1,p}(\Omega)} \leq \|\boldsymbol{u}(t)\|_{\boldsymbol{L}^p(\Omega)} + \|\mathbb{D}\boldsymbol{u}(t)\|_{\boldsymbol{L}^p(\Omega)}$$

and $\boldsymbol{u}(t)$ satisfies the estimates (1.6.4) and (1.6.6), we get,

$$\|\boldsymbol{u}(t)\|^q_{\boldsymbol{W}^{1,p}(\Omega)} \leq C(1 + t^{-q/2})\|\boldsymbol{u}_0\|_{\boldsymbol{L}^p(\Omega)}$$

which implies $\boldsymbol{u} \in L^q(0,T;\boldsymbol{W}^{1,p}(\Omega))$ only for $1 \leq q < 2$ and for all $0 < T < \infty$.

Moreover, as the operator $B_{p,\alpha} : \boldsymbol{D}(B_{p,\alpha}) \to [\boldsymbol{H}^{p'}_0(\text{div},\Omega)]'$ is an isomorphism, we have the equivalence of norm, for any $\boldsymbol{v} \in \boldsymbol{D}(B_{p,\alpha})$, $\|B_{p,\alpha}\boldsymbol{v}\|_{[\boldsymbol{H}^{p'}_0(\text{div},\Omega)]'}$ $\simeq \|\boldsymbol{v}\|_{\boldsymbol{D}(B_{p,\alpha})}$ and here $B_{p,\alpha}\boldsymbol{u} = \frac{\partial \boldsymbol{u}}{\partial t}$. Thus $\frac{\partial \boldsymbol{u}}{\partial t} \in L^q(0,T;[\boldsymbol{H}^{p'}_0(\text{div},\Omega)]')$.

Finally from the equation $\nabla\pi = \Delta\boldsymbol{u} - \frac{\partial \boldsymbol{u}}{\partial t}$, the regularity of π follows. ■

Theorem 1.6.2. Let α satisfy (1.3.5). Then for all $p \leq q < \infty$ and $\boldsymbol{u}_0 \in \boldsymbol{L}^p_{\sigma,\boldsymbol{\tau}}(\Omega)$, there exists $\delta > 0$ such that the unique solution $\boldsymbol{u}(t)$ of the problem (1.6.1) belongs to $\boldsymbol{L}^q(\Omega)$ and satisfies, for all $t > 0$:

$$\|\boldsymbol{u}(t)\|_{\boldsymbol{L}^q(\Omega)} \leq C\, e^{-\delta t} t^{-3/2(1/p-1/q)}\|\boldsymbol{u}_0\|_{\boldsymbol{L}^p(\Omega)}. \tag{1.6.7}$$

Moreover, the following estimates also hold:

$$\|\mathbb{D}\boldsymbol{u}(t)\|_{\boldsymbol{L}^q(\Omega)} \leq C\, e^{-\delta t} t^{-3/2(1/p-1/q)-1/2}\|\boldsymbol{u}_0\|_{\boldsymbol{L}^p(\Omega)}, \tag{1.6.8}$$

$$\forall\, m,n \in \mathbb{N}, \quad \|\frac{\partial^m}{\partial t^m} A^n_{p,\alpha}\boldsymbol{u}(t)\|_{\boldsymbol{L}^q(\Omega)} \leq C\, e^{-\delta t} t^{-(m+n)-3/2(1/p-1/q)}\|\boldsymbol{u}_0\|_{\boldsymbol{L}^p(\Omega)}. \tag{1.6.9}$$

All the above constants $C = C(\Omega, \alpha, p)$ are independent of t and δ.

Proof. First observe that in the case of $p = q$, the estimates (1.6.7)–(1.6.9) follow from the classical semigroup theory and the result that $\|T(t)\| \leq Ce^{-\delta t}$ [33, Theorem 6.13, Chapter 2].

Suppose that $p \neq q$. Let $s \in \mathbb{R}$ such that $\frac{3}{2}(\frac{1}{p} - \frac{1}{q}) < s < \frac{3}{2p}$ and set $\frac{1}{p_0} = \frac{1}{p} - \frac{2s}{3}$. It is clear that $p < q < p_0$. Since for all $t > 0$ and for all $l \in \mathbb{R}^+$, $\boldsymbol{u}(t) \in \mathbf{D}(A^l_{p,\alpha})$, thanks to Theorem 1.5.3, $\boldsymbol{u}(t) \in \mathbf{D}(A^s_{p,\alpha}) \hookrightarrow \boldsymbol{L}^{p_0}(\Omega)$. Now $\frac{1}{q} = \frac{\theta}{p_0} + \frac{1-\theta}{p}$ for $\theta = \frac{1/p-1/q}{1/p-1/p_0} \in (0,1)$. Thus $\boldsymbol{u}(t) \in \boldsymbol{L}^q(\Omega)$ and

$$\begin{aligned}\|\boldsymbol{u}(t)\|_{\boldsymbol{L}^q(\Omega)} \leq C\|\boldsymbol{u}(t)\|^{\theta}_{\boldsymbol{L}^{p_0}(\Omega)}\|\boldsymbol{u}(t)\|^{1-\theta}_{\boldsymbol{L}^p(\Omega)} &\leq C\|A^s_{p,\alpha}T(t)\boldsymbol{u}_0\|^{\theta}_{\boldsymbol{L}^p(\Omega)}\|T(t)\boldsymbol{u}_0\|^{1-\theta}_{\boldsymbol{L}^p(\Omega)}\\ &\leq C\,e^{-\delta t}t^{-\theta s}\|\boldsymbol{u}_0\|_{\boldsymbol{L}^p(\Omega)},\end{aligned}$$

where the last estimate follows from [33, Chapter 2, Theorem 6.13].

In order to prove (1.6.9) we first obtain from (1.6.2): $\frac{\partial^m}{\partial t^m}A^n_{p,\alpha}\boldsymbol{u}(t) \in \boldsymbol{L}^q(\Omega)$ for any $m, n \in \mathbb{N}$ and then

$$\|\frac{\partial^m}{\partial t^m}A^n_{p,\alpha}\boldsymbol{u}(t)\|_{\boldsymbol{L}^q(\Omega)} = \|A^{(m+n)}_{p,\alpha}T(t)\boldsymbol{u}_0\|_{\boldsymbol{L}^q(\Omega)} \leq Ce^{-\delta t}t^{-(m+n)-3/2(1/p-1/q)}\|\boldsymbol{u}_0\|_{\boldsymbol{L}^p(\Omega)}.$$

To prove estimate (1.6.8), we first deduce from (1.6.9) for $m = 1$ and $n = 0$:

$$\|A_{p,\alpha}\boldsymbol{u}(t)\|_{\boldsymbol{L}^q(\Omega)} \leq Ce^{-\delta t}t^{-1-\frac{3}{2}(\frac{1}{p}-\frac{1}{q})}. \tag{1.6.10}$$

Then, from Gagliardo Nirenberg's inequality, we obtain

$$\begin{aligned}\|\mathbb{D}\boldsymbol{u}(t)\|_{\boldsymbol{L}^q(\Omega)} &\leq C\ \|\boldsymbol{u}(t)\|_{\boldsymbol{W}^{1,q}(\Omega)} \leq C\ \|\boldsymbol{u}(t)\|^{1/2}_{\boldsymbol{W}^{2,q}(\Omega)}\|\boldsymbol{u}(t)\|^{1/2}_{\boldsymbol{L}^q(\Omega)}\\ &\leq C\ \|\boldsymbol{u}(t)\|^{1/2}_{\mathbf{D}(A_{q,\alpha})}\|\boldsymbol{u}(t)\|^{1/2}_{\boldsymbol{L}^q(\Omega)} \leq C\ \|A_{q,\alpha}\boldsymbol{u}(t)\|^{1/2}_{\boldsymbol{L}^q(\Omega)}\|\boldsymbol{u}(t)\|^{1/2}_{\boldsymbol{L}^q(\Omega)}.\end{aligned}$$

Thus (1.6.8) follows from (1.6.7) and (1.6.10). ∎

Proof of Theorem 1.1.2. This essentially follows from Theorems 1.6.1 and 1.6.2. ∎

1.6.2 Weak Solution

The following result says that if the initial data is in $[\boldsymbol{H}^{p'}_0(\mathrm{div},\Omega)]'$, we have the weak solution for the homogeneous problem (1.6.1). Here, as in Theorem 1.6.1, we use the analyticity of the semigroup generated by the operator $B_{p,\alpha}$ and the fact that $\|T(t)\| \leq Ce^{-\delta t}$ for some $C > 0$ and $\delta > 0$.

Theorem 1.6.3. Let $1 < p < \infty$ and $\alpha \in L^{t(p)}(\Gamma)$ where $t(p)$ defined in (1.3.1). Then, for all $\boldsymbol{u}_0 \in [\boldsymbol{H}^{p'}_0(\mathrm{div},\Omega)]'$, the problem (1.6.1) has a unique solution $\boldsymbol{u}(t)$ with the regularity

$$\boldsymbol{u} \in C([0,\infty), [\boldsymbol{H}^{p'}_0(\mathrm{div},\Omega)]') \cap C((0,\infty), \mathbf{D}(B_{p,\alpha})) \cap C^1((0,\infty), [\boldsymbol{H}^{p'}_0(\mathrm{div},\Omega)]')$$

and

$$\boldsymbol{u} \in C^k((0,\infty), \mathbf{D}(B^\ell_{p,\alpha})) \quad \forall\, k \in \mathbb{N},\ \forall\, \ell \in \mathbb{N}\backslash\{0\}.$$

Also there exist constants $C = C(\Omega, \alpha, p) > 0$, independent of $\delta > 0$ such that for all $t > 0$,

$$\|\boldsymbol{u}(t)\|_{[\boldsymbol{H}_0^{p'}(\mathrm{div},\Omega)]'} \leq Ce^{-\delta t}\|\boldsymbol{u}_0\|_{[\boldsymbol{H}_0^{p'}(\mathrm{div},\Omega)]'},$$

and

$$\left\|\frac{\partial \boldsymbol{u}(t)}{\partial t}\right\|_{[\boldsymbol{H}_0^{p'}(\mathrm{div},\Omega)]'} + \|\boldsymbol{u}\|_{\boldsymbol{W}^{1,p}(\Omega)} + \|\nabla\pi\|_{[\boldsymbol{H}_0^{p'}(\mathrm{div},\Omega)]'} \leq C\,\frac{e^{-\delta t}}{t}\|\boldsymbol{u}_0\|_{[\boldsymbol{H}_0^{p'}(\mathrm{div},\Omega)]'}.$$

In the same way as we deduced in Corollary 1.6.1, we can have the following regularity result from Theorem 1.6.3.

Corollary 1.6.2. Let $p \in (1,\infty)$ and $\alpha \in L^{t(p)}(\Gamma)$ where $t(p)$ defined in (1.3.1). Moreover, suppose $\boldsymbol{u}_0 \in [\boldsymbol{H}_0^{p'}(\mathrm{div},\Omega)]', 0 < T < \infty$ and $(\boldsymbol{u},\pi)$ be the unique solution of problem (1.6.1) given by Theorem 1.6.3. Then, for all $1 \leq q < 2$, we have

$$\boldsymbol{u} \in L^q(0,T; \boldsymbol{L}^p(\Omega)).$$

Proof. We know the interpolation inequality

$$\|\boldsymbol{u}(t)\|_{\boldsymbol{L}^p(\Omega)} \leq \|\boldsymbol{u}(t)\|^{1/2}_{\boldsymbol{W}^{1,p}(\Omega)}\|\boldsymbol{u}(t)\|^{1/2}_{\boldsymbol{W}^{-1,p}(\Omega)}.$$

Now using the estimates in Theorem 1.6.3 and the fact that $[\boldsymbol{H}_0^{p'}(\mathrm{div},\Omega)]' \hookrightarrow \boldsymbol{W}^{-1,p}(\Omega)$ in the above inequality, we get the result. ■

1.7 The Non-homogeneous Stokes Problem

Here we discuss the non-homogeneous Stokes problem:

$$\begin{cases} \dfrac{\partial \boldsymbol{u}}{\partial t} - \Delta\boldsymbol{u} + \nabla\pi = \boldsymbol{f}, \quad \mathrm{div}\ \boldsymbol{u} = 0 & \text{in } \Omega\times(0,T), \\ \boldsymbol{u}\cdot\boldsymbol{n} = 0, \quad 2[(\mathbb{D}\boldsymbol{u})\boldsymbol{n}]_{\boldsymbol{\tau}} + \alpha\boldsymbol{u}_{\boldsymbol{\tau}} = \boldsymbol{0} & \text{on } \Gamma\times(0,T), \\ \boldsymbol{u}(0) = \boldsymbol{u}_0 & \text{in } \Omega. \end{cases} \tag{1.7.1}$$

It is known that if $-\mathcal{A}$ generates a bounded analytic semigroup on a Banach space X, then we can construct a strong solution of

$$u' + \mathcal{A}u = f \ \text{ for a.e. } t \in (0,T), \quad u(0) = a \tag{1.7.2}$$

if f is Hölder continuous in time with values in X. But the analyticity of $e^{-t\mathcal{A}}$ is not sufficient to deduce the existence of solutions of (1.7.2) for general $f \in L^p(0,T;X)$ unless X is a Hilbert space. Therefore, we use the result on abstract Cauchy problem by Giga and Sohr [20, Theorem 2.3] which used the notion of ζ-convexity. For completeness, we recall the definition of ζ-convexity (see [11], also refer to [35]):

A Banach space X is said to be ζ-convex if there exists a symmetric biconvex function ζ on $X \times X$ such that $\zeta(0,0) > 0$ and

$$\zeta(x,y) \leq \|x+y\| \quad \text{if} \quad \|x\| \leq 1 \leq \|y\|.$$

The concept of ζ-convexity is stronger than that of reflexivity. For application purpose, it is important to recall [20] that X is ζ-convex iff for some $1 < s < \infty$, the truncated Hilbert transform

$$(H_\varepsilon f)(t) = \frac{1}{\pi}\int_{|\tau|>\varepsilon} \frac{f(t-\tau)}{\tau}\, d\tau, \quad f \in L^s(\mathbb{R}, X)$$

converges as $\varepsilon \to 0$ for almost all $t \in \mathbb{R}$ and there is a constant $C = C(s,x)$ independent of f such that

$$\|Hf\|_{L^s(\mathbb{R},X)} \leq C\|f\|_{L^s(\mathbb{R},X)},$$

where $(Hf)(t) = \lim_{\varepsilon\to 0}(H_\varepsilon f)(t)$.

The result in [20, Theorem 2.3] is useful in two senses: (i) it can be used even when $\mathcal{A}$ does not have a bounded inverse (though in our case, both A_p and B_p have bounded inverse) and (ii) the constant in the estimate is independent of time T, hence gives global in time results.

Here we introduce the notation for the space, for any $1 < p, q < \infty$,

$$D_{\mathcal{A}}^{\frac{1}{q},p} = \left\{ v \in X : \|v\|_{D_{\mathcal{A}}^{\frac{1}{q},p}} = \|v\|_X + \left(\int_0^\infty \|t^{1-\frac{1}{q}} \mathcal{A} e^{-t\mathcal{A}} v\|_X^p \frac{dt}{t} \right)^{1/p} < \infty \right\}$$

which actually agrees with the real interpolation space $(D(\mathcal{A}), X)_{1-1/q,p}$ when $e^{-t\mathcal{A}}$ is an analytic semigroup. First we deduce the strong solution of the Stokes system (1.7.1) and obtain $L^p - L^q$ estimates.

Proof of Theorem 1.1.3. Since $\boldsymbol{L}^p_{\sigma,\boldsymbol{\tau}}(\Omega)$ is ζ-convex [20, page 81] and $A_{p,\alpha}$ satisfies the estimate (1.5.1), all the assumptions of [20, Theorem 2.3] are fulfilled with $\mathcal{A} = A_{p,\alpha}$ and $X = \boldsymbol{L}^p_{\sigma,\boldsymbol{\tau}}(\Omega)$. As a result, the regularity of $\boldsymbol{u}$ and $\frac{\partial \boldsymbol{u}}{\partial t}$ follow. The regularity of π comes from the fact that

$$\nabla\pi = \boldsymbol{f} - \frac{\partial \boldsymbol{u}}{\partial t} + \Delta \boldsymbol{u}. \tag{1.7.3}$$

Also we get the estimate

$$\int_0^T \left\| \frac{\partial \boldsymbol{u}}{\partial t} \right\|^q_{\boldsymbol{L}^p(\Omega)} dt + \int_0^T \|A_{p,\alpha}\boldsymbol{u}\|^q_{\boldsymbol{L}^p(\Omega)} dt \leq C \left(\int_0^T \|\boldsymbol{f}(t)\|^q_{\boldsymbol{L}^p(\Omega)} dt + \|\boldsymbol{u}_0\|^q_{D_{A_{p,\alpha}}^{1-\frac{1}{q},q}} \right)$$

which yields (1.1.7) using the fact that $A_{p,\alpha}\boldsymbol{u} = -\Delta\boldsymbol{u} + \nabla\pi$. ■

In the same way, using $\mathcal{A} = B_{p,\alpha}$ and $X = [\boldsymbol{H}_0^{p'}(\mathrm{div}, \Omega)]'_{\sigma,\boldsymbol{\tau}}$ in [20, Theorem 2.3], since $[\boldsymbol{H}_0^{p'}(\mathrm{div}, \Omega)]'_{\sigma,\boldsymbol{\tau}}$ is ζ-convex [3, Proposition 2.16] and $B_{p,\alpha}$ satisfies the estimate on the pure imaginary power (1.5.2), we get the weak solution of the problem (1.7.1) with corresponding estimates as follows:

Theorem 1.7.1. Let $0 < T \leq \infty, 1 < p, q < \infty$ and $\alpha \in L^{t(p)}(\Gamma)$ where $t(p)$ be defined in (1.3.1). Then for every $\boldsymbol{f} \in L^q(0,T; [\boldsymbol{H}_0^{p'}(\mathrm{div}, \Omega)]'_{\sigma,\boldsymbol{\tau}})$ and $\boldsymbol{u}_0 \in D_{B_{p,\alpha}}^{1-\frac{1}{q},q}$ there exists a unique solution $(\boldsymbol{u}, \pi)$ of (1.7.1) satisfying the properties:

$$\boldsymbol{u} \in L^q(0, T_0; \boldsymbol{W}^{1,p}(\Omega)) \text{ for all } T_0 \leq T \text{ if } T < \infty \text{ and } T_0 < \infty \text{ if } T = \infty,$$

$$\pi \in L^q(0,T; L_0^p(\Omega)), \quad \frac{\partial \boldsymbol{u}}{\partial t} \in L^q(0, T, [\boldsymbol{H}_0^{p'}(\mathrm{div}, \Omega)]'_{\sigma,\boldsymbol{\tau}}),$$

$$\int_0^T \left\| \frac{\partial \boldsymbol{u}}{\partial t} \right\|^q_{[\boldsymbol{H}_0^{p'}(\mathrm{div},\Omega)]'} dt + \int_0^T \|\boldsymbol{u}\|^q_{\boldsymbol{W}^{1,p}(\Omega)} dt + \int_0^T \|\pi\|^q_{L^p(\Omega)/\mathbb{R}} dt$$
$$\leq C \left(\int_0^T \|\boldsymbol{f}\|^q_{[\boldsymbol{H}_0^{p'}(\mathrm{div},\Omega)]'} dt + \|\boldsymbol{u}_0\|^q_{D_{B_{p,\alpha}}^{1-\frac{1}{q},q}} \right).$$

1.8 Nonlinear Problem

In this section, we consider the initial value problem for the Navier–Stokes system with Navier boundary condition:

$$\begin{cases} \dfrac{\partial \boldsymbol{u}}{\partial t} - \Delta \boldsymbol{u} + (\boldsymbol{u} \cdot \nabla)\boldsymbol{u} + \nabla \pi = \boldsymbol{0}, \quad \mathrm{div}\, \boldsymbol{u} = 0 & \text{in } \Omega \times (0,T), \\ \boldsymbol{u} \cdot \boldsymbol{n} = 0, \quad 2[(\mathbb{D}\boldsymbol{u})\boldsymbol{n}]_{\boldsymbol{\tau}} + \alpha \boldsymbol{u}_{\boldsymbol{\tau}} = \boldsymbol{0} & \text{on } \Gamma \times (0,T), \\ \boldsymbol{u}(0) = \boldsymbol{u}_0 & \text{in } \Omega. \end{cases} \tag{1.8.1}$$

The semigroup theory formulated in Sects. 1.3–1.5 for the Stokes operator provides us the necessary properties with which we can obtain some existence, uniqueness, and regularity result for the nonlinear problem as well. Here we want to employ the results of [18] for the abstract semi linear parabolic equation of the form

$$u_t + \mathcal{A}u = Fu, \quad u(0) = a, \tag{1.8.2}$$

where Fu represents the nonlinear part and $\mathcal{A}$ is an elliptic operator. This abstract theory gives the existence of a local solution $u(t)$ for certain class of Fu. The solution can be extended globally also, provided norm of the initial data is sufficiently small. Moreover, this solution belongs to $L^q(0,T; L^p)$ with suitably chosen p, q. Since, $u \in L^q(0,T; L^p)$ is equivalent of saying $\|u(t)\|_{L^p(\Omega)} \in L^q(0,T)$, this gives the asymptotic behavior of $\|u(t)\|_{L^p(\Omega)}$ as $t \to 0$ and $t \to \infty$.

To apply [18, Theorem 1 and Theorem 2], we need to verify the hypothesis therein, which we state below for convenience:

For a closed subspace E^p of $L^p(\Omega)$, let $P : L^p(\Omega) \to E^p$ be a continuous projection for $p \in (1, \infty)$ such that the restriction of P on $C_c(\Omega)$, the space of continuous functions with compact support, is independent of p and $C_c(\Omega) \cap E^p$ be dense in E^p. Let $e^{-t\mathcal{A}}$ be a strongly continuous operator on E^p for all $p \in (1, \infty)$. Also, there exist constants $n, m \geq 1$ such that for a fixed $T \in (0, \infty)$, the estimate

$$\|e^{-t\mathcal{A}} f\|_{L^p(\Omega)} \leq M \|f\|_{L^s(\Omega)}/t^\sigma, \quad f \in E^s, \quad t \in (0, T) \qquad \textbf{(A)}$$

holds with $\sigma = (\frac{1}{s} - \frac{1}{p})\frac{n}{m}$ for $p \geq s > 1$ and constant M depending only on p, s, T.

Moreover, let Fu be written as

$$Fu = LGu,$$

where L is a closed, linear operator, densely defined from $L^p(\Omega)$ to E^q for some $q > 1$ such that for some $\gamma, 0 \leq \gamma < m$, the estimate

$$\|e^{-t\mathcal{A}} Lf\|_{L^p(\Omega)} \leq N_1 \|f\|_{L^p(\Omega)}/t^{\gamma/m}, \quad f \in E^p, \quad t \in (0, T) \qquad \textbf{(N1)}$$

holds with N_1 depending only on T and p, for all $p \in (1, \infty)$ and G is a nonlinear mapping from E^p to $L^h(\Omega)$ such that for some $\beta > 0$, the estimate

$$\|Gv - Gw\|_{L^h(\Omega)} \leq N_2 \|v - w\|_{L^p(\Omega)} \left(\|v\|^\beta_{L^p(\Omega)} + \|w\|^\beta_{L^p(\Omega)} \right), \quad G(0) = 0 \quad \textbf{(N2)}$$

holds with $1 \leq h = p/(1+\beta)$ and N_2 depending only on p, for all $p \in (1, \infty)$.

With these assumptions, the next Theorem follows directly from [18, Theorem 1 and Theorem 2].

Theorem 1.8.1. For $\boldsymbol{u}_0 \in \boldsymbol{L}^r_{\sigma,\boldsymbol{\tau}}(\Omega)$ and $\alpha \in W^{1-\frac{1}{r},r}(\Gamma), r \geq 3, \alpha \geq 0$, there exist $T_0 > 0$ and a unique solution $\boldsymbol{u}(t)$ of (1.8.1) on $[0, T_0)$ such that

$$\boldsymbol{u} \in C([0, T_0); \boldsymbol{L}^r_{\sigma,\boldsymbol{\tau}}(\Omega)) \cap L^q(0, T_0; \boldsymbol{L}^p_{\sigma,\boldsymbol{\tau}}(\Omega)) \tag{1.8.3}$$

$$t^{1/q}\boldsymbol{u} \in C([0, T_0); \boldsymbol{L}^p_{\sigma,\boldsymbol{\tau}}(\Omega)) \quad \text{and} \quad t^{1/q}\|\boldsymbol{u}\|_{\boldsymbol{L}^p(\Omega)} \to 0 \text{ as } t \to 0$$

with $\frac{2}{q} = \frac{3}{r} - \frac{3}{p}, p, q > r$. Moreover, there exists a constant $\delta > 0$ such that if $\|\boldsymbol{u}_0\|_{\boldsymbol{L}^r(\Omega)} < \delta$, then T_0 can be taken as infinity for $r = 3$.

Let $(0, T_\star)$ be the maximal interval such that $\boldsymbol{u}$ solves (1.8.1) in $C((0, T_\star); \boldsymbol{L}^r_{\sigma,\boldsymbol{\tau}}(\Omega)), r > 3$. Then

$$\|u(t)\|_{\boldsymbol{L}^r(\Omega)} \geq C(T_\star - t)^{(3-r)/2r}$$

where C is independent of $T_\star$ and t.

Proof. As our Stokes operator has all the same properties and estimates satisfied by the Stokes operator with Dirichlet boundary condition, we are exactly in the same setup as in [18] and hence the proof goes similar to that. However, we briefly review it for completeness.

Let E^p be $\boldsymbol{L}^p_{\sigma,\boldsymbol{\tau}}(\Omega)$ and $P : \boldsymbol{L}^p(\Omega) \to \boldsymbol{L}^p_{\sigma,\boldsymbol{\tau}}(\Omega)$ be the Helmholtz projection, defined in (1.3.4). It is trivial to see that P is independent of $p \in (1,\infty)$ on $C_c(\Omega)$ and $C_c(\Omega) \cap \boldsymbol{L}^p_{\sigma,\boldsymbol{\tau}}(\Omega)$ is dense in $\boldsymbol{L}^p_{\sigma,\boldsymbol{\tau}}(\Omega)$. The Stokes operator $A_{p,\alpha}$ on $\boldsymbol{L}^p_{\sigma,\boldsymbol{\tau}}(\Omega)$ is defined in (1.3.6)–(1.3.7) with dense domain and $-A_{p,\alpha}$ generates bounded analytic semigroup on $\boldsymbol{L}^p_{\sigma,\boldsymbol{\tau}}(\Omega)$ for all $p \in (1,\infty)$ also (cf. Theorem 1.3.6). Applying P on both sides of the Navier–Stokes system (1.8.1) gives

$$\boldsymbol{u}_t + A_{p,\alpha}\boldsymbol{u} = -P(\boldsymbol{u}\cdot\nabla)\boldsymbol{u}, \qquad \boldsymbol{u}(0) = \boldsymbol{u}_0$$

which is obviously in the form (1.8.2) with $Fu = -P(\boldsymbol{u}\cdot\nabla)\boldsymbol{u}$.

We now need to verify the assumptions (**A**), (**N1**), and (**N2**). Since $-A_{p,\alpha}$ generates a bounded analytic semigroup with bounded inverse, we have (cf. [33, Chapter 2, Theorem 6.13])

$$\forall \boldsymbol{f} \in \boldsymbol{L}^s_{\sigma,\boldsymbol{\tau}}(\Omega), \qquad \|A^\sigma_{s,\alpha} e^{-tA_{s,\alpha}}\boldsymbol{f}\|_{\boldsymbol{L}^s(\Omega)} \leq M\|\boldsymbol{f}\|_{\boldsymbol{L}^s(\Omega)}/t^\sigma.$$

As $\boldsymbol{D}(A^\sigma_{s,\alpha})$ is continuously embedded in $\boldsymbol{W}^{2\sigma,s}(\Omega)$, this together with the Sobolev embedding theorem yields (**A**) with $m = 2, n = 3$.

Next we want to write the nonlinear term Fu. Since div $\boldsymbol{u} = 0$, we have $(\boldsymbol{u}\cdot\nabla)\boldsymbol{u}_i = \sum_{j=1}^3 \partial_j(u_j u_i)$. If we define $g : \mathbb{R}^3 \to \mathbb{R}^9$ by

$$(g(x))_{ij} = -x_i x_j$$

and $G : \boldsymbol{L}^p_{\sigma,\boldsymbol{\tau}}(\Omega) \to (L^p(\Omega))^9$ by

$$G\boldsymbol{u}(x) = g(\boldsymbol{u}(x))$$

and $L : (L^p(\Omega))^9 \to \boldsymbol{L}^q_{\sigma,\boldsymbol{\tau}}(\Omega)$ by

$$Lg_{ij} = \sum_{j=1}^3 P\nabla_j g_{ij}$$

which is a linear operator, it implies $Fu = LG\boldsymbol{u}$. Also it is easy to see from Hölder inequality that

$$|g(y) - g(z)| \leq N_2|y - z|(|y| + |z|), \quad g(0) = 0$$

which gives in turn (**N2**) with $\beta = 1$.

Finally

$$\|e^{-tA_{p,\alpha}}L\boldsymbol{f}\|_{\boldsymbol{L}^p(\Omega)} = \|A^{1/2}_{p,\alpha}e^{-tA_p}A^{-1/2}_{p,\alpha}L\boldsymbol{f}\|_{\boldsymbol{L}^p(\Omega)} \leq \frac{C}{t^{1/2}}\|A^{-1/2}_{p,\alpha}L\boldsymbol{f}\|_{\boldsymbol{L}^p(\Omega)}$$

and since $A_{p,\alpha}^{-1/2}L$ is bounded in $\boldsymbol{L}^p(\Omega)$ (cf. [19, Lemma 2.1]), the assumption (**N1**) is verified for $\gamma = 1$.

Thus applying the abstract results by Giga, we get the existence of a unique solution $\boldsymbol{u}$ of the form

$$\boldsymbol{u}(t) = e^{-tA_{p,\alpha}}\boldsymbol{u}_0 - \int_0^t e^{-(t-\tau)A_{p,\alpha}} P(\boldsymbol{u}(\tau)) \cdot \nabla)\boldsymbol{u}(\tau)\, \mathrm{d}\tau.$$

This completes the proof. ■

Next we show that the solution of (1.8.1) given in the above Theorem in the integral form is actually regular enough and satisfies (1.8.1).

Theorem 1.8.2. Let $\boldsymbol{u}_0 \in \boldsymbol{L}^r_{\sigma,\boldsymbol{\tau}}(\Omega), r \geq 3$ and $\boldsymbol{u}(t)$ be the unique solution of (1.8.1) given by Theorem 1.8.1. Then

$$\boldsymbol{u} \in C((0,T_*], \mathbf{D}(A_{p,\alpha})) \cap C^1((0,T_*]; \boldsymbol{L}^r_{\sigma,\boldsymbol{\tau}}(\Omega))$$

Proof. As in the previous Theorem, the proof follows exactly the same way as in the case of the Dirichlet boundary condition in [19]. Since the Stokes operator with Navier boundary condition, defined in (1.3.6)–(1.3.7), have all the same properties as for the Stokes operator with Dirichlet boundary condition, [19, Theorem 2.5] gives (with $f = 0$) that $\boldsymbol{u} \in C((0,T_*], \mathbf{D}(A_{p,\alpha}))$. And $\boldsymbol{u} \in C^1((0,T_*]; \boldsymbol{L}^r_{\sigma,\boldsymbol{\tau}}(\Omega))$ follows from [14, Lemma 2.14] (with $f = 0$). ■

Next we show that regular solutions satisfy energy inequality provided the initial condition is in $\boldsymbol{L}^2_{\sigma,\boldsymbol{\tau}}(\Omega)$.

Proposition 1.8.1. Let $\boldsymbol{u}$ be the regular solution of (1.8.1) on $(0,T_0)(T_0 < \infty)$ satisfying (1.8.3). Suppose $\boldsymbol{u}_0 \in \boldsymbol{L}^2_{\sigma,\boldsymbol{\tau}}(\Omega)$. Then

$$\boldsymbol{u} \in L^\infty(0,T_0; \boldsymbol{L}^2_{\sigma,\boldsymbol{\tau}}(\Omega)) \cap L^2(0,T_0; \boldsymbol{H}^1(\Omega))$$

and satisfies the energy equality

$$\frac{1}{2}\int_\Omega |\boldsymbol{u}(t)|^2 + 2\int_0^t\int_\Omega |\mathbb{D}\boldsymbol{u}|^2 + \int_0^t\int_\Gamma \alpha|\boldsymbol{u}_{\boldsymbol{\tau}}|^2 = \frac{1}{2}\int_\Omega |\boldsymbol{u}_0|^2.$$

Proof. pt The proof follows the same reasoning as in [18, Proposition 1, Section 5]. ■

1.9 The Limit as $\alpha \to \infty$

Let us denote now u_α the solutions of the unsteady Stokes or Navier–Stokes equation with NBC for a given slip coefficient $\alpha \geq 0$ and a fixed initial data u_0. A very formal argument suggests that when $\alpha \to \infty$, we may expect that $u_\alpha \to u_\infty$

in some sense, where u_∞ is the solution of the same equation, with the same initial data, but with Dirichlet boundary condition. The existence of solutions u_∞ for such a problem, for a suitable set of initial data has been proved for example in [38, Theorem 1.1, Chapter III] for the Stokes equation and [38, Theorem 3.1, Chapter III] for the Navier–Stokes equation.

This question has already been considered in [24] for Ω a two dimensional domain and $\frac{1}{\alpha} \in L^\infty(\Gamma)$. The author proves in Theorem 9.2 that when $\|\frac{1}{\alpha}\|_{L^\infty(\Gamma)} \to 0$ and $\boldsymbol{u}_0 \in \boldsymbol{H}^3(\Omega) \cap \boldsymbol{H}_0^1(\Omega) \cap \boldsymbol{L}^2_{\sigma,\boldsymbol{\tau}}(\Omega)$, the solution of problem (1.1.1)–(1.1.5) converges to the solution of the Navier–Stokes problem with Dirichlet boundary condition in $L^\infty(0,T;L^2(\Omega)) \cap L^2(0,T;\dot{H}^1(\Omega)) \cap L^2(0,T;L^2(\Gamma))$ [24, Theorem 9.2].

Our results on this problem are based on the uniform estimates of the solutions with NBC with respect to the parameter α proved in the previous Sections. Then, we may only consider the case where the function α is a non-negative constant. But, on the other hand, our convergence result in the Hilbert case, with the same rate of convergence as in [24] only needs the initial data to satisfy $\boldsymbol{u}_0 \in \boldsymbol{L}^2_{\sigma,\boldsymbol{\tau}}(\Omega)$ and we also obtain convergence results for the non-Hilbert cases.

In the first two results of this Section, we prove that when α is a constant and the initial data is such that $\boldsymbol{u}_0 \in \mathbf{D}(A_{p,\alpha})$ for all α sufficiently large, the solutions of the Stokes equation with Navier boundary conditions (1.6.1) converge in the energy space to the solutions of the Stokes equation with Dirichlet boundary condition obtained in [17]. Moreover, we also obtain estimates on the rates of convergence.

Theorem 1.9.1. Let $\boldsymbol{u}_0 \in \boldsymbol{L}^2_{\sigma,\boldsymbol{\tau}}(\Omega)$, α be a constant, and $T_\alpha(t) : \boldsymbol{L}^p_{\sigma,\boldsymbol{\tau}}(\Omega) \to \boldsymbol{L}^p_{\sigma,\boldsymbol{\tau}}(\Omega)$ the semigroup generated by the Stokes operator $A_{p,\alpha}$, defined in (1.3.2)–(1.3.3). Then for any $T < \infty$,

$$T_\alpha(t)\boldsymbol{u}_0 \to T_\infty(t)\boldsymbol{u}_0 \quad \text{in} \quad L^2(0,T;\boldsymbol{H}^1(\Omega)) \quad \text{as} \quad \alpha \to \infty, \tag{1.9.1}$$

where $T_\infty(t)$ is the semigroup generated by the Stokes operator with Dirichlet boundary condition [17]. Also we have

$$\int_0^T \int_\Gamma |T_\alpha(t)\boldsymbol{u}_0 - T_\infty(t)\boldsymbol{u}_0|^2 \leq \frac{C}{\alpha}. \tag{1.9.2}$$

Moreover, if $\boldsymbol{u}_0 \in \boldsymbol{H}_0^1(\Omega)$ with div $\boldsymbol{u}_0 = 0$ in Ω, we further obtain

$$\int_0^T \int_\Omega |\mathbb{D}(T_\alpha(t)\boldsymbol{u}_0 - T_\infty(t)\boldsymbol{u}_0)|^2 + \int_0^T \int_\Gamma |T_\alpha(t)\boldsymbol{u}_0 - T_\infty(t)\boldsymbol{u}_0|^2 \leq \frac{C}{\alpha}. \tag{1.9.3}$$

Proof. **(i)** Let us denote $\boldsymbol{u}_\alpha := T_\alpha(t)\boldsymbol{u}_0$. Then $\boldsymbol{u}_\alpha$ is the solution of the Stokes problem with Navier boundary condition (1.6.1), given by Theorem 1.6.1 where

π_α is the associated pressure. So $\boldsymbol{u}_\alpha$ satisfies the following energy equality:

$$\frac{1}{2}\int_\Omega |\boldsymbol{u}_\alpha(T)|^2 + 2\int_0^T\int_\Omega |\mathbb{D}\boldsymbol{u}_\alpha|^2 + \alpha\int_0^T\int_\Gamma |\boldsymbol{u}_{\alpha\boldsymbol{\tau}}|^2 = \frac{1}{2}\int_\Omega |\boldsymbol{u}_0|^2$$

which shows that as $\alpha \to \infty$,

$$\mathbb{D}\boldsymbol{u}_\alpha \text{ is bounded in } L^2(0,T;\boldsymbol{L}^2(\Omega))$$

and

$$\boldsymbol{u}_{\alpha\boldsymbol{\tau}} \text{ is bounded in } L^2(0,T;\boldsymbol{L}^2(\Gamma)).$$

Therefore, $\boldsymbol{u}_\alpha$ is bounded in $L^2\ (0,T;\boldsymbol{H}^1(\Omega))$. Hence π_α is also bounded in $L^2(0,T;\boldsymbol{L}^2(\Omega))$ since $\|B_2\boldsymbol{v}\|_{[\boldsymbol{H}_0^2(\mathrm{div},\Omega)]'} \simeq \|\boldsymbol{v}\|_{\boldsymbol{H}^1(\Omega)}$ for all $\boldsymbol{v} \in \mathbf{D}(B_2)$. This deduces that $\frac{\partial \boldsymbol{u}_\alpha}{\partial t}$ is as well bounded in $L^2(0,T;\boldsymbol{H}^{-1}(\Omega))$. So there exists $(\boldsymbol{u}_\infty, \pi_\infty) \in L^2(0,T;\boldsymbol{H}^1(\Omega)) \times L^2(0,T;L^2(\Omega))$ with $\frac{\partial \boldsymbol{u}_\infty}{\partial t} \in L^2(0,T;\boldsymbol{H}^{-1}(\Omega))$ such that up to a subsequence,

$$(\boldsymbol{u}_\alpha, \pi_\alpha) \rightharpoonup (\boldsymbol{u}_\infty, \pi_\infty) \text{ weakly in } L^2(0,T;\boldsymbol{H}^1(\Omega)) \times L^2(0,T;L^2(\Omega)) \quad \text{as } \alpha \to \infty$$

and

$$\frac{\partial \boldsymbol{u}_\alpha}{\partial t} \rightharpoonup \frac{\partial \boldsymbol{u}_\infty}{\partial t} \text{ weakly in } L^2(0,T;\boldsymbol{H}^{-1}(\Omega)).$$

Also, by Aubin–Lions Lemma, we have

$$\boldsymbol{u}_\alpha \to \boldsymbol{u}_\infty \quad \text{in} \quad L^2(0,T;\boldsymbol{L}^{6-\varepsilon}(\Omega)) \text{ for any } \varepsilon > 0.$$

Next we claim that $(\boldsymbol{u}_\infty, \pi_\infty)$ satisfies the following Dirichlet problem:

$$\begin{cases} \dfrac{\partial \boldsymbol{u}_\infty}{\partial t} - \Delta \boldsymbol{u}_\infty + \nabla \pi_\infty = \boldsymbol{0}, \quad \operatorname{div} \boldsymbol{u}_\infty = 0 & \text{in } \Omega \times (0,T) \\ \boldsymbol{u}_\infty = \boldsymbol{0} & \text{on } \Gamma \times (0,T) \\ \boldsymbol{u}_\infty(0) = \boldsymbol{u}_0 & \text{in } \Omega. \end{cases} \tag{1.9.4}$$

Indeed, for any $\boldsymbol{v} \in C^1([0,T];\boldsymbol{H}^1_{0,\sigma}(\Omega))$ where we denote $\boldsymbol{H}^1_{0,\sigma}(\Omega) = \boldsymbol{H}^1_0(\Omega) \cap \boldsymbol{L}^2_{\sigma,\boldsymbol{\tau}}(\Omega)$, the weak formulation satisfied by $(\boldsymbol{u}_\alpha, \pi_\alpha)$ is

$$\int_0^T \left\langle \frac{\partial \boldsymbol{u}_\alpha}{\partial t}, \boldsymbol{v} \right\rangle_{\boldsymbol{H}^{-1}(\Omega)\times \boldsymbol{H}^1_0(\Omega)} + 2\int_0^T\int_\Omega \mathbb{D}\boldsymbol{u}_\alpha : \mathbb{D}\boldsymbol{v} = 0. \tag{1.9.5}$$

Then passing limit as $\alpha \to \infty$, we obtain

$$\int_0^T \left\langle \frac{\partial \boldsymbol{u}_\infty}{\partial t}, \boldsymbol{v} \right\rangle_{\boldsymbol{H}^{-1}(\Omega)\times \boldsymbol{H}^1_0(\Omega)} + 2\int_0^T\int_\Omega \mathbb{D}\boldsymbol{u}_\infty : \mathbb{D}\boldsymbol{v} = 0. \tag{1.9.6}$$

Hence,

$$\left\langle \frac{\partial \boldsymbol{u}_\infty}{\partial t}, \boldsymbol{v} \right\rangle_{\boldsymbol{H}^{-1}(\Omega)\times \boldsymbol{H}_0^1(\Omega)} + 2\int_\Omega \mathbb{D}\boldsymbol{u}_\infty : \mathbb{D}\boldsymbol{v} = 0$$

for any $\boldsymbol{v} \in \boldsymbol{H}^1_{0,\sigma}(\Omega)$ and a.e. $0 \le t \le T$. Also we have, $\boldsymbol{u}_\infty \in C([0,T]; \boldsymbol{L}^2(\Omega))$.

In order to show that $\boldsymbol{u}_\infty(0) = \boldsymbol{u}_0$, we can write from (1.9.5), for any $\boldsymbol{v} \in C^1([0,T]; \boldsymbol{H}^1_{0,\sigma}(\Omega))$ with $\boldsymbol{v}(T) = 0$,

$$-\int_0^T\int_\Omega \boldsymbol{u}_\alpha \cdot \frac{\partial \boldsymbol{v}}{\partial t} + 2\int_0^T\int_\Omega \mathbb{D}\boldsymbol{u}_\alpha : \mathbb{D}\boldsymbol{v} = \int_\Omega \boldsymbol{u}_0 \cdot \boldsymbol{v}(0)$$

and similarly from (1.9.6),

$$-\int_0^T\int_\Omega \boldsymbol{u}_\infty \cdot \frac{\partial \boldsymbol{v}}{\partial t} + 2\int_0^T\int_\Omega \mathbb{D}\boldsymbol{u}_\infty : \mathbb{D}\boldsymbol{v} = \int_\Omega \boldsymbol{u}_\infty(0) \cdot \boldsymbol{v}(0).$$

(i) As $\boldsymbol{v}(0)$ is arbitrary, we thus conclude that $\boldsymbol{u}_\infty(0) = \boldsymbol{u}_0$.

(ii) It remains to prove the strong convergence of $(\boldsymbol{u}_\alpha, \pi_\alpha)$ to $(\boldsymbol{u}_\infty, \pi_\infty)$. Note that $(\boldsymbol{v}_\alpha, p_\alpha) := (\boldsymbol{u}_\alpha - \boldsymbol{u}_\infty, \pi_\alpha - \pi_\infty)$ satisfies the following problem:

$$\begin{cases} \dfrac{\partial \boldsymbol{v}_\alpha}{\partial t} - \Delta \boldsymbol{v}_\alpha + \nabla p_\alpha = \boldsymbol{0}, \quad \operatorname{div} \boldsymbol{v}_\alpha = 0 & \text{in } \Omega \times (0,T) \\ \boldsymbol{v}_\alpha \cdot \boldsymbol{n} = 0, \quad 2[(\mathbb{D}\boldsymbol{v}_\alpha)\boldsymbol{n}]_{\boldsymbol{\tau}} + \alpha \boldsymbol{v}_{\alpha\boldsymbol{\tau}} = -2[(\mathbb{D}\boldsymbol{u}_\infty)\boldsymbol{n}]_{\boldsymbol{\tau}} & \text{on } \Gamma \times (0,T) \\ \boldsymbol{v}_\alpha(0) = \boldsymbol{0} & \text{in } \Omega. \end{cases}$$

Multiplying the above system by $\boldsymbol{v}_\alpha$, we obtain the following energy estimate:

$$\frac{1}{2}\int_\Omega |\boldsymbol{v}_\alpha(t)|^2 + 2\int_0^T\int_\Omega |\mathbb{D}\boldsymbol{v}_\alpha|^2 + \int_0^T\int_\Gamma \alpha|\boldsymbol{v}_{\alpha\boldsymbol{\tau}}|^2 = 2\int_0^T \langle [(\mathbb{D}\boldsymbol{u}_\infty)\boldsymbol{n}]_{\boldsymbol{\tau}}, \boldsymbol{v}_{\alpha\boldsymbol{\tau}}\rangle_\Gamma \quad (1.9.7)$$

which shows that $\boldsymbol{v}_\alpha \to \boldsymbol{0}$ in $L^2(0,T; \boldsymbol{L}^2(\Gamma))$ as $\alpha \to \infty$ and thus (1.9.2) follows. Also since $\boldsymbol{v}_{\alpha\boldsymbol{\tau}} \rightharpoonup \boldsymbol{0}$ weakly in $L^2(0,T; \boldsymbol{H}^{\frac{1}{2}}(\Gamma))$ and $[(\mathbb{D}\boldsymbol{u}_\infty)\boldsymbol{n}]_{\boldsymbol{\tau}} \in L^2(0,T; \boldsymbol{H}^{-\frac{1}{2}}(\Gamma))$, the right hand side in the above relation goes to 0 as $\alpha \to \infty$. So $\mathbb{D}\boldsymbol{v}_\alpha \to 0$ in $L^2(0,T; \boldsymbol{L}^2(\Omega))$. This proves (1.9.1).

(iii) If we assume $\boldsymbol{u}_0 \in \boldsymbol{H}^1_0(\Omega) \cap \boldsymbol{L}^2_{\sigma,\boldsymbol{\tau}}(\Omega)$, then $\boldsymbol{u}_\infty \in L^2(0,T; \boldsymbol{H}^2(\Omega))$ (cf. [38, Proposition 1.2, Chapter III]) and thus the energy equality (1.9.7) can be estimated as

$$\int_0^T\int_\Omega |\mathbb{D}\boldsymbol{v}_\alpha|^2 \le \int_0^T\int_\Gamma [(\mathbb{D}\boldsymbol{u}_\infty)\boldsymbol{n}]_{\boldsymbol{\tau}} \cdot \boldsymbol{v}_{\alpha\boldsymbol{\tau}} \le \|\boldsymbol{u}_\infty\|_{L^2(0,T;\boldsymbol{H}^2(\Omega))} \|\boldsymbol{v}_\alpha\|_{L^2(0,T;\boldsymbol{L}^2(\Gamma))}.$$

This along with the estimate (1.9.2) gives (1.9.3). ∎

We prove now the convergence result as $\alpha \to \infty$ of the solutions to the Navier–Stokes equation.

Proof of Theorem 1.1.4. We proceed as in the linear case.

(i) From Proposition 1.8.1 we can see that as $\alpha \to \infty$,

$$\mathbb{D}\boldsymbol{u}_\alpha \text{ is bounded in } L^2(0,T;\boldsymbol{L}^2(\Omega))$$

and

$$\boldsymbol{u}_{\alpha\boldsymbol{\tau}} \text{ is bounded in } L^2(0,T;\boldsymbol{L}^2(\Gamma)).$$

Therefore, $\boldsymbol{u}_\alpha$ is bounded in $L^2(0,T;\boldsymbol{H}^1(\Omega))$. And since $\|B_{2,\alpha}\boldsymbol{v}\|_{[\boldsymbol{H}_0^2(\mathrm{div},\Omega)]'} \simeq \|\boldsymbol{v}\|_{\boldsymbol{H}^1(\Omega)}$ for any $\boldsymbol{v} \in \mathbf{D}(B_{2,\alpha})$, π_α is as well bounded in $L^2(0,T;\boldsymbol{L}^2(\Omega))$. This implies $\frac{\partial \boldsymbol{u}_\alpha}{\partial t}$ is also bounded in $L^2(0,T;\boldsymbol{H}^{-1}(\Omega))$. Hence, there exists $(\boldsymbol{u}_\infty, \pi_\infty) \in L^2(0,T;\boldsymbol{H}^1(\Omega)) \times L^2(0,T;L^2(\Omega))$ with $\frac{\partial \boldsymbol{u}_\infty}{\partial t} \in L^2(0,T;\boldsymbol{H}^{-1}(\Omega))$ such that up to a subsequence,

$$(\boldsymbol{u}_\alpha, \pi_\alpha) \rightharpoonup (\boldsymbol{u}_\infty, \pi_\infty) \text{ weakly in } L^2(0,T;\boldsymbol{H}^1(\Omega)) \times L^2(0,T;L^2(\Omega)) \quad \text{as } \alpha \to \infty$$

and

$$\frac{\partial \boldsymbol{u}_\alpha}{\partial t} \rightharpoonup \frac{\partial \boldsymbol{u}_\infty}{\partial t} \quad \text{weakly in} \quad L^2(0,T;\boldsymbol{H}^{-1}(\Omega)).$$

Also, by Aubin–Lions Lemma,

$$\boldsymbol{u}_\alpha \to \boldsymbol{u}_\infty \quad \text{in} \quad L^2(0,T;\boldsymbol{L}^{6-\varepsilon}(\Omega)) \quad \text{for any} \quad \varepsilon > 0.$$

Next we show that $(\boldsymbol{u}_\infty, \pi_\infty)$ satisfies the Dirichlet problem (1.1.8). Indeed, for any $\boldsymbol{v} \in C^1([0,T];\boldsymbol{H}^1_{0,\sigma}(\Omega))$, the weak formulation of the problem (1.8.1) is,

$$\int_0^T \left\langle \frac{\partial \boldsymbol{u}_\alpha}{\partial t}, \boldsymbol{v} \right\rangle_{\boldsymbol{H}^{-1}(\Omega)\times \boldsymbol{H}_0^1(\Omega)} + 2\int_0^T\int_\Omega \mathbb{D}\boldsymbol{u}_\alpha : \mathbb{D}\boldsymbol{v} + \int_0^T\int_\Omega (\boldsymbol{u}_\alpha \cdot \nabla)\boldsymbol{u}_\alpha \cdot \boldsymbol{v} = 0. \quad (1.9.8)$$

Then passing limit as $\alpha \to \infty$, we obtain

$$\int_0^T \left\langle \frac{\partial \boldsymbol{u}_\infty}{\partial t}, \boldsymbol{v} \right\rangle_{\boldsymbol{H}^{-1}(\Omega)\times \boldsymbol{H}_0^1(\Omega)} + 2\int_0^T\int_\Omega \mathbb{D}\boldsymbol{u}_\infty : \mathbb{D}\boldsymbol{v} + \int_0^T\int_\Omega (\boldsymbol{u}_\infty \cdot \nabla)\boldsymbol{u}_\infty \cdot \boldsymbol{v} = 0. \tag{1.9.9}$$

To pass to the limit in the nonlinear term, we used the standard relation

$$\int_\Omega (\boldsymbol{u}_\alpha \cdot \nabla)\boldsymbol{u}_\alpha \cdot \boldsymbol{v} = -\int_\Omega (\boldsymbol{u}_\alpha \cdot \nabla)\boldsymbol{v} \cdot \boldsymbol{u}_\alpha.$$

Also we have, $\boldsymbol{u}_\infty \in C([0,T];\boldsymbol{L}^2(\Omega))$.

In order to prove $\boldsymbol{u}_\infty(0) = \boldsymbol{u}_0$, we write from (1.9.8), for any $\boldsymbol{v} \in C^1([0,T]; \boldsymbol{H}^1_{0,\sigma}(\Omega))$ with $\boldsymbol{v}(T) = 0$,

$$-\int_0^T\int_\Omega \boldsymbol{u}_\alpha \cdot \frac{\partial \boldsymbol{v}}{\partial t} + 2\int_0^T\int_\Omega \mathbb{D}\boldsymbol{u}_\alpha : \mathbb{D}\boldsymbol{v} + \int_0^T\int_\Omega (\boldsymbol{u}_\alpha \cdot \nabla)\boldsymbol{u}_\alpha \cdot \boldsymbol{v} = \int_\Omega \boldsymbol{u}_0 \cdot \boldsymbol{v}(0)$$

and similarly from (1.9.9),

$$-\int_0^T\int_\Omega \boldsymbol{u}_\infty \cdot \frac{\partial \boldsymbol{v}}{\partial t} + 2\int_0^T\int_\Omega \mathbb{D}\boldsymbol{u}_\infty : \mathbb{D}\boldsymbol{v} + \int_0^T\int_\Omega (\boldsymbol{u}_\infty \cdot \nabla)\boldsymbol{u}_\infty \cdot \boldsymbol{v} = \int_\Omega \boldsymbol{u}_\infty(0) \cdot \boldsymbol{v}(0).$$

As $\boldsymbol{v}(0)$ is arbitrary, we thus conclude $\boldsymbol{u}_\infty(0) = \boldsymbol{u}_0$.

(ii) Finally to show the strong convergence of $(\boldsymbol{u}_\alpha, \pi_\alpha)$ to $(\boldsymbol{u}_\infty, \pi_\infty)$, setting $\boldsymbol{v}_\alpha = \boldsymbol{u}_\alpha - \boldsymbol{u}_\infty$ and $p_\alpha = \pi_\alpha - \pi_\infty$, it solves the following problem

$$\begin{cases} \dfrac{\partial \boldsymbol{v}_\alpha}{\partial t} - \Delta \boldsymbol{v}_\alpha + (\boldsymbol{u}_\alpha \cdot \nabla)\boldsymbol{u}_\alpha - (\boldsymbol{u}_\infty \cdot \nabla)\boldsymbol{u}_\infty + \nabla p_\alpha = \boldsymbol{0}, \quad \operatorname{div} \boldsymbol{v}_\alpha = 0 & \text{in } \Omega \times (0,T) \\ \boldsymbol{v}_\alpha \cdot \boldsymbol{n} = 0, \quad 2[(\mathbb{D}\boldsymbol{v}_\alpha)\boldsymbol{n}]_{\boldsymbol{\tau}} + \alpha \boldsymbol{v}_{\alpha\boldsymbol{\tau}} = -2[(\mathbb{D}\boldsymbol{u}_\infty)\boldsymbol{n}]_{\boldsymbol{\tau}} & \text{on } \Gamma \times (0,T) \\ \boldsymbol{v}_\alpha(0) = 0 & \text{in } \Omega. \end{cases}$$

Multiplying the above system by $\boldsymbol{v}_\alpha$ and integrating by parts over $\Omega \times (0,T)$, we deduce

$$\begin{aligned} &\frac{1}{2}\|\boldsymbol{v}_\alpha(T)\|^2_{\boldsymbol{L}^2(\Omega)} + 2\int_0^T \|\mathbb{D}\boldsymbol{v}_\alpha\|^2_{\mathbb{L}^2(\Omega)} + \alpha\int_0^T \|\boldsymbol{v}_{\alpha\boldsymbol{\tau}}\|^2_{\boldsymbol{L}^2(\Gamma)} \\ &= -2\int_0^T \langle[(\mathbb{D}\boldsymbol{v}_\infty)\boldsymbol{n}]_{\boldsymbol{\tau}}, \boldsymbol{v}_{\alpha\boldsymbol{\tau}}\rangle_\Gamma - \int_0^T \langle(\boldsymbol{u}_\alpha \cdot \nabla)\boldsymbol{u}_\alpha - (\boldsymbol{u}_\infty \cdot \nabla)\boldsymbol{u}_\infty, \boldsymbol{v}_\alpha\rangle_\Omega. \end{aligned} \tag{1.9.10}$$

This shows that $\boldsymbol{v}_\alpha \to \boldsymbol{0}$ in $L^2(0,T;\boldsymbol{L}^2(\Gamma))$ proving (1.1.9). As $\boldsymbol{v}_\alpha \to \boldsymbol{0}$ in $L^2(0,T;\boldsymbol{L}^4(\Omega))$,

$$\int_0^T \langle(\boldsymbol{u}_\alpha \cdot \nabla)\boldsymbol{u}_\alpha - (\boldsymbol{u}_\infty \cdot \nabla)\boldsymbol{u}_\infty, \boldsymbol{v}_\alpha\rangle_\Omega = -\int_0^T\int_\Omega (\boldsymbol{v}_\alpha \cdot \nabla)\boldsymbol{u}_\infty \cdot \boldsymbol{v}_\alpha \le \|\boldsymbol{v}_\alpha\|^2_{\boldsymbol{L}^4(\Omega)}\|\nabla \boldsymbol{u}_\infty\|_{\boldsymbol{L}^2(\Omega)} \to 0.$$

Also since $\boldsymbol{v}_\alpha \rightharpoonup \boldsymbol{0}$ weakly in $L^2(0,T;\boldsymbol{H}^{\frac{1}{2}}(\Gamma))$ and $[(\mathbb{D}\boldsymbol{u}_\infty)\boldsymbol{n}]_{\boldsymbol{\tau}} \in L^2(0,T;\boldsymbol{H}^{-\frac{1}{2}}(\Gamma))$, it implies,

$$\langle 2[(\mathbb{D}\boldsymbol{u}_\infty)\boldsymbol{n}]_{\boldsymbol{\tau}}, \boldsymbol{v}_\alpha\rangle_{\boldsymbol{H}^{-\frac{1}{2}}(\Gamma)\times\boldsymbol{H}^{\frac{1}{2}}(\Gamma)} \to 0.$$

Therefore, $\mathbb{D}\boldsymbol{v}_\alpha \to \boldsymbol{0}$ in $L^2(0,T;\boldsymbol{L}^2(\Omega))$. The strong convergence for the pressure term follows from the equation.

(iii) Now to obtain (1.1.10), we estimate suitably the right hand side of (1.9.10). Because $\boldsymbol{u}_0 \in \boldsymbol{H}^1_{0,\sigma}(\Omega)$, we have $\boldsymbol{u}_\infty \in L^2(0,T;\boldsymbol{H}^2(\Omega))$ and thus

$$\int_0^T \langle [(\mathbb{D}\boldsymbol{v}_\infty)\boldsymbol{n}]_{\boldsymbol{\tau}}, \boldsymbol{v}_{\alpha\boldsymbol{\tau}}\rangle_\Gamma \leq C\|\boldsymbol{u}_\infty\|_{L^2(0,T;\boldsymbol{H}^2(\Omega))}\|\boldsymbol{u}_\alpha\|_{L^2(0,T;\boldsymbol{L}^2(\Gamma))}.$$

Similarly Γ is $\mathcal{C}^{2,1}$ and $\boldsymbol{u}_0 \in \boldsymbol{H}^2(\Omega) \cap \boldsymbol{H}^1_{0,\sigma}(\Omega)$ implies the regularity $\boldsymbol{u}_\infty \in L^2(0,T;\boldsymbol{H}^3(\Omega))$ by the same argument as [38, Theorem 3.10, Chapter III]. Hence $\nabla\boldsymbol{u}_\infty \in L^\infty(0,T;\boldsymbol{H}^2(\Omega)) \subset L^\infty(0,T;\boldsymbol{L}^\infty(\Omega))$ and thus

$$\left|\int_0^T\int_\Omega (\boldsymbol{v}_\alpha\cdot\nabla)\boldsymbol{u}_\infty\cdot\boldsymbol{v}_\alpha\right| \leq C\|\boldsymbol{v}_\alpha\|^2_{L^2(0,T;\boldsymbol{L}^2(\Omega))}\|\nabla\boldsymbol{u}_\infty\|_{L^\infty(0,T;\boldsymbol{L}^\infty(\Omega))}.$$

Therefore, combining these estimates, we get

$$\frac{1}{2}\|\boldsymbol{v}_\alpha(T)\|^2_{\boldsymbol{L}^2(\Omega)} + 2\int_0^T \|\mathbb{D}\boldsymbol{v}_\alpha\|^2_{\mathbb{L}^2(\Omega)} \leq C\|\boldsymbol{u}_\alpha\|_{L^2(0,T;\boldsymbol{L}^2(\Gamma))} + C\int_0^T \|\boldsymbol{v}_\alpha\|^2_{\boldsymbol{L}^2(\Omega)}$$

which yields by applying Gronwall's lemma,

$$\|\boldsymbol{v}_\alpha(T)\|^2_{\boldsymbol{L}^2(\Omega)} \leq C\|\boldsymbol{u}_\alpha\|_{L^2(0,T;\boldsymbol{L}^2(\Gamma))}e^{Ct} \leq \frac{C}{\alpha}.$$

The convergence in $L^\infty(0,T;\boldsymbol{L}^2(\Omega))$ and thus also in $L^2(0,T;\boldsymbol{H}^1(\Omega))$ follow immediately. ■

References

1. R. A. Adams. *Sobolev spaces.* Academic Press [A subsidiary of Harcourt Brace Jovanovich, Publishers], New York-London, 1975. Pure and Applied Mathematics, Vol. 65.

2. H. Al Baba, C. Amrouche, and M. Escobedo. Analyticity of the semi-group generated by the Stokes operator with Navier-type boundary conditions on L_p-spaces. In *Recent advances in partial differential equations and applications*, volume 666 of *Contemp. Math.*, pages 23–40. Amer. Math. Soc., Providence, RI, 2016.

3. H. Al Baba, C. Amrouche, and M. Escobedo. Semi-group theory for the Stokes operator with Navier-type boundary conditions on L^p-spaces. *Arch. Ration. Mech. Anal.*, 223(2):881–940, 2017.

4. H. Al Baba, C. Amrouche, and A. Rejaiba. The time-dependent Stokes problem with Navier slip boundary conditions on L^p-spaces. *Analysis (Berlin)*, 36(4):269–285, 2016.

5. H. Amann. *Linear and quasilinear parabolic problems. Vol. I*, volume 89 of *Monographs in Mathematics*. Birkhäuser Boston, Inc., Boston, MA, 1995. Abstract linear theory.

6. C. Amrouche, A. Ghosh, C. Conca, and P. Acevedo. Stokes and Navier-Stokes equations with Navier boundary condition. arXiv: https://arxiv.org/abs/1805.07760.

7. C. Amrouche and A. Rejaiba. L^p-theory for Stokes and Navier-Stokes equations with Navier boundary condition. *J. Differential Equations*, 256(4):1515–1547, 2014.

8. V. Barbu. *Nonlinear semigroups and differential equations in Banach spaces*. Editura Academiei Republicii Socialiste România, Bucharest; Noordhoff International Publishing, Leiden, 1976. Translated from the Romanian.

9. H. Beirão da Veiga. Remarks on the Navier-Stokes evolution equations under slip type boundary conditions with linear friction. *Port. Math. (N.S.)*, 64(4):377–387, 2007.

10. M. Beneš. Mixed initial-boundary value problem for the three-dimensional Navier-Stokes equations in polyhedral domains. *Discrete Contin. Dyn. Syst.*, (Dynamical systems, differential equations and applications. 8th AIMS Conference. Suppl. Vol. I):135–144, 2011.

11. J. Bourgain. Some remarks on Banach spaces in which martingale difference sequences are unconditional. *Ark. Mat.*, 21(2):163–168, 1983.

12. N. V. Chemetov and F. Cipriano. Inviscid limit for Navier-Stokes equations in domains with permeable boundaries. *Appl. Math. Lett.*, 33:6–11, 2014.

13. T. Clopeau, A. Mikelić, and R. Robert. On the vanishing viscosity limit for the 2D incompressible Navier-Stokes equations with the friction type boundary conditions. *Nonlinearity*, 11(6):1625–1636, 1998.

14. H Fujita and T Kato. On the Navier-Stokes initial value problem. I. *Arch. Rational Mech. Anal.*, 16:269–315, 1964.

15. D. Gérard-Varet and N. Masmoudi. Relevance of the slip condition for fluid flows near an irregular boundary. *Communications in Mathematical Physics*, 295(1):99–137, Apr 2010.

16. Y. Giga. The Stokes operator in L_r spaces. *Proc. Japan Acad. Ser. A Math. Sci.*, 57(2):85–89, 1981.

17. Y. Giga. Weak and strong solutions of the Navier-Stokes initial value problem. *Publ. Res. Inst. Math. Sci.*, 19(3):887–910, 1983.

18. Y. Giga. Solutions for semilinear parabolic equations in L^p and regularity of weak solutions of the Navier-Stokes system. *J. Differential Equations*, 62(2):186–212, 1986.

19. Y. Giga and T. Miyakawa. Solutions in L_r of the Navier-Stokes initial value problem. *Arch. Rational Mech. Anal.*, 89(3):267–281, 1985.

20. Y. Giga and H. Sohr. Abstract L^p estimates for the Cauchy problem with applications to the Navier-Stokes equations in exterior domains. *J. Funct. Anal.*, 102(1):72–94, 1991.

21. P. Grisvard. *Elliptic problems in nonsmooth domains*, volume 24 of *Monographs and Studies in Mathematics*. Pitman (Advanced Publishing Program), Boston, MA, 1985.

22. D. Iftimie and F. Sueur. Viscous boundary layers for the Navier-Stokes equations with the Navier slip conditions. *Arch. Ration. Mech. Anal.*, 199(1):145–175, 2011.

23. Sh. Itoh, N. Tanaka, and A. Tani. Stability of steady-states solution to Navier-Stokes equations with general Navier slip boundary condition. *Zap. Nauchn. Sem. S.-Peterburg. Otdel. Mat. Inst. Steklov. (POMI)*, 362(Kraevye Zacachi Matematicheskoĭ Fiziki i Smezhnye Voprosy Teorii Funktsiĭ. 39):153–175, 366, 2008.

24. J. P. Kelliher. Navier-Stokes equations with Navier boundary conditions for a bounded domain in the plane. *SIAM J. Math. Anal.*, 38(1):210–232, 2006.

25. H. Komatsu. Fractional powers of operators. *Pacific J. Math.*, 19:285–346, 1966.

26. P. Kučera and J. Neustupa. On L^3-stability of strong solutions of the Navier-Stokes equations with the Navier-type boundary conditions. *J. Math. Anal. Appl.*, 405(2):731–737, 2013.

27. N. Masmoudi and F. Rousset. Uniform regularity for the Navier-Stokes equation with Navier boundary condition. *Arch. Ration. Mech. Anal.*, 203(2):529–575, 2012.

28. N. Masmoudi and L. Saint-Raymond. From the Boltzmann equation to the Stokes-Fourier system in a bounded domain. *Communications on Pure and Applied Mathematics*, 56(9):1263–1293.

29. J. C. Maxwell. On stresses in rarified gases arising from inequalities of temperature. *Philosophical Transactions of the Royal Society of London*, 170:231–256, 1879.

30. S. Monniaux and E. M. Ouhabaz. The incompressible Navier-Stokes system with time-dependent Robin-type boundary conditions. *J. Math. Fluid Mech.*, 17(4):707–722, 2015.

31. C.L.M.H. Navier. Mémoire sur les lois du mouvement des fluides. *Mém. Acad. Sci. Inst. de France (2)*, pages 389–440, 1823.

32. D. Pal, N. Rudraiah, and R. Devanathan. The effects of slip velocity at a membrane surface on blood flow in the microcirculation. *Journal of Mathematical Biology*, 26(6):705–712, Dec 1988.

33. A. Pazy. *Semigroups of linear operators and applications to partial differential equations*, volume 44 of *Applied Mathematical Sciences*. Springer-Verlag, New York, 1983.

34. J. Prüss and G. Simonett. *Moving interfaces and quasilinear parabolic evolution equations*, volume 105 of *Monographs in Mathematics*. Birkhäuser/Springer, [Cham], 2016.

35. J. L. Rubio De Francia. Martingale and integral transforms of Banach space valued functions. In *Probability and Banach spaces (Zaragoza, 1985)*, volume 1221 of *Lecture Notes in Math.*, pages 195–222. Springer, Berlin, 1986.

36. N. Seloula. Mathematical analysis and numerical approximation of the stokes and Navier-Stokes equations with non standard boundary conditions. PhD Thesis, Université de Pau et des Pays de l'Adour, 2010, HAl.

37. C. G. Simader and H. Sohr. A new approach to the Helmholtz decomposition and the Neumann problem in L^q-spaces for bounded and exterior domains. In *Mathematical problems relating to the Navier-Stokes equation*, volume 11 of *Ser. Adv. Math. Appl. Sci.*, pages 1–35. World Sci. Publ., River Edge, NJ, 1992.

38. R. Temam. *Navier-Stokes equations. Theory and numerical analysis.* North-Holland Publishing Co., Amsterdam-New York-Oxford, 1977. Studies in Mathematics and its Applications, Vol. 2.

39. H. Triebel. *Interpolation theory, function spaces, differential operators.* VEB Deutscher Verlag der Wissenschaften, Berlin, 1978.

40. Y. Xiao and Z. Xin. On 3D Lagrangian Navier-Stokes α model with a class of vorticity-slip boundary conditions. *J. Math. Fluid Mech.*, 15(2):215–247, 2013.

Chapter 2

Theoretical and Numerical Results for a Chemorepulsion Model with Non-constant Diffusion Coefficients

Francisco Guillén-González and María Ángeles Rodríguez-Bellido (✉)
Universidad de Sevilla, Facultad de Matemáticas, Departamento de Ecuaciones Diferenciales y Análisis Numérico, Campus de Reina Mercedes, C/ Tarfia, Sevilla Spain
e-mail: guillen@us.es, angeles@us.es

Diego Armando Rueda-Gómez
Universidad Industrial de Santander Escuela de Matemáticas, Bucaramanga-Colombia
e-mail: diaruego@uis.edu.co

2.1 Introduction

Chemotaxis is the biological process of the movement of living organisms in response to a chemical stimulus that can be given toward a higher (attractive) or lower (repulsive) concentration of a chemical substance. At the same time, the presence of living organisms can produce or consume chemical substance.

T. Bodnár et al. (eds.), *Waves in Flows*, Advances in Mathematical Fluid Mechanics, https://doi.org/10.1007/978-3-030-68144-9_2

The classical chemotaxis system was written by Keller–Segel (1970–1971) using the following system of partial differential equations:

$$\begin{cases} \partial_t u - D_u \Delta u + \chi \nabla \cdot (u \nabla v) = 0 \ \text{ in } \Omega, \ t > 0, \\ \partial_t v - D_v \Delta v + \beta v = \alpha u \text{ in } \ \Omega, \ t > 0, \\ \frac{\partial v}{\partial \mathbf{n}} = \frac{\partial u}{\partial \mathbf{n}} = 0 \ \text{ on } \partial\Omega, \ t > 0, \\ v(\mathbf{x}, 0) = v_0(\mathbf{x}) \geq 0, \ u(\mathbf{x}, 0) = u_0(\mathbf{x}) \geq 0 \ \text{ in } \Omega, \end{cases} \tag{2.1}$$

where the unknowns for this model are u and v denoting the cell density and the chemical concentration, respectively. Constants D_u and D_v denote the (positive) self-diffusion coefficients for u and v, respectively; $\chi \in \mathbb{R}$ is a constant representing the strength of the chemotaxis process, $\alpha > 0$ is a constant related to the influence of the cell density in the production of chemical concentration, and $\beta > 0$ is the coefficient that measures the self-degradation of the chemical substance. The term $\chi \nabla \cdot (u \nabla v)$, also called **chemotaxis term**, models the transport of cells toward the higher concentrations of chemical signal if $\chi > 0$ (which is known as **chemoattraction**), and toward the lower concentrations of chemical signal if $\chi < 0$ (which is known as **chemorepulsion**).

Nowadays, several variants of model (2.1) have been analyzed in the literature. The most common ones modify the chemotaxis term. In this work, we consider the chemorepulsion case with a modification of the production term appearing in the equation for v: in fact the linear term αu in $(2.1)_2$ is replaced by $\alpha\, u^2$. The case of αu^p for $p \in (1, 2]$ is analyzed for a parabolic (in u) and elliptic (in v) system in [21] (and the references therein) and justified when the process of signal production through cells needs no longer dependence on the population density in a linear manner, for instance, when saturation effects at large (or short) densities are taken into account. The analysis of $p \in (1, 2)$ for (2.1) with $\chi < 0$ can be seen in [12].

2.2 Chemorepulsion Production System with Quadratic Production Term

Along this work, we will consider (2.1) with $\alpha = \beta = 1$ and $\chi = -1$, and a more general second self-diffusion order operator

$$\begin{cases} \partial_t u - \nabla \cdot (\nu(\boldsymbol{x}) \nabla u) - \nabla \cdot (u \nabla v) = 0 \ \text{ in } \Omega, \ t > 0, \\ \\ \partial_t v - \nabla \cdot (\mu(\boldsymbol{x}) \nabla v) + v = u^2 \text{ in } \ \Omega, \ t > 0, \\ \\ \frac{\partial v}{\partial \mathbf{n}} = \frac{\partial u}{\partial \mathbf{n}} = 0 \ \text{ on } \partial\Omega, \ t > 0, \\ \\ v(\mathbf{x}, 0) = v_0(\mathbf{x}) \geq 0, \ u(\mathbf{x}, 0) = u_0(\mathbf{x}) \geq 0 \ \text{ in } \Omega, \end{cases} \tag{2.2}$$

where the self-diffusion coefficients $\nu(\boldsymbol{x})$ and $\mu(\boldsymbol{x})$ are variable functions depending on the domain.

T. Cieslak et al. (see [2]) proved that model (2.1) with $\chi < 0$ is well-posed. In fact, there exist global in time weak solutions (based on an energy inequality, see Definition 2.3.1), and for $1D$ and $2D$ domains, there exists a unique global in time strong solution, see (2.11) below. Unconditionally energy stable numerical schemes for this model can be seen in [11].

The aim of this work is to compile the main results on existence, regularity, uniqueness, and asymptotic behavior for the chemorepulsion model with quadratic production term given in (2.2). We will use arguments based on some properties of (2.2) like positivity, behavior of the total mass of cells and chemical, and the existence of an energy law. Observe that these results generalize the case of constants $\mu(\boldsymbol{x})$ and $\nu(\boldsymbol{x})$ that were analyzed in [8, 9].

Along this paper, we will consider the usual Sobolev spaces $H^m(\Omega)$ and Lebesgue spaces $L^p(\Omega)$, $1 \leq p \leq \infty$, with norms $\|\cdot\|_m$ and $\|\cdot\|_{L^p}$, respectively. In particular, the $L^2(\Omega)$-norm will be denoted by $\|\cdot\|_0$. We also consider the space

$$\mathbf{H}^1_{\mathbf{n}}(\Omega) := \{\boldsymbol{\sigma} \in \mathbf{H}^1(\Omega) : \boldsymbol{\sigma} \cdot \mathbf{n} = 0 \text{ on } \partial\Omega\},$$

and we will use the following equivalent norms in $H^1(\Omega)$ and $\mathbf{H}^1_{\mathbf{n}}(\Omega)$, respectively (see [15] and [1, Corollary 3.5], respectively):

$$\|u\|_1^2 = \|\nabla u\|_0^2 + \left(\int_\Omega u\right)^2, \quad \forall u \in H^1(\Omega), \tag{2.3}$$

$$\|\boldsymbol{\sigma}\|_1^2 = \|\boldsymbol{\sigma}\|_0^2 + \|\text{rot } \boldsymbol{\sigma}\|_0^2 + \|\nabla \cdot \boldsymbol{\sigma}\|_0^2, \quad \forall \boldsymbol{\sigma} \in \mathbf{H}^1_{\mathbf{n}}(\Omega). \tag{2.4}$$

In particular, (2.4) implies that

$$\|\nabla v\|_1^2 = \|\nabla v\|_0^2 + \|\Delta v\|_0^2, \quad \forall v : \nabla v \in \mathbf{H}^1_{\mathbf{n}}(\Omega) \tag{2.5}$$

is an equivalent norm for ∇v in $\mathbf{H}^1(\Omega)$. If Z is a general Banach space, then its topological dual will be denoted by Z'. Moreover, the letters C_i, K_i will denote different positive constants (independent of discrete parameters).

In general, we will assume at least $\nu \in L^\infty(\Omega)$, $\mu \in C^1(\overline{\Omega})$, and

$$0 < \nu_{min} \leq \nu(\boldsymbol{x}), \quad 0 < \mu_{min} \leq \mu(\boldsymbol{x}) \quad a.e.\ \boldsymbol{x} \in \Omega. \tag{2.6}$$

We are going to consider the regularity of the operator $A_\mu : H^1(\Omega) \to H^1(\Omega)'$ defined by

$$\langle A_\mu w, \bar{w}\rangle = (\mu(\boldsymbol{x})\,\nabla w, \nabla \bar{w}) + (w, \bar{w}), \quad \forall w, \bar{w} \in H^1(\Omega). \tag{2.7}$$

The corresponding differential form of $A_\mu w = g$ is

$$\begin{cases} -\nabla\cdot(\mu(\boldsymbol{x})\,\nabla w) + w &= g \quad \text{in } \Omega, \\ \dfrac{\partial w}{\partial \mathbf{n}} &= 0 \quad \text{on } \partial\Omega. \end{cases} \tag{2.8}$$

Given $g \in H^1(\Omega)'$, the existence of a unique solution $w \in H^1(\Omega)$ can be derived from the Lax–Milgram Lemma. Moreover, for $m > -1$, assuming $g \in H^m(\Omega)$ and $\mu \in \mathcal{C}^{m+1}(\overline{\Omega})$, then the solution of problem (2.8) satisfies $w \in H^{m+2}(\Omega)$, see [3, Theorem 5.50].

2.3 Existence of Weak Solution

Definition 2.3.1. **(Weak Solutions of (2.2))** Let $\nu \in L^\infty(\Omega)$ and $\mu \in C^1(\overline{\Omega})$ satisfying hypothesis (2.6). Given $(u_0, v_0) \in L^2(\Omega) \times H^1(\Omega)$ with $u_0 \geq 0$, $v_0 \geq 0$ a.e. $\boldsymbol{x} \in \Omega$, a pair (u, v) is called weak solution of problem (2.2) in $(0, +\infty)$, if the following properties hold:

(i) $u \geq 0$, $v \geq 0$ a.e. $(t, \boldsymbol{x}) \in (0, +\infty) \times \Omega$;

(ii) (u, v) has the following regularity:

$$(u,v) \in L^\infty(0,T;L^2(\Omega)\times H^1(\Omega)) \cap L^2(0,T;H^1(\Omega)\times H^2(\Omega)), \quad \forall T > 0,$$

$$(\partial_t u, \partial_t v) \in L^{q'}(0,T;H^1(\Omega)'\times L^2(\Omega)), \quad \forall T > 0, \tag{2.9}$$

where $q' = 2$ in the one- and two-dimensional cases ($1D$-$2D$) and $q' = 4/3$ in the three-dimensional case ($3D$) (q' is the conjugate exponent of $q = 2$ in $2D$ and $q = 4$ in $3D$);

(iii) the following variational formulation holds

$$\int_0^T \langle \partial_t u, \bar{u}\rangle + \int_0^T (\nu(\boldsymbol{x})\nabla u, \nabla \bar{u}) + \int_0^T (u\nabla v, \nabla \bar{u}) = 0, \quad \forall \bar{u} \in L^q(0,T;H^1(\Omega)), \ \forall T > 0,$$

where $\langle\cdot,\cdot\rangle$ represents the duality between $H^1(\Omega)'$ and $H^1(\Omega)$, and $(\cdot,\cdot)$ represents the inner product in $L^2(\Omega)$;

(iv) the equation for v holds pointwisely

$$\partial_t v + A_\mu v = u^2 \quad \text{a.e. } (t,\boldsymbol{x}) \in (0,+\infty)\times\Omega;$$

(v) the initial conditions $(2.2)_4$ are satisfied (it has sense because from (2.9) one has in particular $(u, v) \in C([0, T]; H^1(\Omega)' \times L^2(\Omega))$);

(vi) the following energy inequality (in integral version) holds a.e. t_0, t_1 with $t_1 \geq t_0 \geq 0$:

$$
\begin{aligned}
&\mathcal{E}(u(t_1), v(t_1)) - \mathcal{E}(u(t_0), v(t_0)) \\
&+\int_{t_0}^{t_1} \left(\int_\Omega \nu(\boldsymbol{x}) \, |\nabla u(s,\boldsymbol{x})|^2 \, d\boldsymbol{x} + \frac{1}{2} \int_\Omega (\mu(\boldsymbol{x})|\Delta v(s,\boldsymbol{x})|^2 + |\nabla v(s,\boldsymbol{x})|^2) d\boldsymbol{x} \right) ds \\
&+\frac{1}{2} \int_{t_0}^{t_1} \int_\Omega \Delta v \, (\nabla v \cdot \nabla \mu) \, d\boldsymbol{x} \; ds \leq 0,
\end{aligned}
\tag{2.10}
$$

where

$$\mathcal{E}(u, v) = \frac{1}{2}\|u\|_0^2 + \frac{1}{4}\|\nabla v\|_0^2.$$

In this paper, we will see that it is possible to prove the existence of global in time weak solutions of the model (2.2) based on the energy inequality (2.10). Moreover, assuming the H^3-regularity of problem (2.8), there exists a unique local in time (in $(0, T_*)$) strong solution satisfying

$$
\left\{
\begin{array}{rcl}
u & \in & L^\infty(0, T_*; H^1(\Omega)) \cap L^2(0, T_*; H^2(\Omega)), \\
v & \in & L^\infty(0, T_*; H^2(\Omega)) \cap L^2(0, T_*; H^3(\Omega)), \\
\partial_t u & \in & L^\infty(0, T_*; L^2(\Omega)) \cap L^2(0, T_*; H^1(\Omega)), \\
\partial_t v & \in & L^\infty(0, T_*; H^1(\Omega)) \cap L^2(0, T_*; H^2(\Omega)),
\end{array}
\right.
\tag{2.11}
$$

where time T_* is any finite positive time in the one-dimensional and two-dimensional cases. The case of considering $T_* = +\infty$ will be treated more carefully in Sect. 2.8.

2.3.1 First Weak Estimates

In order to prove the existence of weak solutions, we assume that $\mu \in C^1(\overline{\Omega})$, then (2.9) is true but the bounds for $(u(t), v(t)) \in L^2(\Omega) \times H^1(\Omega)$ increase with the time t. Although the proof of it can be explained in more detail, the reason can be seen formally: taking $(u, -\frac{1}{2}\Delta v)$ as test function in $(2.2)_{1,2}$, we obtain

$$\frac{d}{dt}\left(\frac{1}{2}\|u\|_0^2 + \frac{1}{4}\|\nabla v\|_0^2\right) + \nu_{min}\|\nabla u\|_0^2 + \frac{1}{2}\|\nabla v\|_0^2 + \frac{\mu_{min}}{2}\|\Delta v\|_0^2 \leq \frac{1}{2}\int_\Omega |\Delta v||\nabla v||\nabla \mu| \, d\boldsymbol{x} := K(\mu, v).$$

Using the following estimate for the "extra-term":

$$K(\mu, v) \leq \frac{1}{2}\|\Delta v\|_0 \|\nabla v\|_0 \|\nabla \mu\|_{L^\infty} \leq \frac{\mu_{min}}{4}\|\Delta v\|_0^2 + \frac{\|\nabla \mu\|_{L^\infty}^2}{4\mu_{min}}\|\nabla v\|_0^2,$$

we obtain that there exists the constant $C_1 := \frac{\|\nabla\mu\|_{L^\infty}^2}{\mu_{min}} > 0$ such that

$$\frac{d}{dt}\left(\frac{1}{2}\|u\|_0^2 + \frac{1}{4}\|\nabla v\|_0^2\right) + \nu_{min}\|\nabla u\|_0^2 + \frac{1}{2}\|\nabla v\|_0^2 + \frac{\mu_{min}}{4}\|\Delta v\|_0^2 \le C_1\frac{1}{4}\|\nabla v\|_0^2. \tag{2.12}$$

Then, Gronwall Lemma implies

$$\mathcal{E}(u(t), v(t)) \le \mathcal{E}(u_0, v_0)\exp(C_1 t), \quad t > 0. \tag{2.13}$$

Consequently, from (2.12) and (2.13), we deduce that

$$\int_0^t \left(\nu_{min}\|\nabla u(s)\|_0^2 + \frac{1}{2}\min\left\{1, \frac{\mu_{min}}{2}\right\}\|\nabla v(s)\|_1^2\right) ds \le \mathcal{E}(u_0, v_0)(1 + \exp(C_1 t)),\ t > 0. \tag{2.14}$$

From (2.13) and (2.14), we can deduce

$$(u, \nabla v) \in L^\infty(0, T; L^2(\Omega)) \cap L^2(0, T; H^1(\Omega)). \tag{2.15}$$

2.3.2 Some Additional Properties

There are some common properties satisfied for chemotaxis models, which are as follows:

(i) **Conservation of the total mass of u:** The problem is conservative in u, as we can check integrating $(2.2)_1$ in Ω,

$$\frac{d}{dt}\left(\int_\Omega u\right) = 0, \quad \text{i.e.} \quad \int_\Omega u(t) = \int_\Omega u_0, \quad \forall t > 0. \tag{2.16}$$

For the chemical variable, the conservation is not guaranteed, but integrating $(2.2)_2$ in Ω, we deduce the following time ordinary differential equation for $\int_\Omega v$,

$$\frac{d}{dt}\left(\int_\Omega v\right) = \int_\Omega u^2 - \int_\Omega v. \tag{2.17}$$

(ii) **Positivity of the unknowns:** $u \ge 0$ and $v \ge 0$. Assuming $v_0 \ge 0$ in Ω, we can deduce that the unique strong solution $v = v(u^2)$ of problem $(2.2)_2$ satisfies $v \ge 0$ a.e. $(t, \boldsymbol{x}) \in (0, +\infty) \times \Omega$. However, the positivity of any u is not easy to prove only assuming weak regularity of the solution (see Definition 2.3.1) for the chemotaxis model. If higher regularity is proved, maximum principles could conclude the positivity of u. Anyway, we are going to be able to prove that there exists a positive u solving (2.2).

2.3.3 Additional Weak Estimate for v

Assuming $v_0 \geq 0$ in Ω, then $v \geq 0$ in $(0, +\infty) \times \Omega$. Therefore, from (2.17), we observe that the function $y(t) = \int_\Omega v(\boldsymbol{x}, t)\, d\boldsymbol{x} = \|v(t)\|_{L^1}$ satisfies

$$y'(t) + y(t) = z(t), \quad \text{with } z(t) = \int_\Omega u(\boldsymbol{x}, t)^2\, d\boldsymbol{x} = \|u(t)\|_0^2.$$

Therefore,

$$y(t) = y(0)\, e^{-t} + \int_0^t e^{-(t-s)} \|u(s)\|_0^2\, ds,$$

and using (2.13), we obtain

$$\|v(t)\|_{L^1} \leq e^{-t}\|v_0\|_{L^1} + \int_0^t e^{-(t-s)}\|u(s)\|_0^2\, ds \leq e^{-t}\|v_0\|_{L^1} + \mathcal{E}(u_0, v_0)\, t\, e^{C_1 t}, \ \forall t \geq 0. \tag{2.18}$$

Then, from (2.18), we conclude that $v \in L^\infty(0, T; L^1(\Omega))$, which together with (2.15) imply that

$$v \in L^\infty(0, T; H^1(\Omega)) \cap L^2(0, T; H^2(\Omega)), \quad \forall T > 0.$$

2.4 Uniqueness Criteria for Weak Solutions

Theorem 2.4.1. Let $\nu \in L^\infty(\Omega)$, $\mu \in W^{1,\infty}(\Omega)$, and the initial data $(u_0, v_0) \in L^2(\Omega) \times H^1(\Omega)$. Then, the class of weak solutions (u, v) of (2.2) with the additional regularity

$$(u, \nabla v) \in L^4(0, T; L^6(\Omega)) \tag{2.19}$$

has the uniqueness property.

Proof. Let (u_1, v_1) and (u_2, v_2) be two possible solutions of (2.2). We assume that (u_1, v_1) is a weak solution (in the sense of Definition 2.3.1, where (2.10) is satisfied) and (u_2, v_2) is a regular solution.

Denoting $(u, v) = (u_1, v_1) - (u_2, v_2)$, then (u, v) satisfies the following system:

$$\begin{cases} \partial_t u - \nabla \cdot (\nu(\boldsymbol{x})\nabla u) - \nabla \cdot (u_1 \nabla v) - \nabla \cdot (u \nabla v_2) = 0 \ \text{ in } \Omega, \ t > 0, \\ \partial_t v - \nabla \cdot (\mu(\boldsymbol{x})\nabla v) + v = u\,(u_1 + u_2) \text{ in } \ \Omega, \ t > 0, \\ \frac{\partial v}{\partial \mathbf{n}} = \frac{\partial u}{\partial \mathbf{n}} = 0 \ \text{ on } \partial\Omega, \ t > 0, \\ v(\mathbf{x}, 0) = 0, \ u(\mathbf{x}, 0) = 0 \ \text{ in } \Omega. \end{cases} \tag{2.20}$$

Here, we sketch the proof in a formal manner. For a rigorous proof, we use the argument used in [14] for the Navier–Stokes equations, in which it is essential that the weak solution (u_1, v_1) satisfies an energy inequality.

Testing $(2.20)_{1,2}$ by $(u, -\Delta v)$, the terms $\int_\Omega u_1 \nabla v \cdot \nabla u$ cancel, and we obtain

$$\frac{d}{dt}\left(\frac{1}{2}\|u\|_0^2 + \frac{1}{2}\|\nabla v\|_0^2\right) + \nu_{min}\|\nabla u\|_0^2 + \min\{\mu_{min}, 1\}\|\nabla v\|_1^2$$

$$\leq -\int_\Omega \Delta v\,(\nabla v \cdot \nabla \mu)\,d\boldsymbol{x} - \int_\Omega u\,\nabla v_2 \cdot \nabla u\,d\boldsymbol{x} + \int_\Omega u\,\nabla u_1 \cdot \nabla v\,d\boldsymbol{x} - \int_\Omega u\,u_2\,\Delta v\,d\boldsymbol{x}.$$

Therefore,

$$\begin{aligned}\left|\int_\Omega \Delta v\,(\nabla v \cdot \nabla \mu)\,d\boldsymbol{x}\right| &\leq \|\nabla \mu\|_{L^\infty}\|\nabla v\|_0\|\Delta v\|_0 \\ &\leq \frac{\min\{\mu_{min}, 1\}}{8}\|\Delta v\|_0^2 + C(\mu)\|\nabla v\|_0^2,\end{aligned}$$

$$\begin{aligned}\left|-\int_\Omega u\,u_2\,\Delta v\,d\boldsymbol{x}\right| &\leq \|u\|_{L^3}\|u_2\|_{L^6}\|\Delta v\|_0 \leq \|u\|_0^{1/2}\|u\|_1^{1/2}\|u_2\|_{L^6}\|\Delta v\|_0 \\ &\leq \frac{\min\{\mu_{min}, 1\}}{8}\|\Delta v\|_0^2 + \frac{\nu_{min}}{8}\|u\|_1^2 + C(\nu_{min}, \mu_{min})\|u_2\|_{L^6}^4\|u\|_0^2 \\ &\leq \frac{\min\{\mu_{min}, 1\}}{8}\|\Delta v\|_0^2 + \frac{\nu_{min}}{8}\|\nabla u\|_0^2 + C(\nu_{min}, \mu_{min})\left(1 + \|u_2\|_{L^6}^4\right)\|u\|_0^2.\end{aligned}$$

Moreover,

$$\begin{aligned}\left|\int_\Omega u\,\nabla u_1 \cdot \nabla v\,d\boldsymbol{x}\right| &\leq \|u\|_6\|\nabla u_1\|_0\|\nabla v\|_{L^3} \leq C\,\|\nabla u\|_0\|\nabla u_1\|_0\|\nabla v\|_0^{1/2}\|\nabla v\|_1^{1/2} \\ &\leq \frac{\nu_{min}}{8}\|\nabla u\|_0^2 + \frac{\min\{\mu_{min}, 1\}}{8}\|\nabla v\|_1^2 + C(\nu_{min}, \mu_{min})\|\nabla u_1\|_0^4\|\nabla v\|_0^2\end{aligned}$$

and

$$\begin{aligned}\left|\int_\Omega u\,\nabla v_2 \cdot \nabla u\,d\boldsymbol{x}\right| &\leq \|u\|_{L^3}\|\nabla v_2\|_{L^6}\|\nabla u\|_0 \leq \|u\|_0^{1/2}\|u\|_1^{1/2}\|\nabla v_2\|_{L^6}\|\nabla u\|_0 \\ &\leq \frac{\nu_{min}}{8}\|\nabla u\|_0^2 + C(\nu_{min})\left(1 + \|\nabla v_2\|_{L^6}^4\right)\|u\|_0^2.\end{aligned}$$

Bounding conveniently, the following inequality is obtained:

$$\frac{d}{dt}\left(\|u\|_0^2 + \|\nabla v\|_0^2\right) + \nu_{min}\|\nabla u\|_0^2 + \min\{\mu_{min}, 1\}\|\nabla v\|_1^2$$

$$\leq C\left(1 + \|\nabla u_1\|_0^4 + \|u_2\|_{L^6}^4 + \|\nabla v_2\|_{L^6}^4\right)\left(\|u\|_0^2 + \|\nabla v\|_0^2\right) + C\,\|\nabla v\|_0^2.$$

Imposing hypothesis (2.19) for (u_i, v_i), by using Gronwall Lemma, one has the uniqueness of solution for problem (2.2). ■

2.5 Some Constraints Implying Strong Estimates

In a formal manner, we can sketch the proof of the existence of a more regular solution (u, v) satisfying (2.11). A rigorous proof could be done by using Galerkin approximation or the Leray-Schauder fixed point theorem.

Concretely, we can prove some strong and regular estimates in $(0, T_*)$ such that

$$T_* \text{ is small enough, depending on the initial data } (u_0, v_0), \tag{2.21}$$

or $T_* = T$ (for any $T > 0$) if any of the following conditions is satisfied:

$$\begin{cases} (u_0, v_0) \text{ are small initial data such that } \|(u_0, v_0)\|_{H^1(\Omega)\cap H^2(\Omega)} \le C(T), \\ \text{or } \nabla v \in L^4(0,T;\mathbf{H}^1_{\mathbf{n}}(\Omega)), \quad \text{or } \nabla u \in L^4(0,T;\mathbf{L}^2(\Omega)). \end{cases} \tag{2.22}$$

Lemma 2.5.1. Assume $\nu \in W^{1,\infty}(\Omega)$ and $\mu \in W^{2,\infty}(\Omega)$. Then, any regular solution (u, v) of (2.2) satisfies the following inequality:

$$\begin{aligned} \frac{d}{dt}\left(\|u\|_1^2 + \frac{1}{2}\|\nabla v\|_1^2\right) &+ \nu_{min}\,\|\nabla u\|_1^2 + \frac{1}{4}\min\{\mu_{min}, 1\}\,\|\nabla v\|_2^2 \\ &\le C\left(\|\nabla u\|_0^2\|\nabla v\|_1^4 + \|\nabla u\|_0^2 + \|\nabla v\|_1^2\right), \end{aligned} \tag{2.23}$$

where $C = C(\nu_{min}, \mu_{min}, \|\nu\|_{W^{1,\infty}}, \|\mu\|_{W^{2,\infty}})$.

Proof. We first test $(2.2)_{1,2}$ by $(-\Delta u, \frac{1}{2}\Delta^2 v)$, obtaining

$$\begin{aligned} &\frac{d}{dt}\left(\frac{1}{2}\|\nabla u\|_0^2 + \frac{1}{4}\|\Delta v\|_0^2\right) + \int_\Omega \nu(\boldsymbol{x})\,|\Delta u|^2\,d\boldsymbol{x} + \int_\Omega (\nabla u\cdot\nabla\nu)\,\Delta u\,d\boldsymbol{x} \\ &+\frac{1}{2}\int_\Omega \mu(\boldsymbol{x})\,|\nabla(\Delta v)|^2\,d\boldsymbol{x} + \frac{1}{2}\int_\Omega \sum_{j,l=1}^{3} \partial^2_{lj}\mu\,\partial_j v\,\partial_l(\Delta v)\,d\boldsymbol{x} \\ &+\frac{1}{2}\int_\Omega \Big((\nabla(\Delta v)\cdot\nabla\mu)\,\Delta v + \sum_{j=1}^{n}\partial_j(\nabla v)\,\partial_j\mu\cdot\nabla(\Delta v)\Big)\,d\boldsymbol{x} + \frac{1}{2}\|\Delta v\|_0^2 \\ &= \int_\Omega (\nabla u\cdot\nabla)(\nabla v)\,\nabla u\,d\boldsymbol{x} + \frac{1}{2}\int_\Omega |\nabla u|^2\,\Delta v\,d\boldsymbol{x}. \end{aligned} \tag{2.24}$$

Adding (2.12) to (2.24) and using the equivalent norm appearing in (2.5) and the regularity and coercivity hypotheses for ν and μ given in (2.6), we obtain

$$\begin{aligned}
&\frac{d}{dt}\left(\frac{1}{2}\|u\|_1^2+\frac{1}{4}\|\nabla v\|_1^2\right)+\nu_{min}\|\nabla u\|_1^2+\frac{\mu_{min}}{4}\|\Delta v\|_1^2+\frac{1}{2}\|\nabla v\|_1^2\\
&\le\left(\int_\Omega|\nabla\nu||\nabla u||\Delta u|\,d\boldsymbol{x}+\frac{1}{2}\int_\Omega|D^2\mu||\nabla v||\nabla(\Delta v)|\,d\boldsymbol{x}+\frac{3}{2}\int_\Omega|\nabla\mu||\nabla^2 v||\nabla(\Delta v)|\,d\boldsymbol{x}\right.\\
&\left.+\frac{C_1}{4}\|\nabla v\|_0^2+\int_\Omega|\nabla u|^2|\nabla(\nabla v)|\,d\boldsymbol{x}+\int_\Omega|\nabla u|^2|\Delta v|\,d\boldsymbol{x}\right)=\sum_{j=1}^{6}I_j.
\end{aligned}\tag{2.25}$$

Observe that

$$\begin{aligned}
I_1&\le\|\nabla\nu\|_{L^\infty(\Omega)}\|\nabla u\|_0\|\Delta u\|_0\le\frac{\nu_{min}}{6}\|\Delta u\|_0^2+\frac{3\|\nabla\nu\|_{L^\infty(\Omega)}^2}{2\nu_{min}}\|\nabla u\|_0^2,\\
I_2&\le\frac{1}{2}\|D^2\mu\|_{L^\infty(\Omega)}\|\nabla v\|_0\|\nabla(\Delta v)\|_0\le\frac{\mu_{min}}{16}\|\nabla(\Delta v)\|_0^2+\frac{\|D^2\mu\|_{L^\infty(\Omega)}^2}{\mu_{min}}\|\nabla v\|_0^2,\\
I_3&\le\frac{3}{2}\|\nabla\mu\|_{L^\infty(\Omega)}\|\nabla^2 v\|_0\|\nabla(\Delta v)\|_0\le\frac{\mu_{min}}{16}\|\nabla(\Delta v)\|_0^2+\frac{9\|\nabla\mu\|_{L^\infty(\Omega)}^2}{\mu_{min}}\|\nabla v\|_1^2,\\
I_5,\,I_6&\le\|\nabla u\|_{L^3}\|\nabla u\|_{L^6}\|\nabla^2 v\|_0\le C\|\nabla u\|_0^{1/2}\|\nabla u\|_1^{3/2}\|\nabla v\|_1\\
&\le\frac{\nu_{min}}{6}\|\nabla u\|_1^2+\frac{9^3}{4\nu_{min}^3}\|\nabla u\|_0^2\|\nabla v\|_1^4.
\end{aligned}$$

Therefore, we arrive at (2.23). ■

Remark 2.5.2. Note that if

$$\|\nabla\nu\|_{L^\infty(\Omega)},\quad\|\nabla\mu\|_{L^\infty(\Omega)}\quad\text{and}\quad\|D^2\mu\|_{L^\infty(\Omega)}\quad\text{are small enough,}\tag{2.26}$$

then the terms I_1-I_3 appearing in the proof of Lemma 2.5.1 can be directly controlled with the left hand side of (2.23). In particular, it means that if $\nu(\boldsymbol{x})$ and $\mu(\boldsymbol{x})$ are constant functions, then the only terms appearing in the proof of Lemma 2.5.1 are I_5-I_6.

By using Lemma 2.5.1, classical arguments yield to the following.

Corollary 2.5.3. Assume $(u_0,v_0)\in H^1(\Omega)\cap H^2(\Omega)$ and the hypothesis of Lemma 2.5.1. Let $T>0$ be a time satisfying one of the properties (2.21) or (2.22). Then, any weak solution (u,v) of (2.2) also has the following strong regularity:

$$(u,v)\in L^\infty(0,T;H^1(\Omega)\times H^2(\Omega))\cap L^2(0,T;H^2(\Omega)\times H^3(\Omega)).\tag{2.27}$$

Remark 2.5.4 (1D and 2D Cases). In the one- and two-dimensional cases, regularity (2.27) holds without assuming neither (2.21) nor (2.22). Indeed, in this case, the bounds for I_5 and I_6 can be replaced by

$$I_5,\, I_6 \le \|\nabla u\|_{L^4}^2 \|\nabla^2 v\|_0 \le C\|\nabla u\|_0 \|\nabla u\|_1 \|\nabla v\|_1 \le \frac{\nu_{min}}{6}\|\nabla u\|_1^2 + \frac{3}{2\nu_{min}}\|\nabla u\|_0^2 \|\nabla v\|_1^2.$$

Thus, (2.23) becomes

$$\begin{aligned}\frac{d}{dt}\left(\|u\|_1^2 + \frac{1}{2}\|\nabla v\|_1^2\right) &+ \nu_{min}\,\|\nabla u\|_1^2 + \frac{1}{4}\min\{\mu_{min}, 1\}\,\|\nabla v\|_2^2 \\ &\le C_2\left(\|\nabla u\|_0^2 \|\nabla v\|_1^2 + \|\nabla u\|_0^2 + \|\nabla v\|_1^2\right).\end{aligned}$$

Finally, using Gronwall Lemma, (2.15), and estimate (2.14), we deduce the finite time estimate

$$\|u(t)\|_1^2 + \frac{1}{2}\|\nabla v(t)\|_1^2 \le \left(\|u_0\|_1^2 + \frac{1}{2}\|\nabla v_0\|_1^2\right) \exp\left(C_2(t + \mathcal{E}(u_0, v_0)(1 + e^{C_1 t})\right), \quad t > 0,$$

and

$$\begin{aligned}&\int_0^t \left(\|\nabla u(s)\|_1^2 + \|\nabla v(s)\|_2^2\right) ds \\ &\le \left(\|u_0\|_1^2 + \frac{1}{2}\|\nabla v_0\|_1^2\right) \left(1 + \mathcal{E}(u_0, v_0)\, e^{C_1 t}\right) \exp\left(C_2 t + \mathcal{E}(u_0, v_0)(1 + e^{C_1 t})\right),\end{aligned}$$

which directly implies (2.27).

2.6 More Regular Estimates

Now, we will see additional regularity only for u.

Lemma 2.6.1. Assume $\nu, \mu \in W^{2,\infty}(\Omega)$. Any solution (u, v) of (2.2) having the regularity (2.27) satisfies the following inequality:

$$\begin{aligned}\frac{d}{dt}\left(\|u\|_2^2 + \|\nabla v\|_1^2\right) &+ \nu_{min}\,\|\nabla u\|_2^2 + \frac{1}{4}\min\{\mu_{min}, 1\}\,\|\nabla v\|_2^2 \\ &\le C\left(\|\nabla u\|_1^2 \|\nabla v\|_1^4 + \|\nabla v\|_2^2 \|u\|_2^2 + \|\nabla u\|_1^2 + \|\nabla v\|_1^2\right).\end{aligned} \tag{2.28}$$

Proof. Testing now $\Delta(2.2)_1$ by Δu, we obtain

$$\begin{aligned}\frac{1}{2}\frac{d}{dt}\|\Delta u\|_0^2 &+ \int_\Omega \nu(\boldsymbol{x})\,|\nabla(\Delta u)|^2\, d\boldsymbol{x} \le 2\int_\Omega |\nabla \nu(\boldsymbol{x})||\nabla^2 u||\nabla(\Delta u)|\, d\boldsymbol{x} \\ &+ \int_\Omega |\nabla^2 \nu(\boldsymbol{x})||\nabla u||\nabla(\Delta u)|\, d\boldsymbol{x} + \int_\Omega \nabla(\nabla\cdot(u\nabla v))\,\nabla(\Delta u)\, d\boldsymbol{x} := \sum_{j=1}^3 K_j,\end{aligned} \tag{2.29}$$

where

$$K_1 \le 2\|\nabla\nu\|_{L^\infty}\|\nabla u\|_1\|\nabla(\Delta u)\|_0 \le \frac{\nu_{min}}{6}\|\nabla(\Delta u)\|_0^2 + \frac{6\|\nabla\nu\|_{L^\infty}^2}{\nu_{min}}\|\nabla u\|_1^2,$$

$$K_2 \le \|\nabla^2\nu\|_{L^\infty}\|\nabla u\|_0\|\nabla(\Delta u)\|_0 \le \frac{\nu_{min}}{6}\|\nabla(\Delta u)\|_0^2 + \frac{3\|\nabla^2\nu\|_{L^\infty}^2}{2\nu_{min}}\|\nabla u\|_0^2.$$

The bound for K_3 needs more calculations

$$\begin{aligned}
K_3 &\le \int_\Omega \left(|\nabla^2 u||\nabla v| + |\nabla u||\nabla^2 v| + |u||\nabla(\Delta v)|\right)|\nabla(\Delta u)|\,d\boldsymbol{x}\\
&\le \left(\|\nabla^2 u\|_{L^3}\|\nabla v\|_{L^6} + \|\nabla u\|_{L^3}\|\nabla^2 v\|_{L^6} + \|u\|_{L^\infty}\|\nabla(\Delta v)\|_0\right)\|\nabla(\Delta u)\|_0\\
&\le C\left(\|\nabla u\|_1^{1/2}\|\nabla u\|_2^{3/2}\|\nabla v\|_1 + \left(\|\nabla u\|_0^{1/2}\|\nabla u\|_1^{1/2} + \|u\|_1^{1/2}\|u\|_2^{1/2}\right)\|\nabla v\|_2\|\nabla(\Delta u)\|_0\right)\\
&\le \frac{\nu_{min}}{6}\|\nabla u\|_2^2 + C(\nu_{min})\left(\|\nabla u\|_1^2\|\nabla v\|_1^4 + \|\nabla v\|_2^2\|u\|_2^2\right).
\end{aligned}$$

Adding (2.23) to (2.29), and using the estimates for K_1, K_2, and Young inequalities to bound the right hand side of K_3, we obtain (2.28). ■

Using Lemma 2.6.1, the regularity (2.27), and the Gronwall Lemma, we are going to prove additional regularity for u.

Corollary 2.6.2. Assume $u_0, v_0 \in H^2(\Omega)$ and a final time $T > 0$ satisfying hypothesis (2.21) or (2.22). Then, the weak solution (u, v) of (2.2) also has the following additional regularity:

$$u \in L^\infty(0,T;H^2(\Omega)) \cap L^2(0,T;H^3(\Omega)). \tag{2.30}$$

Remark 2.6.3 (No Blow-Up). In particular, (2.27) and (2.30) imply that

$$(u,v) \in L^\infty(0,T;L^\infty(\Omega)\times L^\infty(\Omega)) \quad \forall T > 0.$$

That means that there is no blow-up in $(0, T)$. Moreover, when a 1D or 2D domain is considered, hypothesis $(2.22)_2$ is always satisfied because (2.27) is always true (see Remark 2.5.4). This implies that the solution (u, v) of (2.2) does not blow up at finite time.

Looking at the model (2.2), and using the regularity given in (2.27) and (2.30), one can deduce

$$\partial_t u,\ \partial_t v \in L^\infty(0,T;L^2(\Omega)) \cap L^2(0,T;H^1(\Omega)).$$

Our last attempt of obtaining more regularity for (u, v) will pass by increasing the regularity satisfied by

$$(\widetilde{u}, \widetilde{v}) := (\partial_t u, \partial_t v).$$

Lemma 2.6.4. Assume $\nu \in L^\infty(\Omega)$ and $\mu \in W^{1,\infty}(\Omega)$. Any solution (u,v) of (2.2) having the regularity (2.27) satisfies the following inequality:

$$\frac{d}{dt}\left(\|\widetilde{u}\|_0^2 + \frac{1}{2}\|\nabla\widetilde{v}\|_0^2\right) + \nu_{min}\|\widetilde{u}\|_1^2 + \frac{1}{2}\min\{\mu_{min},1\}\|\nabla\widetilde{v}\|_1^2 \le C\left(\|\nabla\widetilde{v}\|_0^2 + \|\nabla v\|_1^4\|\widetilde{u}\|_0^2 + \|\nabla u\|_0^4\|\widetilde{u}\|_0^2\right).$$

Proof. By deriving (2.2) in time,

$$\begin{cases} \partial_t\widetilde{u} - \nabla\cdot(\nu(\boldsymbol{x})\,\nabla\widetilde{u}) - \nabla\cdot(\widetilde{u}\nabla v) - \nabla\cdot(u\nabla\widetilde{v}) = 0, \\ \partial_t\widetilde{v} - \nabla\cdot(\mu(\boldsymbol{x})\,\nabla\widetilde{v}) + \widetilde{v} = 2u\widetilde{u}. \end{cases} \tag{2.31}$$

Testing by $\widetilde{u}$ in $(2.31)_1$ and $-\frac{1}{2}\Delta\widetilde{v}$ in $(2.31)_2$, observing that the term $(u\,\nabla\tilde{u}, \nabla\widetilde{v})$ cancels (it appears with different sign due to the chemotaxis and production terms), and taking into account that $\int_\Omega \widetilde{u} = 0$ and using classical $3D$ interpolation inequalities and (2.3), we deduce

$$\frac{d}{dt}\left(\frac{1}{2}\|\widetilde{u}\|_0^2 + \frac{1}{4}\|\nabla\widetilde{v}\|_0^2\right) + \nu_{min}\|\widetilde{u}\|_1^2 + \frac{\min\{\mu_{min},1\}}{2}\|\nabla\widetilde{v}\|_1^2$$

$$\le -\int_\Omega \nabla\mu(\boldsymbol{x})\cdot\nabla\widetilde{v}\,\Delta\widetilde{v}\,d\boldsymbol{x} - \int_\Omega \widetilde{u}\nabla v\,\nabla\widetilde{u}\,d\boldsymbol{x} + \int_\Omega \widetilde{u}\nabla u\,\nabla\widetilde{v}\,d\boldsymbol{x}$$

$$\le \|\nabla\mu\|_{L^\infty}\,\|\nabla\widetilde{v}\|_0\|\Delta\widetilde{v}\|_0 + \|\widetilde{u}\|_{L^3}\left(\|\nabla v\|_{L^6}\|\nabla\widetilde{u}\|_0 + \|\nabla\widetilde{v}\|_{L^6}\|\nabla u\|_0\right)$$

$$\le \|\nabla\mu\|_{L^\infty}\|\nabla\widetilde{v}\|_0\|\nabla\widetilde{v}\|_1 + C\,\|\widetilde{u}\|_0^{1/2}\|\nabla\widetilde{u}\|_0^{1/2}\left(\|\nabla v\|_1\|\nabla\widetilde{u}\|_0 + \|\nabla u\|_0\|\nabla\widetilde{v}\|_1\right)$$

$$\le \frac{\nu_{min}}{2}\|\widetilde{u}\|_1^2 + \frac{\min\{\mu_{min},1\}}{2}\|\nabla\widetilde{v}\|_1^2 + C(\nu_{min},\|\nabla\mu\|_{L^\infty})\left(\|\nabla\widetilde{v}\|_0^2 + \|\nabla v\|_1^4\|\widetilde{u}\|_0^2 + \|\nabla u\|_0^4\|\widetilde{u}\|_0^2\right);$$

hence, the lemma is proved. ∎

Therefore, since $\|\nabla v\|_1^4$ and $\|\nabla u\|_0^4 \in L^\infty(0,T)$, we can deduce the following.

Corollary 2.6.5. Under conditions of Corollary 2.5.3, and assuming $(\partial_t u, \partial_t v)(0,\boldsymbol{x}) \in L^2(\Omega)\times H^1(\Omega)$, then, the weak solution (u,v) of (2.2) satisfying the regularity (2.27) also has the following regularity:

$$(\partial_t u, \partial_t v) \in L^\infty(0,T; L^2(\Omega)\times H^1(\Omega)) \cap L^2(0,T; H^1(\Omega)\times H^2(\Omega)).$$

Attending the sense of Definition 2.3.1, we could say that $(\partial_t u, \partial_t v)$ also has weak regularity. Looking at system (2.31), it is possible to prove the following regularity:

$$(\partial_{tt}u, \partial_{tt}v) \in L^2(0,T; H^1(\Omega)' \times L^2(\Omega)).$$

Moreover, by following a bootstrap argument, it would also be possible to obtain more regularity for (u, v). However, we stop here because the regularity obtained so far is sufficient to guarantee the hypothesis required later to prove error estimates (see (2.66)).

2.7 An Equivalent Problem ($\boldsymbol{\sigma} = \nabla v$)

In the work of Zhang et al. (see [22]), model (2.2) for $\nu(\boldsymbol{x}) = 1$ and $\mu(\boldsymbol{x}) = 1$ is analyzed by introducing the auxiliary variable $\boldsymbol{\sigma} = \nabla v$. Here, we modify the problem satisfied by $\boldsymbol{\sigma}$ that appears in [22], and we study the regularity of the solution of problem (2.2) by using such auxiliary variable. Concretely, applying the gradient to Eq. $(2.2)_2$ and adding the term $\frac{1}{\mu}\mathrm{rot}(\mathrm{rot}\ \boldsymbol{\sigma})$ (that vanishes because $rot\boldsymbol{\sigma} = rot(\nabla v) = 0$), we obtain

$$\begin{cases} \partial_t u - \nabla \cdot (\nu(\boldsymbol{x})\, \nabla u) = \nabla \cdot (u\boldsymbol{\sigma}) \ \text{ in } \Omega,\ t > 0, \\ \partial_t \boldsymbol{\sigma} - \nabla\left(\nabla \cdot (\mu(\boldsymbol{x})\, \boldsymbol{\sigma})\right) + \dfrac{1}{\mu(\boldsymbol{x})}\,\mathrm{rot}(\mathrm{rot}\ \boldsymbol{\sigma}) + \boldsymbol{\sigma} = \nabla(u^2) \text{ in } \ \Omega,\ t > 0, \\ \dfrac{\partial u}{\partial \mathbf{n}} = 0 \ \text{ on } \partial\Omega,\ t > 0, \\ \boldsymbol{\sigma} \cdot \mathbf{n} = 0, \ \ [\mathrm{rot}\ \boldsymbol{\sigma} \times \mathbf{n}]_{tang} = 0 \ \text{ on } \partial\Omega,\ t > 0, \\ u(x, 0) = u_0(x) \geq 0,\ \boldsymbol{\sigma}(x, 0) = \nabla v_0(x) \ \text{ in } \Omega. \end{cases} \tag{2.32}$$

Here, the operator *rot* depends on the spatial dimension (it is a scalar operator in dimension 2, and a vector operator in dimension 3).

Once (2.32) is solved, we can recover v from u^2 solving

$$\begin{cases} \partial_t v - \nabla \cdot (\mu(\boldsymbol{x})\, \nabla v) + v = u^2 \text{ in } \ \Omega,\ t > 0, \\ \frac{\partial v}{\partial \mathbf{n}} = 0 \ \text{ on } \partial\Omega,\ t > 0, \\ v(x, 0) = v_0(x) \geq 0 \ \text{ in } \Omega. \end{cases} \tag{2.33}$$

2.7.1 The Stationary Problem for $\boldsymbol{\sigma}$

In order to simplify the notation, we are going to use the following linear elliptic operator:

$$B_\mu \boldsymbol{\sigma} = \boldsymbol{f} \quad \Leftrightarrow \quad \begin{cases} -\nabla(\nabla \cdot (\mu\, \boldsymbol{\sigma})) + \dfrac{1}{\mu}\ \mathrm{rot}\ (\mathrm{rot}\ \boldsymbol{\sigma}) + \boldsymbol{\sigma} = \boldsymbol{f} \ \text{ in } \Omega, \\ \boldsymbol{\sigma} \cdot \mathbf{n} = 0, \ \ [\mathrm{rot}\ \boldsymbol{\sigma} \times \mathbf{n}]_{tang} = \mathbf{0} \ \text{ on } \partial\Omega. \end{cases} \tag{2.34}$$

The corresponding variational form is given by $B_\mu : \mathbf{H}^1_{\mathbf{n}}(\Omega) \to \mathbf{H}^1_{\mathbf{n}}(\Omega)'$ such that

$$\langle B_\mu \boldsymbol{\sigma}, \bar{\boldsymbol{\sigma}} \rangle = (\boldsymbol{\sigma}, \bar{\boldsymbol{\sigma}}) + (\nabla \cdot (\mu \boldsymbol{\sigma}), \nabla \cdot \bar{\boldsymbol{\sigma}}) + \left(\text{rot } \boldsymbol{\sigma}, \text{rot } \left(\frac{1}{\mu} \bar{\boldsymbol{\sigma}} \right) \right), \quad \forall \boldsymbol{\sigma}, \bar{\boldsymbol{\sigma}} \in \mathbf{H}^1_{\mathbf{n}}(\Omega). \tag{2.35}$$

For the results below, it is easier to work in the space $\mathbf{H}^1_{\mathbf{n}}(\Omega)$ endowed with the norm

$$\|\boldsymbol{\sigma}\|_{1,\mu} = \left(\|\nabla \cdot (\mu \boldsymbol{\sigma})\|_0^2 + \|rot\, \boldsymbol{\sigma}\|_0^2 + \|\boldsymbol{\sigma}\|_0^2 \right)^{1/2}. \tag{2.36}$$

First, we will prove that the norm $\|\cdot\|_{1,\mu}$ is equivalents to the $\mathbf{H}^1_{\mathbf{n}}(\Omega)$-norm defined in (2.4).

Lemma 2.7.1. Let $\mu \in W^{1,\infty}(\Omega)$. The norms (2.36) and (2.4) are equivalent in $\mathbf{H}^1_{\mathbf{n}}(\Omega)$. In fact, there are $C_4, C_5 > 0$ such that

$$\|\boldsymbol{\sigma}\|_{1,\mu} \le C_4 \|\boldsymbol{\sigma}\|_1, \tag{2.37}$$

$$\|\boldsymbol{\sigma}\|_1 \le C_5 \|\boldsymbol{\sigma}\|_{1,\mu}. \tag{2.38}$$

Proof. Take $\boldsymbol{\sigma} \in \mathbf{H}^1_{\mathbf{n}}(\Omega)$ and consider $\|\boldsymbol{\sigma}\|_{1,\mu}$. Observe that

$$\nabla \cdot (\mu \boldsymbol{\sigma}) = \mu \nabla \cdot \boldsymbol{\sigma} + \nabla \mu \cdot \boldsymbol{\sigma}, \tag{2.39}$$

which implies that

$$\|\nabla \cdot (\mu \boldsymbol{\sigma})\|_0^2 \le \|\mu\|^2_{L^\infty(\Omega)} \|\nabla \cdot \boldsymbol{\sigma}\|_0^2 + \|\nabla \mu\|^2_{L^\infty(\Omega)} \|\boldsymbol{\sigma}\|_0^2.$$

Therefore,

$$\|\boldsymbol{\sigma}\|^2_{1,\mu} \le \|\mu\|^2_{L^\infty(\Omega)} \|\nabla \cdot \boldsymbol{\sigma}\|_0^2 + \|rot\, \boldsymbol{\sigma}\|_0^2 + \left(1 + \|\nabla \mu\|^2_{L^\infty(\Omega)}\right) \|\boldsymbol{\sigma}\|_0^2.$$

Since $\mu \in W^{1,\infty}(\Omega)$, then there exists a constant $C_4 > 0$ such that (2.37) holds.

On the other hand, using (2.39), we deduce that

$$\nabla \cdot \boldsymbol{\sigma} = \frac{1}{\mu} \left(\nabla \cdot (\mu \boldsymbol{\sigma}) - \nabla \mu \cdot \boldsymbol{\sigma} \right),$$

which, using $(2.6)_2$, implies that

$$\|\nabla \cdot \boldsymbol{\sigma}\|_0^2 \le \frac{1}{\mu_{min}} \left(\|\nabla \cdot (\mu \boldsymbol{\sigma})\|_0^2 + \|\nabla \mu\|^2_{L^\infty(\Omega)} \|\boldsymbol{\sigma}\|_0^2 \right). \tag{2.40}$$

In this case, using (2.4) and (2.40),

$$\|\boldsymbol{\sigma}\|_1^2 \le \frac{1}{\mu_{min}} \|\nabla \cdot (\mu \boldsymbol{\sigma})\|_0^2 + \|rot\, \boldsymbol{\sigma}\|_0^2 + \left(1 + \frac{1}{\mu_{min}} \|\nabla \mu\|^2_{L^\infty(\Omega)}\right) \|\boldsymbol{\sigma}\|_0^2,$$

and again we can deduce the existence of a constant $C_5 > 0$ such that (2.38) holds. ∎

In order to prove the existence of a unique solution $\boldsymbol{\sigma} \in \mathbf{H}^1_{\mathbf{n}}(\Omega)$ of problem (2.34), we will use the following generalization of the Lax–Milgram theorem.

Theorem 2.7.1 (Nečas Theorem). (see [5]) Let U and V be two Hilbert spaces, let $a : U \times V \to \mathbb{R}$ be a bilinear form, and $f \in V'$. The problem

$$\text{Find } u \in U \text{ such that } a(u,v) = f(v), \quad \forall v \in V$$

is well-posed if and only if

(n1) there exists a constant $\alpha > 0$ such that

$$\inf_{u\in U} \sup_{v\in V} \frac{a(u,v)}{\|u\|_U \|v\|_V} \geq \alpha > 0, \tag{2.41}$$

(n2) it holds

$$\text{if } \ a(u,v) = 0, \ \forall\, u \in U \quad \Rightarrow \quad v = 0. \tag{2.42}$$

Under the previous conditions, the following estimate holds:

$$\|u\|_U \leq \frac{1}{\alpha} \|f\|_{V'}, \quad \forall f \in V'.$$

Theorem 2.7.2. Let $\mu \in W^{1,\infty}(\Omega)$ and $\boldsymbol{f} \in \mathbf{H}^1_{\mathbf{n}}(\Omega)'$. Then, there exists a unique solution $\boldsymbol{\sigma} \in \mathbf{H}^1_{\mathbf{n}}(\Omega)$ of problem (2.34). Moreover, the solution depends continuously on data, satisfying

$$\|\boldsymbol{\sigma}\|_1 \leq \frac{1}{\alpha} \|\boldsymbol{f}\|_{\mathbf{H}^1_{\mathbf{n}}(\Omega)'}$$

for a constant $\alpha > 0$ defined as

$$\alpha = \min\{1, \mu_{min}\} \frac{C_6}{C_5^2\, C_7\, \|\mu\|_{W^{1,\infty}(\Omega)}}, \tag{2.43}$$

with C_6, $C_7 > 0$ such that

$$C_6\, \|\boldsymbol{\sigma}\|_1 \leq \|\boldsymbol{\sigma}\|_{H^1(\Omega)} \leq C_7\, \|\boldsymbol{\sigma}\|_1 \tag{2.44}$$

(recall that the $\mathbf{H}^1_{\mathbf{n}}(\Omega)$-norm $\|\boldsymbol{\sigma}\|_1$ defined in (2.4) is an equivalent norm in $\mathbf{H}^1_{\mathbf{n}}(\Omega)$ to $\|\boldsymbol{\sigma}\|_{H^1(\Omega)}$ defined as $\|\boldsymbol{\sigma}\|^2_{H^1(\Omega)} = \|\boldsymbol{\sigma}\|_0^2 + \|\nabla \boldsymbol{\sigma}\|_0^2$).

Proof. In order to prove this result, we use Theorem 2.7.1 for the bilinear operator $\mathcal{B}_\mu : \mathbf{H}^1_{\mathbf{n}}(\Omega) \times \mathbf{H}^1_{\mathbf{n}}(\Omega) \to \mathbb{R}$ defined in (2.35), which is clearly a continuous and bilinear operator. Observe that, thanks to Lemma 2.7.1, the space $\mathbf{H}^1_{\mathbf{n}}(\Omega)$ is a Hilbert space endowed with either the norm defined in (2.4) or the norm defined in (2.36).

Taking $\bar{\boldsymbol{\sigma}} = \mu\,\boldsymbol{\sigma}$ in (2.35), we obtain

$$\mathcal{B}_\mu(\boldsymbol{\sigma}, \mu\,\boldsymbol{\sigma}) = \|\nabla\cdot(\mu\,\boldsymbol{\sigma})\|_0^2 + \|rot\,\boldsymbol{\sigma}\|_0^2 + \|\sqrt{\mu}\,\boldsymbol{\sigma}\|_0^2 \geq \min\{1, \mu_{min}\}\,\|\boldsymbol{\sigma}\|_{1,\mu}^2.$$

Thus,

$$\|\mu\,\boldsymbol{\sigma}\|_1 \leq C_6^{-1}\,\|\mu\,\boldsymbol{\sigma}\|_{H^1(\Omega)} \leq C_6^{-1}\left(\|\mu\|_{L^\infty(\Omega)}^2\|\nabla\boldsymbol{\sigma}\|_0^2 + \|\nabla\mu\|_{L^\infty(\Omega)}^2\|\boldsymbol{\sigma}\|_0^2 + \|\mu\|_{L^\infty(\Omega)}^2\|\boldsymbol{\sigma}\|_0^2\right)^{1/2}$$
$$\leq C_6^{-1}\,\|\mu\|_{W^{1,\infty}(\Omega)}\|\boldsymbol{\sigma}\|_{H^1(\Omega)} \leq \frac{C_7}{C_6}\,\|\mu\|_{W^{1,\infty}(\Omega)}\|\boldsymbol{\sigma}\|_1. \tag{2.45}$$

Therefore, using (2.38), (2.44), and (2.45),

$$\sup_{\boldsymbol{\varphi}\in\mathbf{H}_\mathbf{n}^1(\Omega)} \frac{\mathcal{B}_\mu(\boldsymbol{\sigma},\boldsymbol{\varphi})}{\|\boldsymbol{\varphi}\|_1} \geq \frac{\mathcal{B}_\mu(\boldsymbol{\sigma},\mu\,\boldsymbol{\sigma})}{\|\mu\,\boldsymbol{\sigma}\|_1} \geq \min\{1,\mu_{min}\}\frac{\|\boldsymbol{\sigma}\|_{1,\mu}^2}{\|\mu\,\boldsymbol{\sigma}\|_1} \geq \min\{1,\mu_{min}\}\frac{C_6\,\|\boldsymbol{\sigma}\|_1}{C_5^2\,C_7\,\|\mu\|_{W^{1,\infty}(\Omega)}}$$

and (2.41) is true for α defined in (2.43).

In order to satisfy (2.42), assume that $\mathcal{B}_\mu(\boldsymbol{\sigma},\boldsymbol{\varphi}) = 0$, $\forall\boldsymbol{\sigma}\in\mathbf{H}_\mathbf{n}^1(\Omega)$. In particular,

$$\mathcal{B}_\mu\left(\frac{1}{\mu}\,\boldsymbol{\varphi},\boldsymbol{\varphi}\right) = 0,$$

and therefore,

$$\mathcal{B}_\mu(\frac{1}{\mu}\,\boldsymbol{\varphi},\boldsymbol{\varphi}) = \|\nabla\cdot\boldsymbol{\varphi}\|_0^2 + \left\|rot\left(\frac{1}{\mu}\boldsymbol{\varphi}\right)\right\|_0^2 + \left\|\frac{1}{\sqrt{\mu}}\boldsymbol{\varphi}\right\|_0^2 = 0,$$

implying that

$$\|\mu\|_{L^\infty(\Omega)}^{-1}\,\|\boldsymbol{\varphi}\|_0^2 \leq \left\|\frac{1}{\sqrt{\mu}}\boldsymbol{\varphi}\right\|_0^2 = 0 \quad\Rightarrow\quad \boldsymbol{\varphi} = \mathbf{0}.$$

Condition (2.42) is therefore true. ■

Observe that, when we consider that the right hand side of problem (2.34) is given by $\boldsymbol{f} = \nabla g$ with $g \in H^1(\Omega)$, then, taking $\boldsymbol{\sigma} = \nabla w$ with w the solution of the problem (2.8), the following H^2-regularity for problem (2.34) can be proved.

Corollary 2.7.2. (see [8]) If $\boldsymbol{f} = \nabla g$ with $g \in H^1(\Omega)$, then the solution $\boldsymbol{\sigma}$ of problem (2.34) belongs to $\mathbf{H}^2(\Omega)$. Moreover,

$$\|\boldsymbol{\sigma}\|_2 \leq C\,\|\nabla g\|_0. \tag{2.46}$$

The proof is based on [8, Lemma 2.1], and it is a consequence of Theorem 2.7.2 together with the H^3-regularity of problem (2.8).

2.7.2 Weak Regularity of the $(u, \boldsymbol{\sigma})$ Problem (2.32)

With the same spirit of Definition 2.3.1 for the (u, v) problem (2.2), we state the following.

Definition 2.7.3. **(Weak Solutions of (2.32))** Given $(u_0, \boldsymbol{\sigma}_0) \in L^2(\Omega) \times \mathbf{L}^2(\Omega)$ with $u_0 \geq 0$ a.e. $\boldsymbol{x} \in \Omega$, a pair $(u, \boldsymbol{\sigma})$ is called weak solution of problem (2.32) in $(0, +\infty)$, if $u \geq 0$ a.e. $(t, \boldsymbol{x}) \in (0, +\infty) \times \Omega$,

$$(u, \boldsymbol{\sigma}) \in L^\infty(0, T; L^2(\Omega) \times \mathbf{L}^2(\Omega)) \cap L^2(0, T; H^1(\Omega) \times \mathbf{H}^1(\Omega)), \quad \forall T > 0,$$

$$(\partial_t u, \partial_t \boldsymbol{\sigma}) \in L^{q'}(0, T; H^1(\Omega)' \times \mathbf{H}^1(\Omega)'), \quad \forall T > 0,$$

where q is as in Definition 2.3.1; the following variational formulations hold

$$\int_0^T \langle \partial_t u, \bar{u} \rangle + \int_0^T (\nu(\boldsymbol{x}) \nabla u, \nabla \bar{u}) + \int_0^T (u\boldsymbol{\sigma}, \nabla \bar{u}) = 0, \ \forall \bar{u} \in L^q(0, T; H^1(\Omega)), \ \forall T > 0,$$

$$\int_0^T \langle \partial_t \boldsymbol{\sigma}, \bar{\boldsymbol{\sigma}} \rangle + \int_0^T \langle B_\mu \boldsymbol{\sigma}, \bar{\boldsymbol{\sigma}} \rangle = 2 \int_0^T (u \nabla u, \bar{\boldsymbol{\sigma}}), \quad \forall \bar{\boldsymbol{\sigma}} \in L^q(0, T; \mathbf{H}^1(\Omega)), \quad \forall T > 0,$$

the initial conditions $(2.32)_5$ are satisfied, and the following energy inequality (in integral version) holds a.e. t_0, t_1 with $t_1 \geq t_0 \geq 0$:

$$\begin{aligned}
&\mathcal{E}(u(t_1), \boldsymbol{\sigma}(t_1)) - \mathcal{E}(u(t_0), \boldsymbol{\sigma}(t_0)) + \int_{t_0}^{t_1} \int_\Omega \nu(\boldsymbol{x}) |\nabla u(s, \boldsymbol{x})|^2 \, d\boldsymbol{x} \, ds \\
&+ \frac{1}{2} \int_{t_0}^{t_1} \int_\Omega \left(\mu(\boldsymbol{x}) \, |\nabla \cdot \boldsymbol{\sigma}(s, \boldsymbol{x})|^2 + \frac{1}{\mu(\boldsymbol{x})} \, |rot \, \boldsymbol{\sigma}(s, \boldsymbol{x})|^2 + |\boldsymbol{\sigma}(s, \boldsymbol{x})|^2 \right) d\boldsymbol{x} \, ds \\
&+ \frac{1}{2} \int_{t_0}^{t_1} \int_\Omega (\nabla \mu(\boldsymbol{x}) \cdot \boldsymbol{\sigma}(s, \boldsymbol{x})) \, (\nabla \cdot \boldsymbol{\sigma}(s, \boldsymbol{x})) \, d\boldsymbol{x} \, ds \\
&- \frac{1}{2} \int_{t_0}^{t_1} \int_\Omega \frac{1}{\mu^2(\boldsymbol{x})} \, rot \, \boldsymbol{\sigma}(s, \boldsymbol{x}) \cdot (\nabla \mu(\boldsymbol{x}) \times \boldsymbol{\sigma}(s, \boldsymbol{x})) \, d\boldsymbol{x} \, ds \leq 0,
\end{aligned}$$

where

$$\mathcal{E}(u, \boldsymbol{\sigma}) = \frac{1}{2} \|u\|_0^2 + \frac{1}{4} \|\boldsymbol{\sigma}\|_0^2. \tag{2.47}$$

Assuming that $\mu \in W^{1,\infty}(\Omega)$, we can explain formally how to obtain the weak regularity for $(u, \boldsymbol{\sigma})$: we consider $(u, \frac{1}{2}\boldsymbol{\sigma})$ as test functions in $(2.32)_{1,2}$, obtaining

$$\begin{aligned}
\frac{d}{dt} (\mathcal{E}(u, \boldsymbol{\sigma})) \quad &+ \quad \nu_{min} \, \|\nabla u\|_0^2 + \frac{\mu_{min}}{2} \, \|\nabla \cdot \boldsymbol{\sigma}\|_0^2 + \frac{1}{2\|\mu\|_{L^\infty(\Omega)}} \, \|rot \, \boldsymbol{\sigma}\|_0^2 + \|\boldsymbol{\sigma}\|_0^2 \\
&+ \quad \frac{1}{2} \int_\Omega \left((\nabla \mu \cdot \boldsymbol{\sigma}) \, (\nabla \cdot \boldsymbol{\sigma}) - \left(\frac{\nabla \mu}{\mu^2} \times \boldsymbol{\sigma} \right) \cdot rot \, \boldsymbol{\sigma} \right) d\boldsymbol{x} \leq 0.
\end{aligned} \tag{2.48}$$

Observe that

$$\int_\Omega (\nabla\mu\cdot\boldsymbol{\sigma})\,(\nabla\cdot\boldsymbol{\sigma})\,d\boldsymbol{x} \le \|\nabla\mu\|_{L^\infty(\Omega)}\|\boldsymbol{\sigma}\|_0\|\nabla\cdot\boldsymbol{\sigma}\|_0^2 \le \frac{\mu_{min}}{2}\,\|\nabla\cdot\boldsymbol{\sigma}\|_0^2 + \frac{\|\nabla\mu\|^2_{L^\infty(\Omega)}}{2\mu_{min}}\,\|\boldsymbol{\sigma}\|_0^2 \tag{2.49}$$

and

$$\begin{aligned}\left|-\int_\Omega\left(\frac{\nabla\mu}{\mu^2}\times\boldsymbol{\sigma}\right)\cdot rot\,\boldsymbol{\sigma}\,d\boldsymbol{x}\right| &\le \frac{\|\nabla\mu\|_{L^\infty(\Omega)}}{\mu^2_{min}}\,\|\boldsymbol{\sigma}\|_0\|rot\,\boldsymbol{\sigma}\|_0\\ &\le \frac{1}{2\|\mu\|_{L^\infty(\Omega)}}\,\|rot\,\boldsymbol{\sigma}\|_0^2 + \frac{\|\mu\|_{L^\infty(\Omega)}\|\nabla\mu\|^2_{L^\infty(\Omega)}}{2\mu^4_{min}}\,\|\boldsymbol{\sigma}\|_0^2.\end{aligned} \tag{2.50}$$

Putting (2.49)–(2.50) into (2.48), we obtain

$$\begin{aligned}\frac{d}{dt}\mathcal{E}(u,\boldsymbol{\sigma}) &+ \nu_{min}\,\|\nabla u\|_0^2 + \frac{\mu_{min}}{4}\,\|\nabla\cdot\boldsymbol{\sigma}\|_0^2 + \frac{1}{4\|\mu\|_{L^\infty(\Omega)}}\,\|rot\,\boldsymbol{\sigma}\|_0^2 + \|\boldsymbol{\sigma}\|_0^2\\ &\le \frac{\|\nabla\mu\|^2_{L^\infty(\Omega)}}{4\mu_{min}}\left(1+\frac{\|\mu\|_{L^\infty(\Omega)}}{\mu^3_{min}}\right)\,\|\boldsymbol{\sigma}\|_0^2.\end{aligned}$$

Therefore, we can deduce that

$$\mathcal{E}(u(t),\boldsymbol{\sigma}(t)) \le \mathcal{E}(u_0,\boldsymbol{\sigma}_0)\exp(C_8 t)$$

for $C_8 = \dfrac{\|\nabla\mu\|^2_{L^\infty(\Omega)}}{\mu_{min}}\left(1+\dfrac{\|\mu\|_{L^\infty(\Omega)}}{\mu^3_{min}}\right)$, and consequently,

$$\int_0^t\left(\nu_{min}\,\|\nabla u(s)\|_0^2 + \min\left\{\frac{\mu_{min}}{4},\frac{1}{4\|\mu\|_{L^\infty(\Omega)}},1\right\}\,\|\boldsymbol{\sigma}(s)\|_1^2\right)\,ds \le \mathcal{E}(u_0,\boldsymbol{\sigma}_0)\,(1+\exp(C_8 t)),\ t>0,$$

which, in particular, implies that $(u,\boldsymbol{\sigma}) \in L^\infty(0,T;\mathbf{L}^2(\Omega)) \cap L^2(0,T;\mathbf{H}^1(\Omega))$.

2.7.3 Equivalence of Problems (u,v) and $(u,\boldsymbol{\sigma})$

In order to prove that to solve (2.32)–(2.33) is equivalent to solve (2.2), the following result is essential.

Lemma 2.7.3. If $\boldsymbol{\sigma}_0 = \nabla v_0$, problems (2.2) and (2.32)-(2.33) are equivalent in the following sense: If (u,v) is a weak solution of (2.2), then $(u,\boldsymbol{\sigma})$ with $\boldsymbol{\sigma} = \nabla v$ is a weak solution of (2.32); and reciprocally, if $(u,\boldsymbol{\sigma})$ is a weak solution of (2.32) and $v = v(u^2)$ is the unique strong solution of problem (2.33) (i.e., $v \in L^p(0,T;W^{2,p}(\Omega)) \cap L^\infty(0,T;W^{1,p}(\Omega)) \cap L^{q'}(0,T;H^2(\Omega))$ since $u^2 \in L^p(0,T;L^p(\Omega)) \cap L^{q'}(0,T;L^2(\Omega))$ for $p = 5/3$ in 3D, $p = 2$ in 2D and q' is as in Definition 2.3.1, see also [6, Theorem 10.22]), then $\boldsymbol{\sigma} = \nabla v$ and (u,v) is a weak solution of (2.2).

Proof. Consider (u,v) a weak solution of (2.2). Testing the equation for v by $-\nabla\cdot\bar{\mathbf{w}}$, with $\bar{\mathbf{w}}\in L^q(0,T;\mathbf{H}^1(\Omega))$ (q as in Definition 2.3.1), integrating by parts, we have

$$(\nabla(\partial_t v)-\nabla(\nabla\cdot(\mu\,\nabla v))+\nabla v,\bar{\mathbf{w}})=(\nabla(u^2),\bar{\mathbf{w}}). \tag{2.51}$$

Taking into account that $rot(\nabla v)=0$, we can write (2.51) as

$$(\partial_t(\nabla v)+B_\mu(\nabla v),\bar{\mathbf{w}})=(\nabla(u^2),\bar{\mathbf{w}}),\quad \forall\bar{\mathbf{w}}\in L^q(0,T;\mathbf{H}^1(\Omega)) \tag{2.52}$$

for $B_\mu(\cdot)$ defined in (2.34). On the other hand, if $(u,\boldsymbol{\sigma})$ is a weak solution of (2.32), $(u,\boldsymbol{\sigma})$ satisfies

$$(\partial_t\boldsymbol{\sigma}+B_\mu(\boldsymbol{\sigma}),\bar{\mathbf{w}})=(\nabla(u^2),\bar{\mathbf{w}}),\quad \forall\bar{\mathbf{w}}\in L^q(0,T;\mathbf{H}^1(\Omega)), \tag{2.53}$$

and $v=v(u^2)$ is the unique weak solution of problem (2.33), and ∇v satisfies (2.52). The statement of Lemma 2.7.3 will be proved if we show that $\boldsymbol{\sigma}=\nabla v$.

Indeed, as a consequence of (2.52)–(2.53),

$$(\partial_t(\boldsymbol{\sigma}-\nabla v),\bar{\mathbf{w}})+(B_\mu(\boldsymbol{\sigma}-\nabla v),\bar{\mathbf{w}})=0,\quad \forall\bar{\mathbf{w}}\in L^q(0,T;\mathbf{H}^1(\Omega)). \tag{2.54}$$

We cannot take $\bar{\mathbf{w}}=\boldsymbol{\sigma}-\nabla v$ in (2.54) because $\boldsymbol{\sigma}-\nabla v\notin L^q(0,T;\mathbf{H}^1(\Omega))$. However, $\boldsymbol{\sigma}-\nabla v\in L^\infty(0,T;\mathbf{L}^2(\Omega)$, and thus, $B_\mu^{-1}(\boldsymbol{\sigma}-\nabla v)\in L^\infty(0,T;\mathbf{H}^2(\Omega)\hookrightarrow L^q(0,T;\mathbf{H}^1(\Omega))$.

Therefore, considering $\bar{\mathbf{w}}=\mu\,B_\mu^{-1}(\boldsymbol{\sigma}-\nabla v)$ in (2.54), we obtain

$$(\partial_t(\boldsymbol{\sigma}-\nabla v),\mu\,B_\mu^{-1}(\boldsymbol{\sigma}-\nabla v))+(B_\mu(\boldsymbol{\sigma}-\nabla v),\mu\,B_\mu^{-1}(\boldsymbol{\sigma}-\nabla v))=0. \tag{2.55}$$

Observe that denoting $\boldsymbol{\Phi}=B_\mu^{-1}(\boldsymbol{\sigma}-\nabla v)$, we have

$$\begin{aligned}(\partial_t(\boldsymbol{\sigma}-\nabla v),\mu\,B_\mu^{-1}(\boldsymbol{\sigma}-\nabla v))&=(\partial_t(B_\mu\boldsymbol{\Phi}),\mu\,\boldsymbol{\Phi})\\ &=(\partial_t(-\nabla(\nabla\cdot(\mu\,\boldsymbol{\Phi}))+\tfrac{1}{\mu}rot\,(rot\,\boldsymbol{\Phi})+\boldsymbol{\Phi},\mu\,\boldsymbol{\Phi})\\ &=\frac{1}{2}\frac{d}{dt}\left(\|\nabla\cdot(\mu\,\boldsymbol{\Phi})\|_0^2+\|rot\,\boldsymbol{\Phi}\|_0^2+\|\mu^{1/2}\,\boldsymbol{\Phi}\|_0^2\right)\end{aligned}$$

and

$$\begin{aligned}&(B_\mu(\boldsymbol{\sigma}-\nabla v),\mu\,B_\mu^{-1}(\boldsymbol{\sigma}-\nabla v))=(B_\mu(\boldsymbol{\sigma}-\nabla v),\mu\,\boldsymbol{\Phi})\\ &\quad=(-\nabla(\nabla\cdot(\mu\,(\boldsymbol{\sigma}-\nabla v)))+\tfrac{1}{\mu}rot\,(rot\,(\boldsymbol{\sigma}-\nabla v))+(\boldsymbol{\sigma}-\nabla v),\mu\,\boldsymbol{\Phi})\\ &\quad=(\nabla\cdot(\mu\,(\boldsymbol{\sigma}-\nabla v)),\nabla\cdot(\mu\,\boldsymbol{\Phi}))+(rot\,(\boldsymbol{\sigma}-\nabla v),rot\,(\boldsymbol{\Phi}))+(\boldsymbol{\sigma}-\nabla v,\mu\,\boldsymbol{\Phi})\\ &\quad=(\mu\,(\boldsymbol{\sigma}-\nabla v),-\nabla(\nabla\cdot(\mu\,\boldsymbol{\Phi}))+\tfrac{1}{\mu}rot\,(rot\,(\boldsymbol{\Phi}))+\boldsymbol{\Phi})\\ &\quad=(\mu\,(\boldsymbol{\sigma}-\nabla v),B_\mu\boldsymbol{\Phi})=(\mu\,(\boldsymbol{\sigma}-\nabla v),B_\mu B_\mu^{-1}(\boldsymbol{\sigma}-\nabla v))=\|\mu^{1/2}\,(\boldsymbol{\sigma}-\nabla v)\|_0^2.\end{aligned} \tag{2.56}$$

From (2.55)–(2.56), we deduce

$$\frac{1}{2}\frac{d}{dt}\left(\|\nabla\cdot(\mu\,\boldsymbol{\Phi})\|_0^2+\|rot\,\boldsymbol{\Phi}\|_0^2+\|\mu^{1/2}\,\boldsymbol{\Phi}\|_0^2\right)+\|\mu^{1/2}\,(\boldsymbol{\sigma}-\nabla v)\|_0^2=0,$$

which together with the fact that $\boldsymbol{\sigma}_0-\nabla v_0=\mathbf{0}$ and $\mu\geq\mu_{min}>0$ imply that

$$\boldsymbol{\sigma}-\nabla v=\mathbf{0}\quad\text{in } L^2(0,T;\mathbf{L}^2(\Omega)),$$

that is $\boldsymbol{\sigma}=\nabla v$ and the proof finished. ■

2.8 Asymptotic Analysis

One of the main characteristics of chemoattractant models is the fact that the solutions can blow up at finite time in space dimension greater or equal to 2. In chemorepulsion models, this phenomenon is not expected. In the literature, we can find some related works. Osaki and Yagi [16] studied the convergence of the solution of the Keller–Segel model to a stationary solution in the one-dimensional case; the convergence of the Keller–Segel model with an additional cross-diffusion term to a steady state is made in Hittmeir and Jungel [13]; the convergence of a chemorepulsion model with linear production is made by Cieslak et al. in [2].

For model (2.2), when constant diffusion coefficients (for instance, $\nu(\boldsymbol{x})=1$ and $\mu(\boldsymbol{x})=1$) are considered, the exponential convergence of any weak solution (u,v) to a constant solution can be proved using Galerkin approximations (see [10]). The main idea is the obtention of a "clean" energy inequality of type

$$\mathcal{E}(u(t_1),v(t_1))+\int_{t_0}^{t_1}\left(\|\nabla u(s)\|_0^2+\frac{1}{2}\|\nabla v(s)\|_0^2+\frac{1}{2}\|\Delta v(s)\|_0^2\right)ds\leq\mathcal{E}(u(t_0),v(t_0)),\tag{2.57}$$

which gives dissipation of the energy $\mathcal{E}(u(t),v(t))$ allowing to obtain uniform bounds for any time $t\in(0,+\infty)$ for the solution $(u(t),v(t))\in L^2(\Omega)\times H^1(\Omega)$.

However, the case $\nu(\boldsymbol{x})=\mu(\boldsymbol{x})=1$ (or constant) is not the only structure for $\nu(\boldsymbol{x})$ and $\mu(\boldsymbol{x})$ that implies an energy inequality of type (2.57). In order to describe possible forms for $\nu(\boldsymbol{x})$ and $\mu(\boldsymbol{x})$ having a clean energy inequality, we focus on the integral version of the energy inequality (2.10). Manipulating the last term of the general energy inequality (2.10) as follows:

$$\int_\Omega\Delta v\,(\nabla v\cdot\nabla\mu)\,d\boldsymbol{x}\leq\|\Delta v\|_0\|\nabla v\|_0\|\nabla\mu\|_{L^\infty(\Omega)}\leq\frac{\mu_{min}}{2}\,\|\Delta v\|_0^2+\frac{\|\nabla\mu\|_{L^\infty(\Omega)}^2}{2\,\mu_{min}}\,\|\nabla v\|_0^2,\tag{2.58}$$

then from (2.10) and (2.58), using (2.6), we can deduce

$$\mathcal{E}(u(t_1),v(t_1))-\mathcal{E}(u(t_0),v(t_0))$$

$$+\int_{t_0}^{t_1}\left(\nu_{min}\,\|\nabla u(s)\|_0^2+\frac{\mu_{min}}{4}\,\|\Delta v(s)\|_0^2+\frac{1}{2}\left(1-\frac{\|\nabla\mu\|_{L^\infty(\Omega)}^2}{\mu_{min}}\right)\|\nabla v(s)\|_0^2\right)ds\leq 0.$$

Therefore, if the following constraint is assumed,

$$\|\nabla\mu\|^2_{L^\infty} < \mu_{min}, \tag{2.59}$$

we can arrive at the following energy inequality (similar to (2.57)):

$$\frac{d}{dt}\left(\frac{1}{2}\|u\|_0^2 + \frac{1}{4}\|\nabla v\|_0^2\right) + \nu_{min}\,\|\nabla u\|_0^2 + \frac{1}{4}\|\nabla v\|_0^2 + \frac{\mu_{min}}{4}\|\Delta v\|_0^2 \leq 0.$$

Starting from this inequality, one can obtain uniform in time weak regularity and to study large-time behavior of the model.

Remark 2.8.1 (Higher Regularity and No Blow-Up). When $\nu(\boldsymbol{x})$ and $\mu(\boldsymbol{x})$ satisfy smallness constraints (2.26) and (2.59), assuming $(2.22)_2$, then the regularities given in (2.27) and (2.30) hold for any time in $(0,+\infty)$, and Corollary 2.6.2 is true for $T = +\infty$. In the explanation, it is necessary to consider Remark 2.5.2. In particular, one has

$$(u, v) \in L^\infty(0,+\infty; L^\infty(\Omega) \times L^\infty(\Omega)),$$

which means that, in this case, any global in time weak solution of (2.2) satisfying $(2.22)_2$ does not blow up, neither at finite nor infinite time.

Remark 2.8.2 (1D and 2D Cases). When a 1D or 2D domain is considered and $\nu(\boldsymbol{x})$ and $\mu(\boldsymbol{x})$ satisfying (2.26) and (2.59), then hypothesis $(2.22)_2$ for $T = +\infty$ is always satisfied. In order to prove it, the bounds for the terms $I_1 - I_3$ in the proof of Lemma 2.5.1 must be done using (2.26). In this case, instead of obtaining estimate (2.23), we arrive at

$$\frac{d}{dt}\left(\|u\|_1^2+\frac{1}{2}\|\nabla v\|_1^2\right) +\nu_{min}\,\|\nabla u\|_1^2+ \min\{\mu_{min},1\}\,\|\nabla v\|_2^2 \leq C\,\left(\|\nabla u\|_0^2\|\nabla v\|_1^2+\|\nabla v\|_0^2\right).$$

In particular, this implies that the weak solution (u, v) of (2.2) does not blow up, neither at finite nor infinite time in the norms associated to the regularity (2.27).

Centering in the case when $\nu(\boldsymbol{x})$ and $\mu(\boldsymbol{x})$ are constant, it is possible to prove the following asymptotic results (see [10]).

In a first step, we obtain exponential bounds for weak solutions a.e. $t \geq 0$. Concretely, the result reads as follows:

Theorem 2.8.1 ([10]). Let (u, v) be any weak–strong solution of problem (2.2) with $\nu(\boldsymbol{x}) = \mu(\boldsymbol{x}) = 1$, obtained by Galerkin approximations. Then, the following estimates hold

$$\|(u(t) - m_0, \nabla v(t))\|_0^2 \leq C_0 \exp(-2t), \quad \text{a.e. } t \geq 0,$$

$$\|v(t) - (m_0)^2\|_0^2 \leq C_0 \exp(-t), \quad \forall t \geq 0,$$

where $m_0 := \dfrac{1}{|\Omega|} \displaystyle\int_\Omega u_0$ and C_0 is a positive constant depending on the data (u_0, v_0), but independent of t.

When stronger norms are used, the following result is also shown.

Theorem 2.8.2 ([10]). For any $\varepsilon > 0$ small enough, it holds

$$\|(u(t) - m_0, \nabla v(t))\|_1^2 \leq \varepsilon \exp\left(-\frac{1}{2}(t - t_2)\right), \quad \text{a.e. } t \geq t_2(\varepsilon),$$

with $t_2 := t_2(\varepsilon) \geq 0$ a large enough time (increasing respect to ε).

Remark 2.8.3. Note that, if the general model with non-constant diffusion coefficients (2.2) is considered, we can guarantee the existence of weak solutions for any finite $T > 0$. However, for any $\mu(\boldsymbol{x})$, we cannot guarantee the exponential decreasing of such solution to a constant function due to the exponential bounds (2.13) and (2.14). In this sense, the fact of introducing a variable $\mu(\boldsymbol{x})$ in Eq. $(2.2)_2$ can modify the asymptotic behavior at infinite time.

2.9 Time-Discrete Schemes

Once the analysis of the continuous model has been made, an important question is how to discretize the problem in order to reproduce numerically the properties observed in the continuous behavior. First, we design time-discrete schemes for both continuous models: (2.2) and (2.32)–(2.33).

In general, we consider $t_n = n\,k$ a uniform partition of $[0, T]$, with $k = T/N$ the time step and $\delta_t a_n = \dfrac{a_n - a_{n-1}}{k}$ a time-discrete derivative.

For the first one, using a backward Euler approximation for the derivative in time, we consider the following first order, nonlinear, and coupled scheme.

- <u>*Scheme* ***UV***</u>:

  Initialization: Start from $u_0 = u(0)$ and $v_0 = v(0)$.

  Time step n: Given $(u_{n-1}, v_{n-1}) \in H^1(\Omega) \times H^2(\Omega)$ with $u_{n-1} \geq 0$ and $v_{n-1} \geq 0$, compute $(u_n, v_n) \in H^1(\Omega) \times H^2(\Omega)$ with $u_n \geq 0$ and $v_n \geq 0$ and solving

$$\begin{cases} (\delta_t u_n, \bar{u}) + (\nu \nabla u_n, \nabla \bar{u}) + (u_n \nabla v_n, \nabla \bar{u}) = 0, & \forall \bar{u} \in H^1(\Omega), \\[2ex] \delta_t v_n + A_\mu v_n - u_n^2 = 0, & \text{a.e. } \boldsymbol{x} \in \Omega, \end{cases}$$

  where the operator A_μ is defined in (2.7).

In the case of considering the reformulation in the $(u, \boldsymbol{\sigma})$-problem (2.32), one also has the following first order, nonlinear, and coupled scheme.

- *Scheme **US***:

  Initialization: Start from $v_0 = v(0)$ and $(u_0, \boldsymbol{\sigma}_0) = (u(0), \boldsymbol{\sigma}(0))$, with $\boldsymbol{\sigma}_0 = \nabla v_0$.

  Time step n: Given $(u_{n-1}, \boldsymbol{\sigma}_{n-1}) \in H^1(\Omega) \times \mathbf{H}^1_{\mathbf{n}}(\Omega)$ with $u_{n-1} \geq 0$, compute $(u_n, \boldsymbol{\sigma}_n) \in H^1(\Omega) \times \mathbf{H}^1_{\mathbf{n}}(\Omega)$ with $u_n \geq 0$ and solving

$$
\begin{cases}
(\delta_t u_n, \bar{u}) + (\nu \nabla u_n, \nabla \bar{u}) + (u_n \boldsymbol{\sigma}_n, \nabla \bar{u}) = 0, \quad \forall \bar{u} \in H^1(\Omega), \\
(\delta_t \boldsymbol{\sigma}_n, \bar{\boldsymbol{\sigma}}) + \langle B_\mu \boldsymbol{\sigma}_n, \bar{\boldsymbol{\sigma}} \rangle - 2(u_n \nabla u_n, \bar{\boldsymbol{\sigma}}) = 0, \quad \forall \bar{\boldsymbol{\sigma}} \in \mathbf{H}^1_{\mathbf{n}}(\Omega),
\end{cases}
$$

  where the operator B_μ is defined in (2.34).

  Once $(u_n, \boldsymbol{\sigma}_n)$ is solved, given $v_{n-1} \in H^1(\Omega)$ with $v_{n-1} \geq 0$, we can recover $v_n = v_n(u_n^2) \in H^2(\Omega)$ (with $v_n \geq 0$) solving

$$
\delta_t v_n + A_\mu v_n = u_n^2, \quad \text{a.e. } \boldsymbol{x} \in \Omega. \tag{2.60}
$$

2.9.1 Main Properties of the Time-Discrete Schemes

As in the continuous case, Scheme **UV** and Scheme **US** satisfy the following properties (see [8] for the proofs).

- Both schemes are **conservative** in u, that is, $\int_\Omega u_n = \int_\Omega u_{n-1} = \cdots = \int_\Omega u_0$. This is the time-discrete version of the continuous total mass conservation property (2.16).
- Scheme **UV** guarantees the positivity of the time-discrete variables: $u_n \geq 0$, $v_n \geq 0$.
- It can prove the **existence of solution** $(u_n, \boldsymbol{\sigma}_n, v_n)$ of scheme **US**, nonnegative for (u_n, v_n). Moreover, under a smallness condition on k, this solution is **unique**.
- Both schemes are equivalents because the following time-discrete version of Lemma 2.7.3 holds.

  Lemma 2.9.1. If $\boldsymbol{\sigma}_{n-1} = \nabla v_{n-1}$, the schemes **UV** and **US** are equivalent in the following sense: If (u_n, v_n) is a solution of scheme **UV**, then $(u_n, \boldsymbol{\sigma}_n)$ with $\boldsymbol{\sigma}_n = \nabla v_n$ solves scheme **US**; and reciprocally, if $(u_n, \boldsymbol{\sigma}_n)$ is a solution of the scheme **US** and $v_n = v_n(u_n^2)$ is the unique solution of (2.60), then $\boldsymbol{\sigma}_n = \nabla v_n$, and therefore, (u_n, v_n) is a solution of the scheme **UV**.

Proof. The proof is similar to that of Lemma 2.7.3.

Let (u_n, v_n) be a solution of the scheme **UV**. Testing the v_n-equation by $-\nabla \cdot \bar{\mathbf{w}}$, with any $\bar{\mathbf{w}} \in \mathbf{H}^1_{\mathbf{n}}(\Omega)$, integrating by parts, and taking into account that $rot(\nabla v_n) = 0$, we obtain

$$(\delta_t(\nabla v_n) + B_\mu(\nabla v_n), \bar{\mathbf{w}}) = (\nabla(u_n^2), \bar{\mathbf{w}}), \quad \forall \bar{\mathbf{w}} \in \mathbf{H}^1_{\mathbf{n}}(\Omega), \tag{2.61}$$

where $B_\mu(\cdot)$ is defined in (2.34).

On the other hand, let $(u_n, \boldsymbol{\sigma}_n)$ be a solution of the scheme **US**

$$(\delta_t \boldsymbol{\sigma}_n + B_\mu(\boldsymbol{\sigma}_n), \bar{\mathbf{w}}) = (\nabla(u_n^2), \bar{\mathbf{w}}), \quad \forall \bar{\mathbf{w}} \in \mathbf{H}^1_{\mathbf{n}}(\Omega), \tag{2.62}$$

and $v_n = v_n(u_n^2)$ satisfies (2.60). It suffices to see that $\boldsymbol{\sigma}_n = \nabla v_n$ (because then the statement of Lemma 2.9.1 is true). Indeed, by comparing (2.61) and (2.62), one has

$$(\delta_t(\boldsymbol{\sigma}_n - \nabla v_n), \bar{\mathbf{w}}) + (B_\mu(\boldsymbol{\sigma}_n - \nabla v_n), \bar{\mathbf{w}}) = 0, \quad \forall \bar{\mathbf{w}} \in \mathbf{H}^1_{\mathbf{n}}(\Omega). \tag{2.63}$$

Now, we can take $\bar{\mathbf{w}} = \boldsymbol{\sigma}_n - \nabla v_n$ in (2.63) because $\boldsymbol{\sigma}_n - \nabla v_n \in \mathbf{H}^1_{\mathbf{n}}(\Omega)$. Therefore, considering $\bar{\mathbf{w}} = \mu\,(\boldsymbol{\sigma}_n - \nabla v_n)$ in (2.63), we obtain

$$(\delta_t(\boldsymbol{\sigma}_n - \nabla v_n), \mu\,(\boldsymbol{\sigma}_n - \nabla v_n)) + (B_\mu(\boldsymbol{\sigma}_n - \nabla v_n), \mu\,(\boldsymbol{\sigma}_n - \nabla v_n)) = 0.$$

Observe that using the formula $a\,(a-b) = \frac{1}{2}\,(a^2 - b^2) + \frac{1}{2}\,(a-b)^2$, we deduce

$$\begin{aligned}\delta_t\left(\frac{1}{2}\|\sqrt{\mu}\,(\boldsymbol{\sigma}_n - \nabla v_n)\|_0^2\right) \quad &+ \quad \frac{k}{2}\|\delta_t\sqrt{\mu}\,(\boldsymbol{\sigma}_n - \nabla v_n)\|_0^2 + \|\nabla\cdot(\mu(\boldsymbol{\sigma}_n - \nabla v_n))\|_0^2 \\ &+ \quad \|rot\,(\boldsymbol{\sigma}_n - \nabla v_n)\|_0^2 + \|\sqrt{\mu}\,(\boldsymbol{\sigma}_n - \nabla v_n)\|_0^2 = 0,\end{aligned}$$

which together with the fact that $\boldsymbol{\sigma}_{n-1} - \nabla v_{n-1} = \mathbf{0}$ and $\mu \geq \mu_{min} > 0$ implies that

$$\boldsymbol{\sigma}_n - \nabla v_n = \mathbf{0} \quad \text{in } \mathbf{H}^1(\Omega),$$

that is $\boldsymbol{\sigma}_n = \nabla v_n$. ■

Attending the equivalence given in Lemma 2.9.1, we focus on the analysis of Scheme **US**.

- Scheme **US** is **unconditionally energy stable** when a smallness condition like (2.59) is assumed (see (2.65) below) because it holds

$$\begin{aligned}&\delta_t\mathcal{E}(u_n, \boldsymbol{\sigma}_n) + \frac{k}{2}\|\delta_t u_n\|_0^2 + \frac{k}{4}\|\delta_t\boldsymbol{\sigma}_n\|_0^2 + \nu_{min}\|\nabla u_n\|_0^2\,d\boldsymbol{x} \\ &+\frac{1}{4}\left(\mu_{min}\,\|\nabla\cdot\boldsymbol{\sigma}_n\|_0^2 + \frac{1}{\|\mu\|_{L^\infty}}\,\|rot\,\boldsymbol{\sigma}_n\|_0^2 + \|\boldsymbol{\sigma}_n\|_0^2\right) \leq 0,\end{aligned} \tag{2.64}$$

where $\mathcal{E}(u_n, \boldsymbol{\sigma}_n)$ is defined in (2.47). Indeed, testing by $(u_n, \boldsymbol{\sigma}_n)$ and canceling the chemotaxis and the production terms, one has

$$\delta_t \mathcal{E}(u_n, \boldsymbol{\sigma}_n) + \frac{k}{2}\|\delta_t u_n\|_0^2 + \frac{k}{4}\|\delta_t \boldsymbol{\sigma}_n\|_0^2 + \int_\Omega \nu |\nabla u_n|^2 \, d\boldsymbol{x}$$

$$+\frac{1}{2}\int_\Omega \left(\mu\,|\nabla\cdot\boldsymbol{\sigma}_n|^2 + \frac{1}{\mu}|rot\,\boldsymbol{\sigma}_n|^2 + |\boldsymbol{\sigma}_n|^2\right) d\boldsymbol{x} = K(\mu, \boldsymbol{\sigma}_n),$$

where

$$K(\mu, \boldsymbol{\sigma}_n) = -\frac{1}{2}\int_\Omega \left((\nabla\mu\cdot\boldsymbol{\sigma}_n)\,(\nabla\cdot\boldsymbol{\sigma}_n) - rot\,\boldsymbol{\sigma}_n \cdot \left(\frac{\nabla\mu}{\mu^2}\times\boldsymbol{\sigma}_n\right)\right) d\boldsymbol{x}.$$

Note that the terms on $K(\mu, \boldsymbol{\sigma}_n)$ satisfy estimates (2.49) and (2.50). Therefore, (2.64) is obtained assuming

$$\frac{1}{2\mu_{min}}\|\nabla\mu\|_{L^\infty}^2 \left(1 + \frac{\|\mu\|_{L^\infty}}{\mu_{min}^3}\right) < 1. \tag{2.65}$$

- As a consequence of (2.64), we have estimates $(u_n, \boldsymbol{\sigma}_n) \in l^\infty L^2 \cap l^2 H^1$, and $\int_\Omega v_n \le C$, implying also estimates for $v_n \in l^\infty H^1 \cap l^2 H^2$. In fact, more regular bounds can also be obtained. Notice that $l^p(H)$ is the space of discrete in time H-valued functions.
- In the case of $\nu(\boldsymbol{x}) = \mu(\boldsymbol{x}) = 1$, the time-discrete solution (u_n, σ_n) of Scheme **US** converges toward a weak solution of the continuous model (2.32).

  Theorem 2.9.1 ([8]). There exists a subsequence (k'), with $k' \to 0$, and a weak solution $(u, \boldsymbol{\sigma})$ of (2.32) in $(0, T)$, such that the sequence $(u_{k'}, \boldsymbol{\sigma}_{k'})$ of solutions of discrete scheme **US** corresponding to k' converges to $(u, \boldsymbol{\sigma})$ weakly-* in $L^\infty(0, T; L^2(\Omega))$, weakly in $L^2(0, T; H^1(\Omega))$, and strongly in $L^2(0, T; L^2(\Omega))$.
- Again in the case of $\nu(\boldsymbol{x}) = \mu(\boldsymbol{x}) = 1$, **error estimates** ($\mathcal{O}(k)$) for $e_u^n = u(t_n) - u_n$, $e_{\boldsymbol{\sigma}}^n = \boldsymbol{\sigma}(t_n) - \boldsymbol{\sigma}_n$, $e_v^n = v(t_n) - v_n$ have been obtained in [8]. In fact, assuming

$$(u, \boldsymbol{\sigma}) \in L^\infty(0, T; H^1(\Omega) \times \mathbf{H}^1(\Omega)) \text{ and } (u_{tt}, \boldsymbol{\sigma}_{tt}) \in L^2(0, T; H^1(\Omega)' \times \mathbf{H}^1_{\mathbf{n}}(\Omega)') \tag{2.66}$$

for the exact solution of (2.32), and

$$k\left(\|(\nabla u, \nabla\cdot\boldsymbol{\sigma})\|_{L^\infty(L^2)}^4\right) \quad \text{is small enough,}$$

then the a priori error estimate

$$\|(e_u^n, e_{\boldsymbol{\sigma}}^n)\|_{l^\infty L^2 \cap l^2 H^1} \le C(T)\,k$$

holds, where $C(T) = K_1 exp(K_2 T)$, with $K_1, K_2 > 0$ independent of k.

Moreover, if the exact solution has the additional regularity

$$v_{tt} \in L^1(0,T; H^1(\Omega)'),$$

then the a priori error estimate holds

$$\|e_v^n\|_{l^\infty H^1 \cap l^2 H^2} \leq C(T)\, k,$$

where $C(T) = K_1 exp(K_2 T)$, with $K_1, K_2 > 0$ independent of k.

2.9.2 A Linear Scheme

As in [8], we propose the following first order in time, linear coupled scheme for model (2.32):

- *Scheme* ***LC***:
  Initialization: Start from $v_0 = v(0)$ and $(u_0, \boldsymbol{\sigma}_0) = (u(0), \boldsymbol{\sigma}(0))$, with $\boldsymbol{\sigma}_0 = \nabla v_0$.
  Time step n: Given $(u_{n-1}, \boldsymbol{\sigma}_{n-1}) \in H^1(\Omega) \times \mathbf{H}^1_{\mathbf{n}}(\Omega)$, compute $(u_n, \boldsymbol{\sigma}_n) \in H^1(\Omega) \times \mathbf{H}^1_{\mathbf{n}}(\Omega)$ solving

$$\begin{cases} (\delta_t u_n, \bar{u}) + (\nu\, \nabla u_n, \nabla \bar{u}) = -(u_{n-1}\boldsymbol{\sigma}_n, \nabla \bar{u}), & \forall \bar{u} \in H^1(\Omega), \\ (\delta_t \boldsymbol{\sigma}_n, \bar{\boldsymbol{\sigma}}) + \langle B_\mu \boldsymbol{\sigma}_n, \bar{\boldsymbol{\sigma}} \rangle = 2(u_{n-1}\nabla u_n, \bar{\boldsymbol{\sigma}}), & \forall \bar{\boldsymbol{\sigma}} \in \mathbf{H}^1_{\mathbf{n}}(\Omega). \end{cases}$$

Again, once solved the scheme **LC**, given $v_{n-1} \in H^1(\Omega)$ with $v_{n-1} \geq 0$, we can recover $v_n = v_n(u_n^2) \in H^2(\Omega)$ solving (2.60).

In the case of $\nu(\boldsymbol{x}) = \mu(\boldsymbol{x}) = 1$, for the scheme **LC**, it is possible to prove the same results obtained for the schemes **US** and **UV** given in Sect. 2.9.1, excepting the following two properties (see [8]):

1. It is not clear how to prove the equivalence of the schemes **LC** and **UV** in the sense of Lemma 2.9.1 because the relation $\boldsymbol{\sigma}_n = \nabla v_n$ it is not clear.
2. Although it is possible to prove unconditional well-posedness of the scheme **LC**, the nonnegativity of u_n does not hold in general.

2.10 Fully Discrete Schemes

The literature on numerical analysis for chemotaxis models is scarce: For the Keller–Segel system (i.e., with chemoattraction and linear production), Filbet studied in [7] the existence of discrete solutions and the convergence of a finite volume scheme. Saito, in [19, 20], proved error estimates for a conservative finite element (FE) approximation. A mixed FE approximation is studied in [17], and in [4], some error estimates are proved for a fully discrete discontinuous FE method.

However, as far as we know, there are not works studying FE schemes satisfying the property of energy stability related to the energy inequality (2.57).

Here, we center on two fully discrete schemes, in time and space. The time discretization is based on Schemes **US** and **UV**, and the space discretization uses FE approximations. For simplicity, we consider $\nu(\boldsymbol{x}) = \mu(\boldsymbol{x}) = 1$ for the design of the fully discrete schemes and their numerical analysis. Then, here we state the main results given in [9, 10].

We consider for simplicity a uniform partition of $[0, T]$ with time step $k = T/N : (t_n = nk)_{n=0}^{n=N})$. For the space discretization, we consider $\{\mathcal{T}_h\}_{h>0}$ be a family of shape-regular and quasi-uniform triangulations of $\overline{\Omega}$ made up of simplexes K (intervals in 1D, triangles in 2D, and tetrahedra in 3D), so that $\overline{\Omega} = \cup_{K\in\mathcal{T}_h} K$, where $h = \max_{K\in\mathcal{T}_h} h_K$, with h_K being the diameter of K. Further, let $\mathcal{N}_h = \{\mathbf{a}_i\}_{i\in\mathcal{I}}$ denote the set of all the nodes of $\mathcal{T}_h$. The election of continuous FE spaces for u, $\boldsymbol{\sigma}$, and v can be the following:

$$(U_h, \boldsymbol{\Sigma}_h, V_h) \subset H^1 \times \mathbf{H}^1_{\mathbf{n}} \times W^{1,6}, \quad \text{generated by } \mathbb{P}_m, \mathbb{P}_r, \mathbb{P}_s \text{ with } m, r, s \geq 1.$$

We also need to use $A_h : U_h \to U_h$, $B_h : \boldsymbol{\Sigma}_h \to \boldsymbol{\Sigma}_h$ and $\widetilde{A}_h : V_h \to V_h$, and the linear operators are defined as follows:

$$\begin{cases} (A_h u_h, \bar{u}_h) = (\nabla u_h, \nabla \bar{u}_h) + (u_h, \bar{u}_h), \quad \forall \bar{u}_h \in U_h, \\ (B_h \boldsymbol{\sigma}_h, \bar{\boldsymbol{\sigma}}_h) = (\nabla\cdot\boldsymbol{\sigma}_h, \nabla\cdot\bar{\boldsymbol{\sigma}}_h) + (\text{rot}\ \boldsymbol{\sigma}_h, \text{rot}\ \bar{\boldsymbol{\sigma}}_h) + (\boldsymbol{\sigma}_h, \bar{\boldsymbol{\sigma}}_h), \quad \forall \bar{\boldsymbol{\sigma}}_h \in \boldsymbol{\Sigma}_h, \\ (\widetilde{A}_h v_h, \bar{v}_h) = (\nabla v_h, \nabla \bar{v}_h) + (v_h, \bar{v}_h), \quad \forall \bar{v}_h \in V_h. \end{cases} \tag{2.67}$$

Moreover, some interpolation operators must be used:

$$\mathcal{R}_h^u : H^1(\Omega) \to U_h, \quad \mathcal{R}_h^{\boldsymbol{\sigma}} : \mathbf{H}^1_{\mathbf{n}}(\Omega) \to \boldsymbol{\Sigma}_h, \quad \mathcal{R}_h^v : H^1(\Omega) \to V_h$$

such that for all $u \in H^1(\Omega)$, $\boldsymbol{\sigma} \in \mathbf{H}^1_{\mathbf{n}}(\Omega)$, and $v \in H^1(\Omega)$, $\mathcal{R}_h^u u \in U_h$, $\mathcal{R}_h^{\boldsymbol{\sigma}} \boldsymbol{\sigma} \in \boldsymbol{\Sigma}_h$ and $\mathcal{R}_h^v v \in V_h$ satisfy

$$(\nabla(\mathcal{R}_h^u u - u), \nabla \bar{u}_h) + (\mathcal{R}_h^u u - u, \bar{u}_h) = 0, \quad \forall \bar{u}_h \in U_h,$$

$$(\nabla\cdot(\mathcal{R}_h^{\boldsymbol{\sigma}}\boldsymbol{\sigma} - \boldsymbol{\sigma}), \nabla\cdot\bar{\boldsymbol{\sigma}}_h) + (\text{rot}(\mathcal{R}_h^{\boldsymbol{\sigma}}\boldsymbol{\sigma} - \boldsymbol{\sigma}), \text{rot}\ \bar{\boldsymbol{\sigma}}_h) + (\mathcal{R}_h^{\boldsymbol{\sigma}}\boldsymbol{\sigma} - \boldsymbol{\sigma}, \bar{\boldsymbol{\sigma}}_h) = 0, \quad \forall \bar{\boldsymbol{\sigma}}_h \in \boldsymbol{\Sigma}_h,$$

$$(\nabla(\mathcal{R}_h^v v - v), \nabla \bar{v}_h) + (\mathcal{R}_h^v v - v, \bar{v}_h) = 0, \quad \forall \bar{v}_h \in V_h.$$

Collecting all the previous information, the **fully discrete schemes US and UV** read as follows:

- *Scheme **UV***:

  Initialization: Take $(u_h^0, v_h^0) = (\mathcal{R}_h^u u_0, \mathcal{R}_h^v v_0) \in U_h \times V_h$. Then, $\int_\Omega u_h^0 = \int_\Omega u_0 = m_0$.

  Time step n: Given $(u_h^{n-1}, v_h^{n-1}) \in U_h \times V_h$, compute $(u_h^n, v_h^n) \in U_h \times V_h$ solving

$$\left\{ \begin{array}{l} (\delta_t u_h^n, \bar{u}_h) + (\nabla u_h^n, \nabla \bar{u}_h) + (u_h^n \nabla v_h^n, \nabla \bar{u}_h) = 0, \quad \forall \bar{u}_h \in U_h, \\ (\delta_t v_h^n, \bar{v}_h) + (\widetilde{A}_h v_h^n, \bar{v}_h) - ((u_h^n)^2, \bar{v}_h) = 0, \quad \forall \bar{v}_h \in V_h, \end{array} \right.$$

  where the linear operator $\widetilde{A}_h$ is defined in (2.67).

- *Scheme **US***:

- **Initialization**: Take $(u_h^0, \boldsymbol{\sigma}_h^0) = (\mathcal{R}_h^u u_0, \mathcal{R}_h^{\boldsymbol{\sigma}} \boldsymbol{\sigma}_0) \in U_h \times \boldsymbol{\Sigma}_h$ and $v_h^0 = \mathcal{R}_h^v v_0 \in V_h$. Then, $\int_\Omega u_h^0 = \int_\Omega u_0 = m_0$.
- **Time step** n: Given $(u_h^{n-1}, \boldsymbol{\sigma}_h^{n-1}) \in U_h \times \boldsymbol{\Sigma}_h$, compute $(u_h^n, \boldsymbol{\sigma}_h^n) \in U_h \times \boldsymbol{\Sigma}_h$ solving

$$\left\{ \begin{array}{l} (\delta_t u_h^n, \bar{u}_h) + (\nabla u_h^n, \nabla \bar{u}_h) + (u_h^n \boldsymbol{\sigma}_h^n, \nabla \bar{u}_h) = 0, \quad \forall \bar{u}_h \in U_h, \\ \\ (\delta_t \boldsymbol{\sigma}_h^n, \bar{\boldsymbol{\sigma}}_h) + (B_h \boldsymbol{\sigma}_h^n, \bar{\boldsymbol{\sigma}}_h) - 2(u_h^n \nabla u_h^n, \bar{\boldsymbol{\sigma}}_h) = 0, \quad \forall \bar{\boldsymbol{\sigma}}_h \in \boldsymbol{\Sigma}_h, \end{array} \right.$$

  where the linear operator B_h is defined in (2.67). Once $(u_h^n, \boldsymbol{\sigma}_h^n)$ is computed, given $v_h^{n-1} \in V_h$, we can recover $v_h^n = v_h^n((u_h^n)^2) \in V_h$ solving

$$(\delta_t v_h^n, \bar{v}_h) + (\widetilde{A}_h v_h^n, \bar{v}_h) = ((u_h^n)^2, \bar{v}_h), \quad \forall \bar{v}_h \in V_h. \tag{2.68}$$

  Given $u_h^n \in U_h$ and $v_h^{n-1} \in V_h$, Lax–Milgram theorem implies that there exists a unique $v_h^n \in V_h$ solution of (2.68).

The main properties satisfied by these fully discrete schemes are summarized below (for a detailed proof of them, see [9, 10]):

- Total mass conservation for u_n, $\int_\Omega u_h^n$, that is,

$$\int_\Omega u_h^n = \int_\Omega u_h^{n-1} = \cdots = \int_\Omega u_h^0,$$

  and the following behavior for $\int_\Omega v_h^n$:

$$\delta_t \left(\int_\Omega v_h^n \right) = \int_\Omega (u_h^n)^2 - \int_\Omega v_h^n.$$

- Existence of solution $(u_n, \boldsymbol{\sigma}_n, v_n)$ of the scheme **US** and (u_n, v_n) of the scheme **UV**, and uniqueness of such solution under a smallness condition on k.

- Unconditional energy stability for the scheme **US**, in the sense

$$\delta_t \mathcal{E}(u_n, \boldsymbol{\sigma}_n) + \frac{k}{2}\|\delta_t u_n\|_0^2 + \frac{k}{4}\|\delta_t \boldsymbol{\sigma}_n\|_0^2 + \|\nabla u_n\|_0^2 + \frac{1}{2}\|\boldsymbol{\sigma}_n\|_1^2 = 0.$$

  In the case of the scheme **UV**, this property can also be proved, if the discrete spaces (U_h, V_h) are generated by $(\mathbb{P}_m, \mathbb{P}_{2m})$-continuous FE, with $m \geq 1$ (see [10]).

- Weak estimates for the discrete solution of the schemes **US** and **UV**, and stronger estimates for the scheme **US** assuming an additional hypothesis (that can be considered as a discrete version of (2.22)). Concretely, one has

$$\|(u_h^n, \boldsymbol{\sigma}_h^n)\|_1^2 \leq K_0 \quad \forall n \geq 0,$$

  with $K_0 > 0$ a constant depending on the initial data, but independent of (k, h) and n.

- Convergence toward weak solutions for both schemes (in a similar sense of the corresponding time-discrete schemes).

- Optimal error estimates for $e_u^n = u(t_n) - u_h^n$, $\mathcal{R}_h^u : H^1(\Omega) \to U_h$, $e_u^n = (\mathcal{I} - \mathcal{R}_h^u)u(t_n) + \mathcal{R}_h^u u(t_n) - u_h^n = e_{u,i}^n + e_{u,h}^n$,

$$\|(e_{u,h}^n, e_{\boldsymbol{\sigma},h}^n)\|_{l^\infty L^2 \cap l^2 H^1}, \|e_{v,h}^n\|_{l^\infty H^1 \cap l^2 W^{1,6}} \leq C(T)(k + h^{m+1}).$$

2.10.1 Large-Time Behavior

It is possible to prove the exponential decreasing of the solution of the schemes **US** and **UV** to the constant solution (m_0, m_0^2). In fact, we have the following results.

Theorem 2.10.1 ([10]). Let (u_h^n, v_h^n) be a solution of the scheme **UV** associated to an initial data $(u_h^0, v_h^0) \in U_h \times V_h$, which is a suitable approximation of $(u_0, v_0) \in L^2(\Omega) \times H^1(\Omega)$, as $h \downarrow 0$, with $\frac{1}{|\Omega|}\int_\Omega u_h^0 = \frac{1}{|\Omega|}\int_\Omega u_0 = m_0$. Then,

$$\|(u_h^n - m_0, \nabla v_h^n)\|_0^2 \leq C_0 \exp\Big(-\frac{2}{1+2k}kn\Big), \quad \forall n \geq 0,$$

$$\|v_h^n - (m_0)^2\|_0^2 \leq C_0 \exp\Big(-\frac{1}{1+k}kn\Big), \quad \forall n \geq 0,$$

$$k\sum_{m>n}\Big(\|\widetilde{u}_h^m\|_1^2 + \frac{1}{2}\|(A_h - I)v_h^n\|_0^2 + \frac{1}{2}\|\nabla v_h^n\|_0^2\Big) \leq C_0 \exp\Big(-\frac{2}{1+2k}kn\Big), \quad \forall n \geq 0,$$

where C_0 is a positive constant depending on the data (u_0, v_0), but independent of (k, h) and n.

Theorem 2.10.2 ([10]). Let $(u_h^n, \boldsymbol{\sigma}_h^n, v_h^n)$ be a solution of the scheme **US** associated to an initial data $(u_h^0, \boldsymbol{\sigma}_h^0, v_h^0) \in U_h \times \boldsymbol{\Sigma}_h \times V_h$, which is a suitable approximation of $(u_0, \boldsymbol{\sigma}_0, v_0) \in L^2(\Omega) \times \mathbf{L}^2(\Omega) \times H^1(\Omega)$, as $h \to 0$, with $\frac{1}{|\Omega|}\int_\Omega u_h^0 = \frac{1}{|\Omega|}\int_\Omega u_0 = m_0$. Then,

$$\|(u_h^n - m_0, \boldsymbol{\sigma}_h^n)\|_0^2 \le C_0 \exp\Big(-\frac{2}{1+2k} kn\Big), \quad \forall n \ge 0,$$

$$\|v_h^n - (m_0)^2\|_0^2 \le C_0 \exp\Big(-\frac{1}{1+k} kn\Big), \quad \forall n \ge 0,$$

$$k \sum_{m>n} \Big(\|\widetilde{u}_h^m\|_1^2 + \frac{1}{2}\|\boldsymbol{\sigma}_h^m\|_1^2 \Big) \le C_0 \exp\Big(-\frac{2}{1+2k} kn\Big), \quad \forall n \ge 0,$$

where C_0 is a positive constant depending on the data $(u_0, \boldsymbol{\sigma}_0)$, but independent of (k, h) and n.

2.11 Numerical Simulations

Some numerical simulations have been made in order to show the well-behavior of the fully discrete schemes. Below, we explain the technical details:

- We use $\mathbb{P}_1 \times \mathbb{P}_1 \times \mathbb{P}_2$-continuous FE approximation for $U_h \times \boldsymbol{\Sigma}_h \times V_h$.
- The domain is $\Omega = [0, 2]^2$, and a structured mesh has been used.
- All the simulations are carried out using **FreeFem++** software.
- The linear iterative method used to approach the nonlinear schemes **US** and **UV** is the Newton Method, and in all the cases, the iterative method stops when the relative error in L^2-norm is less than $tol = 10^{-6}$.

Positivity: Taking into account that the positivity of u_h is not clear for the fully discrete schemes **US** and **UV** (and **LC**); and recalling that for the time-discrete scheme **US** and **UV** the existence of nonnegative solution (u_n, v_n) can be proved (see Sect. 2.9.1), while for the time-discrete scheme **LC**, the nonnegativity of u_n is not clear (see Sect. 2.9.2), some numerical experiments have been performed to quantify in practice the nonnegativity. For this, we have considered the following initial data (see Fig. 2.1):

$$u_0 = -10xy(2-x)(2-y)exp(-10(y-1)^2 - 10(x-1)^2) + 10.0001$$

and

$$v_0 = 200xy(2-x)(2-y)exp(-30(y-1)^2 - 30(x-1)^2) + 0.0001.$$

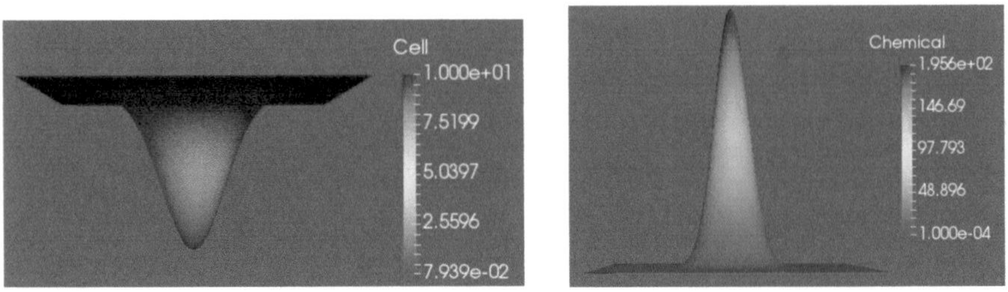

(a) (b)

Figure 2.1: Initial conditions. (**a**) Initial cell density u_0. (**b**) Initial chemical concentration v_0

Taking the time-discrete parameter $k = 10^{-5}$, and meshes in space with increasingly thinner space-discrete parameter ($h = \frac{1}{70}$, $h = \frac{1}{150}$ and $h = \frac{1}{300}$), we observe that the positivity of u_h is not obtained for both schemes **US** and **LC** (see Fig. 2.2a–c); but when the spatial parameter tends to 0, scheme **US** works better than **LC**. Finally, the chemical variable v seems to be positive in all cases, see Fig. 2.2d.

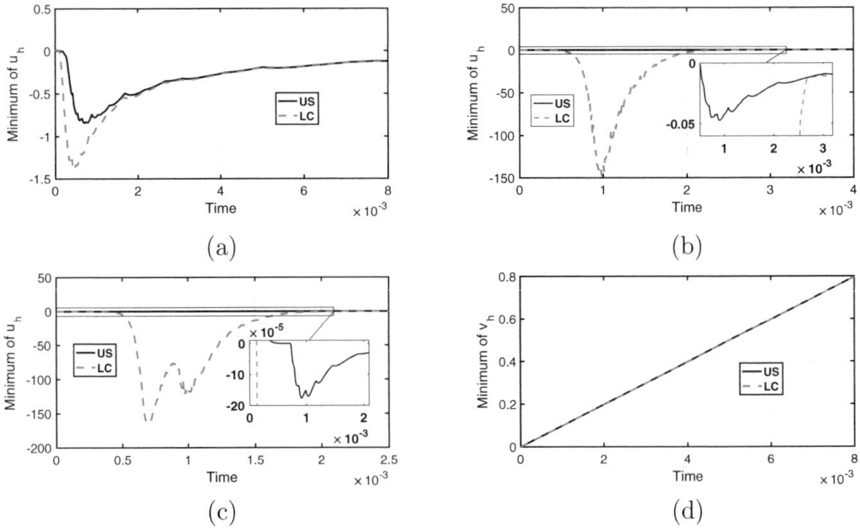

Figure 2.2: Minimum values of u_h and v_h for several values of h. (**a**) Minimum values of u_h, with $h = \frac{1}{35}$. (**b**) Minimum values of u_h, with $h = \frac{1}{75}$. (**c**) Minimum values of u_h, with $h = \frac{1}{150}$. (**d**) Values of the minimum of v_h

Energy Stability: We compare the behavior of the schemes with respect to the following "exact" energy, which comes from the continuous problem:

$$\mathcal{E}(u, v) = \frac{1}{2}\|u\|_0^2 + \frac{1}{4}\|\nabla v\|_0^2.$$

Moreover, we study the behavior of the corresponding discrete residual of the energy law

$$RE^n := \delta_t \mathcal{E}(u_h^n, v_h^n) + \|\nabla u_h^n\|_0^2 + \frac{1}{2}\|\nabla v_h^n\|_0^2 + \frac{1}{2}\|(A_h - I)v_h^n\|_0^2, \quad \forall n.$$

Then, taking $k = 10^{-6}$, $h = \frac{1}{25}$ and similar initial conditions as given in Fig. 2.1, we observe that both schemes **US** and **LC** have decreasing energies and negative RE^n (see Fig. 2.3).

Numerical accuracy: We consider an exact solution for problem (2.2) for $\nu(\boldsymbol{x}) = \mu(\boldsymbol{x}) = 1$, $k = 10^{-5}$ and obtain the spatial numerical error orders for schemes **US** and **UV** that are shown in Tables 2.1 and 2.2. We can see that in both cases, when $h \to 0$, $\|u(t_n) - u_h^n\|_{l^2 H^1}$ is convergent in optimal rate $\mathcal{O}(h)$, and $\|u(t_n) - u_h^n\|_{l^\infty L^2}$ and $\|v(t_n) - v_h^n\|_{l^\infty H^1}$ are convergent in optimal rate $\mathcal{O}(h^2)$.

Asymptotic behavior: We present one numerical experiment in order to illustrate the asymptotic behavior of schemes **UV** and **US**. We consider $k = 10^{-3}$, $h = \frac{1}{25}$, and the initial conditions:

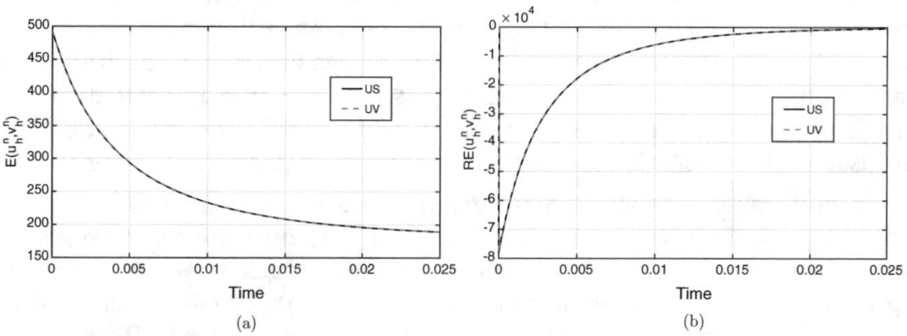

Figure 2.3: Behavior of $\mathcal{E}(u_h^n, v_h^n)$ and RE^n. (**a**) Energy $\mathcal{E}(u_h^n, v_h^n)$ of schemes **UV** and **US**. (**b**) Residual RE^n of schemes **UV** and **US**

Error–h	1/40–1/50	1/50–1/60	1/60–1/70
$\|e_u^n\|_{l^\infty L^2}$	1.9970	1.9980	1.9985
$\|e_u^n\|_{l^2 H^1}$	0.9979	0.9986	0.9989
$\|e_v^n\|_{l^\infty H^1}$	1.9985	1.9990	1.9993

Table 2.1: Error orders in several spaces for scheme **US**

$$u_0 = 5cos(2\pi x)cos(2\pi y) + 5.0001 \quad \text{and} \quad v_0 = -15cos(2\pi x)cos(2\pi y)) + 24.$$

We observe in Fig. 2.4 that $\|(u_h^n - m_0, \nabla v_h^n)\|_0^2$ and $\|u_h^n - m_0\|_1^2$ decrease to 0 faster than $\|v_h^n - (m_0)^2\|_0^2$ and $\|A_h(v_h^n - (m_0)^2)\|_0^2$.

2.12 Conclusions

We have studied the model (2.2) representing a chemorepulsion model with quadratic production and non-constant diffusion coefficients. In the study made for the continuous model, the asymptotic behavior is proved under some particular smallness constraints for the diffusion coefficients $\nu(\boldsymbol{x})$ and $\mu(\boldsymbol{x})$.

With respect to the numerical analysis, we have studied first time-discrete schemes and later fully discrete schemes. For the time-discrete case, we present three unconditionally mass-conservative and energy stable first order time schemes: two (nonlinear) backward Euler type schemes (which conserve positivity) and one linear approximation (where the positivity is not conserved in general). For the fully discrete context, using the auxiliary variable $\boldsymbol{\sigma} = \nabla v$, we design a FE backward Euler scheme that is unconditionally mass-conservative and energy stable.

In both cases, continuous and discrete settings, we analyze positivity, solvability, convergence toward weak solutions, and error estimates of these schemes. In particular, uniqueness of the nonlinear scheme is obtained assuming small time step with respect to a strong norm of the scheme. This hypothesis is avoided in $1D$ and $2D$ domains (the continuous analysis has been made in Remarks 2.5.4 and 2.6.3), where a global in time strong estimate is proved.

With respect to the asymptotic behavior, we analyze the large-time behavior of the global weak solutions under smallness constraints for $\nu(\boldsymbol{x})$ and $\mu(\boldsymbol{x})$, and one has the exponential convergence to a constant state as time goes to infinity, in the case of constant $\nu(\boldsymbol{x})$ and $\mu(\boldsymbol{x})$.

Finally, we provide some numerical results in agreement with our theoretical analysis.

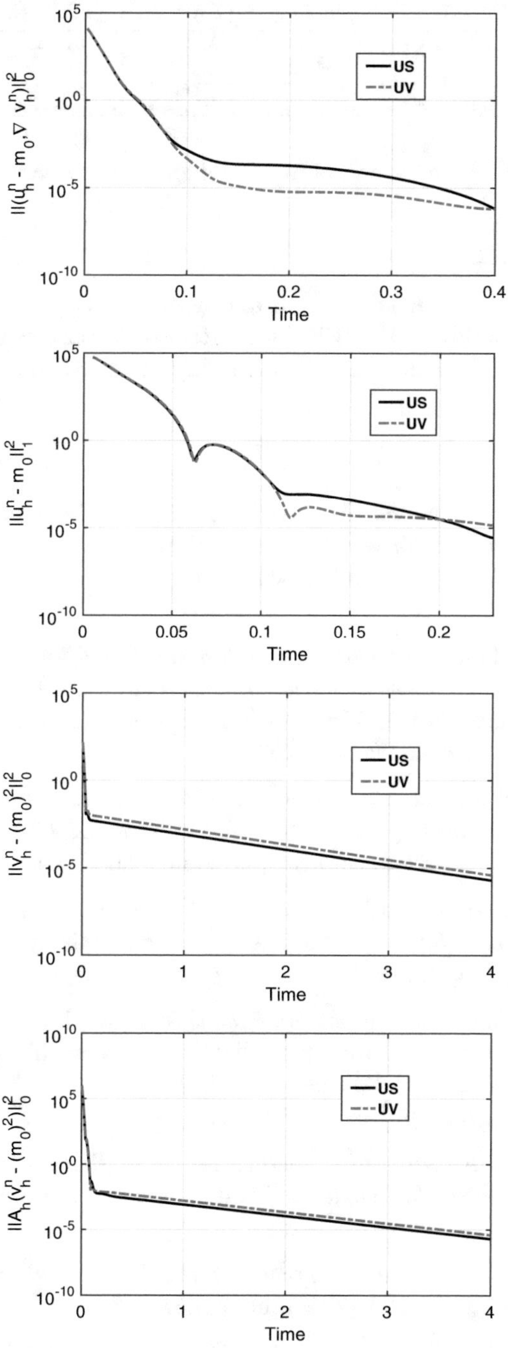

Figure 2.4: Evolution of $\|(u_h^n - m_0, \nabla v_h^n)\|_0^2$, $\|u_h^n - m_0\|_1^2$, $\|v_h^n - (m_0)^2\|_0^2$ and $\|A_h(v_h^n - (m_0)^2)\|_0^2$

Error–h	1/40–1/50	1/50–1/60	1/60–1/70
$\|e_u^n\|_{l^\infty L^2}$	1.9968	1.9979	1.9985
$\|e_u^n\|_{l^2 H^1}$	0.9978	0.9985	0.9989
$\|e_v^n\|_{l^\infty H^1}$	1.9985	1.9990	1.9993

Table 2.2: Error orders in several spaces for scheme **UV**

Acknowledgment

This research was partially supported by MINECO grant MTM2015-69875-P with the participation of FEDER. D. A. Rueda-Gómez has also been partially supported by the VIE of the Universidad Industrial de Santander (Colombia).

References

1. C. Amrouche and N. E. H. Seloula, *Lp-theory for vector potentials and Sobolev's inequalities for vector fields: application to the Stokes equations with pressure boundary conditions.* Math. Models Methods Appl. Sci. 23 (2013), no. 1, 37–92.

2. T. Cieslak, P. Laurençot and C. Morales-Rodrigo, *Global existence and convergence to steady states in a chemorepulsion system.* Parabolic and Navier-Stokes equations. Part 1, 105–117, Banach Center Publ., 81, Part 1, Polish Acad. Sci. Inst. Math., Warsaw, 2008.

3. F. Demengel and G. Demengel, *Functional spaces for the theory of elliptic partial differential equations.* Translated from the 2007 French original by Reinie Erné. Universitext. Springer, London; EDP Sciences, Les Ulis (2012).

4. Y. Epshteyn and A. Izmirlioglu, *Fully discrete analysis of a discontinuous finite element method for the Keller-Segel chemotaxis model.* J. Sci. Comput. **40** (2009), no. 1–3, 211–256.

5. A. Ern, J.-L. Guermond, *Éléments finis: théorie, applications, mise en œuvre.* (French) [Finite elements: theory, applications, implementation] Matématiques & Applications (Berlin) [Mathematics & Applications], 36. Springer-Verlag, Berlin, 2002. x+430 pp. ISBN: 3-540-42615-9

6. E. Feireisl and A. Novotný, *Singular limits in thermodynamics of viscous fluids. Advances in Mathematical Fluid Mechanics.* Birkhäuser Verlag, Basel (2009).

7. F. Filbet, *A finite volume scheme for the Patlak-Keller-Segel chemotaxis model.* Numer. Math. **104** (2006), no. 4, 457–488.

8. F. Guillén-González, M. A. Rodríguez-Bellido, D. A. Rueda-Gómez, *Study of a chemo-repulsion model with quadratic production. Part I: Analysis of the*

continuous problem and time-discrete numerical schemes. ArXiv:1803.02386. *Comput. Math. Appl.* **80** (2020), 692–713. https://doi.org/10.1016/j.camwa.2020.04.009

9. F. Guillén-González, M. A. Rodríguez-Bellido, D. A. Rueda-Gómez, *Study of a chemo-repulsion model with quadratic production. Part II: Analysis of an unconditional energy-stable fully discrete scheme.* ArXiv:1803.02391. *Comput. Math. Appl.* **80** (2020), 636–652. https://doi.org/10.1016/j.camwa.2020.04.010
10. F. Guillén-González, M. A. Rodríguez-Bellido, D. A. Rueda-Gómez, *Asymptotic behaviour for a chemo-repulsion system with quadratic production: The continuous problem and two fully discrete numerical schemes.* ArXiv:1805.00962.
11. F. Guillén-González, M. A. Roríguez-Bellido, D. A. Rueda-Gómez, *Unconditionally energy stable fully discrete schemes for a chemo-repulsion model. Math. Comp.* 88 (2019), no. 319, 2069–2099.
12. F. Guillén-González, M. A. Roríguez-Bellido, D. A. Rueda-Gómez, *Analysis of a chemo-repulsion model with nonlinear production: The continuous problem and unconditionally energy stable fully discrete schemes.* ArXiv:1807.05078v2
13. S. Hittmeir and A. Jüngel, *Cross diffusion preventing blow-up in the two-dimensional Keller-Segel model. SIAM J. Math. Anal.* **43** (2011), no. 2, 997–1022.
14. P.–L. Lions, *Mathematical topics in fluid mechanics. Vol. 1.* Incompressible models, Oxford Science Publications. The Clarendon Press Oxford University Press, New York, 1996.
15. J. Necas, *Les Méthodes Directes en Théorie des Equations Elliptiques.* Editeurs Academia, Prague (1967).
16. K. Osaki and A. Yagi, *Finite dimensional attractors for one-dimensional Keller-Segel equations. Funkcialaj Ekvacioj* **44** (2001), 441–469.
17. A. Marrocco, *Numerical simulation of chemotactic bacteria aggregation via mixed finite elements.* M2AN Math. Model. Numer. Anal. **37** (2003), no. 4, 617–630.
18. D. A. Rueda Gómez, *Análisis teórico y numérico de problemas diferenciales con quimiotaxis repulsiva.* (Ph D Thesis). Universidad de Sevilla, Sevilla (2018). https://idus.us.es/xmlui/handle/11441/79799
19. N. Saito, *Conservative upwind finite-element method for a simplified Keller-Segel system modelling chemotaxis.* IMA J. Numer. Anal. **27** (2007), no. 2, 332–365.
20. N. Saito, *Error analysis of a conservative finite-element approximation for the Keller-Segel system of chemotaxis.* Commun. Pure Appl. Anal. **11** (2012), no. 1, 339–364.

21. M. Winkler, *A critical blow-up exponent in a chemotaxis system with nonlinear signal production*. Nonlinearity 31 (2018), no. 5, 2031–2056.

22. J. Zhang, J. Zhu and R. Zhang, *Characteristic splitting mixed finite element analysis of Keller-Segel chemotaxis models*. Appl. Math. Comput. 278 (2016), 33–44.

Chapter 3

Remarks on the Energy Equality for the 3D Navier-Stokes Equations

Luigi Carlo Berselli and Elisabetta Chiodaroli (✉)
Università di Pisa, Dipartimento di Matematica, Pisa, Italy
e-mail: luigi.carlo.berselli@unipi.it

In a recent paper, jointly with L. C. Berselli, we study the problem of energy conservation for solutions of the initial boundary value problem associated with the 3D Navier-Stokes equations with Dirichlet boundary conditions. While the energy equality is satisfied for strong solutions, the dissipation phenomenon is expected to be connected with the roughness of the solutions. A natural question is, then, which regularity is needed for a weak solution in order to conserve the energy. The importance of this issue was brought out in evidence by Onsager's work.

First, we consider Leray–Hopf weak solutions and we provide some new criteria for energy conservation, involving the gradient of the velocity. Then we compare them with the existing literature in scaling invariant spaces and with the Onsager conjecture.

Finally, we consider the problem of energy conservation for very-weak solutions, showing energy equality for distributional solutions belonging to the so-called Shinbrot class.

T. Bodnár et al. (eds.), *Waves in Flows*, Advances in Mathematical Fluid Mechanics, https://doi.org/10.1007/978-3-030-68144-9_3

3.1 Introduction

We consider the Navier-Stokes equations in a bounded domain $\Omega \subset \mathbb{R}^3$ with smooth boundary $\partial\Omega$, and vanishing Dirichlet boundary conditions

$$\begin{cases} \partial_t u - \Delta u + (u \cdot \nabla)\, u + \nabla p = 0 & \text{in } (0,T) \times \Omega, \\ \nabla \cdot u = 0 & \text{in } (0,T) \times \Omega, \\ u = 0 & \text{on } (0,T) \times \partial\Omega, \\ u(0,\cdot) = u_0 & \text{in } \Omega. \end{cases} \tag{3.1}$$

We study the problem in a bounded domain and for simplicity we treat the problem with unit viscosity and with vanishing external force, but both assumptions are unessential.

For the variational formulation of the Navier-Stokes equations (3.1), we introduce as usual the space $\mathcal{V}$ of smooth and divergence-free vectors fields, with compact support in Ω. We then denote the completion of $\mathcal{V}$ in $L^2(\Omega)$ by H and the completion in $H_0^1(\Omega)$ by V. The Hilbert space H is endowed with the natural L^2-norm $\|\,.\,\|$ and inner product $(\cdot,\cdot)$, while V with the norm $\|\nabla v\|$ and inner product $((u,v)) := (\nabla u, \nabla v)$. As usual, we do not distinguish between scalar and vector valued functions. The dual pairing between V and V' is denoted by $\langle\cdot,\cdot\rangle$, and the dual norm by $\|\,.\,\|_*$.

Since the works of Leray and Hopf [25, 28] for the Cauchy problem and for the initial boundary value problem, it is well-known that for initial data in H, and for all $T > 0$ there exists at least a weak solution in the following sense.

Definition 3.1.1 (Leray–Hopf Weak Solutions). A vector field $u \in L^\infty(0,T;H) \cap L^2(0,T;V)$ is a Leray–Hopf weak solution to the Navier-Stokes equations (3.1) if

(i) u is a solution of (3.1) in the sense of distributions, i.e.,

$$\int_0^T (u, \partial_t\phi) - (\nabla u, \nabla\phi) - ((u \cdot \nabla)\, u, \phi)\, dt = -(u_0, \phi(0)),$$

for all $\phi \in C_0^\infty([0,T[\times\Omega)$ with $\nabla \cdot \phi = 0$;

(ii) u satisfies the global energy inequality

$$\frac{1}{2}\|u(t)\|^2 + \int_0^t \|\nabla u(s)\|^2 \leq \frac{1}{2}\|u_0\|^2 \qquad \forall t \geq 0; \tag{3.2}$$

(iii) the initial datum is attained in the strong sense of $L^2(\Omega)$

$$\|u(t) - u_0\| \to 0 \qquad t \to 0^+.$$

Remark 3.1.1. If u is a weak solution on $[0,T[\times\Omega$ for all $T > 0$, then u will be called a global weak solution.

While existence is well-known in this class, regularity and even uniqueness are still unproved at present. For the construction of Leray–Hopf weak solutions a fundamental role is played by the *kinetic energy*, since its boundedness represents the very basic *a priori* estimate. The work of Leray and Hopf showed that the energy inequality (3.2) (opposed to the equality (3.6)) can be inferred for weak solutions in the class where globally in time existence holds. In the latter inequality the sign "less or equal" is due to the lack of regularity of weak solutions and it comes from a limiting process on smoother functions based on weak convergence. The balance of the kinetic energy instead would formally follow by multiplying the equations by the solutions itself and by performing some integration by parts, only once the needed computations are allowed. The first attempts to determine sufficient conditions implying the validity of the energy equality (3.6) in the class of weak solutions came with a series of papers by Lions [30] and Prodi [34], where the criterion $u \in L^4(0,T;L^4(\Omega))$ has been identified. A few years later, in the wake of the celebrated survey of Serrin [35], Shinbrot [36] derived the criterion

$$u \in L^r(0,T;L^s(\Omega)) \qquad \text{with} \quad \frac{2}{r}+\frac{2}{s}\leq 1 \qquad \text{for} \quad s\geq 4, \tag{3.3}$$

which contains the condition $L^4(0,T;L^4(\Omega))$ as a sub-case, and which extends to a wide range of exponents the condition for energy conservation.

It is extremely relevant to observe that the condition (3.3) does not depend on the space dimension n (in particular it is valid also for $n > 3$) and moreover –very remarkably– it is not scaling invariant. We observe that the results of Ladyžhenskaya, Prodi, and Serrin showed regularity of weak solutions (and hence validity of the energy equality) if the scaling invariant condition

$$u \in L^r(0,T;L^s(\Omega)) \qquad \text{with} \quad \frac{2}{r}+\frac{3}{s}= 1 \qquad \text{for} \quad s> 3 \tag{3.4}$$

holds true (see also Sohr [39]).

The essential difference between the energy equality and the energy inequality is the presence of anomalous dissipation due to the presence of non-linearity. The dissipation phenomenon is expected to be connected with the possible roughness of the solutions. The importance of energy conservation/dissipation (especially in the limit of vanishing viscosity) came from Onsager's work [33]. Moreover, for the Navier-Stokes equations the possible connection between energy conservation and uniqueness of weak solutions still represents an interesting open problem.

The famous Onsager conjecture for the Euler and Navier-Stokes equations predicts the threshold regularity for energy conservation. In this direction, energy conservation for $C^{0,1/3}$-solutions of the Euler equations has been recently addressed in the work of Isett [26] and also Buckmaster, De Lellis, Székelyhidi, and Vicol [9]. On the other hand the role of Besov-Hölder continuous spaces in the energy conservation of the viscous Navier-Stokes equations was studied by Cheskidov, Constantin, Friedlander, and Shvydkoy [10], who proved that weak solutions in the following class

$$u \in L^3(0,T;B^{1/3}_{3,\infty}(\mathbb{R}^3))$$

conserve the energy (Cf. also Constantin, E, and Titi [14] for the Euler equations). The latter result has been recently extended by Cheskidov and Luo [12] to the class

$$u \in L^\beta_w(0,T;B^{\frac{2}{\beta}+\frac{2}{p}-1}_{p,\infty}(\mathbb{R}^3)) \qquad \text{with} \quad \frac{2}{p}+\frac{1}{\beta}<1 \qquad \text{for} \quad 1\leq\beta<p\leq\infty,$$

where L^β_w denotes the weak (Marcinkiewicz) space, while $B^s_{p,q}(\mathbb{R}^3)$ are the standard Besov spaces.

A similar approach, based on Fourier methods, allowed Cheskidov, Friedlander, and Shvydkoy [11] to prove the following sufficient condition for energy conservation (here A denotes the Stokes operator associated with the Dirichlet boundary conditions)

$$A^{5/12}u \in L^3(0,T;L^2(\Omega)),$$

which turns out to be –in terms of scaling– less strict than Shinbrot one (3.3). In fact, the latter criterion by Cheskidov, Friedlander, and Shvydkoy is *equivalent in terms of scaling* to $u \in L^3(0,T;L^{9/2}(\Omega))$. We also recall that Farwig [17] proved the following sufficient condition for energy conservation

$$A^{1/4}u \in L^3(0,T;L^{18/7}(\Omega)),$$

which turns out to have the same scaling.

3.2 Main Results

In the paper [6] the authors looked for regularity requirements sufficient to prove energy conservation and weaker than the existing ones. The improvements coming from the results in [6] can be measured in terms of scaling, see Theorem 3.2.1.

In order to present the first main result of [6] we recall, for the reader not acquainted with all the technicalities necessary when dealing with genuine Leray–Hopf weak solutions of the Navier-Stokes equations, that one main step in the analysis of the energy equality is to rigorously prove the following equality

$$\int_0^T\int_\Omega (u\cdot\nabla)\,u\cdot u\,\mathrm{d}x\mathrm{d}t=0. \tag{3.5}$$

The latter equality is *formally* valid if u is a divergence-free vector field tangential to the boundary. The calculations are justified once u is smooth enough to ensure that the above space-time integral is finite, but this is not the case if u is a weak solution in the 3D case.

The novelty of our approach is to look for conditions involving the gradient of the velocity, instead of the velocity itself, which allow to conclude that the above integral is finite.

The first result of [6] which we present here is the following criterion.

Theorem 3.2.1. Let $u_0 \in H$ and let u be a Leray–Hopf weak solution of (3.1) corresponding to u_0 as initial datum. Let us assume that one the following conditions is satisfied

(i) $\nabla u \in L^{\frac{q}{2q-3}}(0,T;L^q(\Omega))$, for $\frac{3}{2} < q < \frac{9}{5}$;

(ii) $\nabla u \in L^{\frac{5q}{5q-6}}(0,T;L^q(\Omega))$, for $\frac{9}{5} \leq q \leq 3$;

(iii) $\nabla u \in L^{1+\frac{2}{q}}(0,T;L^q(\Omega))$, for $q > 3$.

Then, the velocity u satisfies the energy equality

$$\frac{1}{2}\int_\Omega |u(t,x)|^2 \mathrm{d}x + \int_0^t \int_\Omega |\nabla u(s,x)|^2 \,\mathrm{d}x\mathrm{d}s = \frac{1}{2}\int_\Omega |u_0(x)|^2 \mathrm{d}x \qquad \forall\, t \geq 0. \quad (3.6)$$

The comparison with the previous ones known in literature will be discussed in Sect. 3.3. The overall strategy consists in assuming that $\nabla u \in L^q(\Omega)$ for some fixed q, and in employing all the available regularity coming from the definition of weak solutions, so to find the smaller exponent in the time variable which allows to show (3.5).

Remark 3.2.1. A similar approach had already been implemented also by Farwig and Taniuchi [19], focusing on the degree k of (fractional) smoothness of u and on the correct definition of spaces in the case of general unbounded domains. In particular, their result for $k = 1$ has the same scaling as ours in the case $q = 9/5$. Very recently, our approach of working with levels of regularity for the gradient of the velocity has inspired the paper of Beirão da Veiga and Yang [4] where they treat Newtonian and non-Newtonian fluids and some improvements of the results in (iii) are obtained by using some Gagliardo-Nirenberg inequality.

In the second main result of [6] we deal with another less-restrictive notion of solution, which allows for infinite energy, the so-called *very-weak solutions*.

Definition 3.2.2 (Very-Weak Solutions). A vector field $u \in L^2_{loc}((0,T) \times \Omega)$ is a very-weak solution to the Navier-Stokes equations if

(i) the following identity

$$\int_0^T (u, \partial_t \phi) + (u, \Delta \phi) + (u, (u \cdot \nabla)\,\phi)\,\mathrm{d}t = -(u_0, \phi(0))$$

holds true for all $\phi \in \mathcal{D}_T$, where

$$\mathcal{D}_T := \left\{ \begin{array}{l} \phi \in C^\infty([0,T] \times \overline{\Omega}), \text{with support contained in a compact set of } [0,T] \times \overline{\Omega}, \\ \text{such that } \nabla \cdot \phi = 0 \text{ in } \Omega,\ \phi = 0 \text{ on } \partial\Omega \text{ and } \phi(T) = 0 \text{ in } \Omega \end{array} \right\};$$

(ii) $\nabla \cdot u = 0$ in the sense of $\mathcal{D}'(\Omega)$, i.e., in the sense of distributions in Ω, for a.e. $t \in [0,T]$.

In this class there is not any available regularity on u, apart the solution being in $L^2_{loc}((0,T)\times\Omega)$. The interest for very-weak solutions dates back to Foias [21], who proved their uniqueness under condition (3.4). Later Fabes, Jones, and Rivière [16] proved the existence of very-weak solutions for the Cauchy problem, while the case of the initial boundary value problem has been studied mainly starting from the work of Amann [1]. The notion in the 3D time-evolution case in a bounded domain (especially with analysis of the meaning of boundary conditions) is analyzed, for instance, in Amann [1] and Farwig, Kozono, and Sohr [18], with emphasis on the non-homogeneous problem, too. As usual when dealing with very-weak solutions a duality argument can be employed to show uniqueness, by using properties of the adjoint problem (which in case of the Navier-Stokes equations, is a backward Oseen type-problem). The connection with the energy equality has been very recently developed by Galdi [23], who showed that very-weak solution in the Lions-Prodi class $L^4(0,T;L^4(\Omega))$, and with initial data in $L^2(\Omega)$, satisfy (3.6). It is relevant to observe that the duality argument is used to improve the known regularity of the solution, in order to use the previously established results for usual weak solutions.

The second result contained in [6] that we present here is then the following.

Theorem 3.2.3. Let $u\in L^2_{loc}((0,T)\times\Omega)$ be a distributional solution of (3.1). If the initial datum $u_0\in H$ and if (3.3) holds true, then the energy equality (3.6) is satisfied.

The interpretation of the latter result and its connection with the scaling of space and time variables will be given in Sect. 3.3.3. The techniques employed here are inspired by the approach followed in Galdi [22, 23].

Summarizing the results, the sufficient conditions required for the energy equality are stronger if the requested regularity of the solution u is weaker. In particular, the less strict requirement of the solution (being just a distributional solution) seems to require more restrictive conditions than those in Theorem 3.2.1, which are nevertheless the same as those classically found in [30, 34, 36] for Leray–Hopf weak solutions.

3.3 Comparison with Previous Results

In this section we first introduce some further notation and then compare our results with the ones in the existing literature. We will use the customary Sobolev spaces $(W^{s,p}(\Omega),\|\,.\,\|_{W^{s,p}})$ and we denote the L^p-norm simply by $\|\,.\,\|_p$ and since the Hilbert case plays a special role we denote the $L^2(\Omega)$-norm simply by $\|\,.\,\|$. For X a Banach space we will also denote the usual Bochner spaces of functions defined on $[0,T]$ and with values in X by $(L^p(0,T;X),\|\,.\,\|_{L^p(X)})$. In the case $X=L^q(\Omega)$ we denote the norm of $L^p(0,T;L^q(\Omega))$ simply by $\|\,.\,\|_{p,q}$.

In this section we compare our results with the ones already present in literature. From the physical point of view the most relevant results are those related with large q, especially with $q>3$, being 3 the space dimension.

3.3.1 Energy Conservation and Onsager Conjecture

As mentioned in the introduction, the validity of (3.6) has a strong connection with Onsager conjecture about the threshold regularity of $C^{1/3}$-Hölder continuity of the velocity which allows for energy conservation. While for the 3D Euler equations the conjecture is basically solved in both directions, for the 3D Navier-Stokes equations much has still to be done. This is a strong motivation to investigate further the energy equality for the 3D Navier-Stokes equations and to understand the anomalous energy dissipation phenomenon in this case. Our result involves Sobolev regularity of the velocity, but this can be compared to Hölder regularity as we will explain in the following. In the range $q > 3$, standard Sobolev embedding theorems imply $W^{1,q}(\Omega) \subset C^{0,1-3/q}(\overline{\Omega})$, and $W^{1,q}(\Omega)$ is a *proper* subset of $C^{0,1-3/q}(\overline{\Omega})$. On the other hand, in order to compare our result with previous ones, we can consider the two spaces as if they would be equivalent in a certain sense. Indeed, we can consider $C^{0,1-3/q}(\overline{\Omega})$ as a rough –but meaningful– measure of the classical regularity of $W^{1,q}(\Omega)$ (see [6] for details).

By embedding, results from Theorem 3.2.1 in the range $q > 3$ are very close to the condition

$$u \in L^{1+\frac{2}{q}}(0,T; C^{0,1-\frac{3}{q}}(\overline{\Omega})).$$

In particular, by taking $q = 9/2$ we obtain, as class of solutions conserving the energy, that with scaling comparable to

$$u \in L^{\frac{13}{9}}(0,T; C^{0,1/3}(\overline{\Omega})),$$

which improves the previously cited results

$$u \in L^3(0,T; B^{1/3}_{3,\infty}(\mathbb{R}^3)) \qquad \text{from Ref. [10];}$$

$$u \in L^{\frac{3}{2}}_w(0,T; B^{1/3}_{\infty,\infty}(\mathbb{R}^3)) \qquad \text{from Ref. [12], when } p=\infty, \text{ and } \beta = 3/2.$$

In the space variables the two functional spaces are very close to $C^{0,1/3}(\overline{\Omega})$, but $3 > 3/2 = 1.5 > 13/9 = 1.\overline{4}$.

We also warn again the reader that our results are obtained by embedding, hence they are valid for a proper subset of $C^{0,1/3}(\overline{\Omega})$ and there is not any direct connection between Hölder and Sobolev regularity. Nevertheless, in terms of scaling our results present a better behavior as compared with the previous ones present in literature. This suggests that there could be also room for further improvements.

3.3.2 The Case $q < 3$

In order to explain the comparison (we will make later on) we also wish to recall the notion of scaling invariance for space-time functions. By interpolation one can show that Leray–Hopf weak solution have the following regularity

$$u \in L^r(0,T; L^s(\Omega)) \qquad \text{with} \quad \frac{2}{r} + \frac{3}{s} = \frac{3}{2}, \qquad \text{for} \quad 2 \leq s \leq 6.$$

Several results (starting again from the classical work of Ladyžhenskaya, Prodi, and Serrin) concern uniqueness and regularity with scaling invariant conditions on solutions. In particular, if a weak solution satisfies condition (3.4) (see [35], for example), then it becomes unique, strong, smooth, and it satisfies the energy equality.

Full regularity of weak solutions follows also under alternative assumptions on the gradient of the velocity ∇u. More specifically, if

$$\nabla u \in L^p(0,T;L^q(\Omega)) \qquad \text{with} \quad \frac{2}{p}+\frac{3}{q}=2, \qquad \text{for} \quad q>\frac{3}{2}, \tag{3.7}$$

then weak solutions are regular, see Beirão da Veiga [3] and also [5] for the problem in a bounded domain. In $\mathbb{R}^3$ standard Sobolev embeddings imply that if $\nabla u \in L^p(0,T;L^q(\Omega))$, then $u \in L^p(0,T;L^{q^*}(\Omega))$ where $\frac{1}{q^*}=\frac{1}{q}-\frac{1}{3}$. We recall that for weak solutions ∇u is simply (x,t)-square-integrable and $\frac{2}{2}+\frac{3}{2}=\frac{5}{2}>2$.

The class defined by (3.4) is important from the point of view of the relation between scaling invariance and partial regularity of weak solutions. In fact, if a pair (u,p) solves (3.1), then so does the family $\{(u_\lambda,p_\lambda)\}_{\lambda>0}$ defined by

$$u_\lambda(t,x) := \lambda u(\lambda^2 t, \lambda x) \qquad \text{and} \qquad p_\lambda(x,t) := \lambda^2 p(\lambda^2 t, \lambda x). \tag{3.8}$$

Scaling invariance means that $\|u_\lambda\|_{L^r(0,T/\lambda^2;L^s(\Omega_\lambda))} = \|u\|_{L^r(0,T;L^s(\Omega))}$ and this happens if and only if (r,s) satisfy (3.4). (Here $\Omega_\lambda := \{x/\lambda : \; x\in\Omega\}$.) Likewise, the scaling invariance of ∇u occurs if and only if (3.7) holds true. We will use these notions to compare classical and more recent results with the new ones from Theorem 3.2.1.

If $u \in L^4(0,T;L^4(\Omega))$, then, *in terms of scaling*, this regularity lies in between the class of existence and that of regularity since

$$1=\frac{2}{4}+\frac{2}{4}<\frac{2}{4}+\frac{3}{4}=\frac{5}{4}<\frac{3}{2}.$$

We note that also the class (3.3) identified by Shinbrot shares the same property, even if in terms of scaling

$$1<\frac{2}{r}+\frac{3}{s}=\frac{2}{r}+\frac{2}{s}+\frac{1}{s}=1+\frac{1}{s}<\frac{3}{2},$$

hence, as s increases, this class becomes closer and closer to that of regularity given by (3.4). A natural question is that if we can lower down the level of regularity needed for a weak solution, in order to satisfy the energy equality.

The new results we present are related with the aim of finding sufficient conditions for the energy equality involving the gradient of the velocity.

The fact that condition (3.3) is not scaling invariant makes it possible to conjecture that perhaps some threshold can be broken by a more precise inspection of the calculations. In fact, the aforementioned results in [11, 17] identified the sufficient conditions

$$A^{5/12}u \in L^3(0,T;L^2(\Omega)) \qquad \text{or} \qquad A^{1/4}u \in L^3(0,T;L^{18/7}(\Omega)),$$

which are –in terms of scaling– both comparable with

$$u \in L^3(0,T;L^{9/2}(\Omega)),$$

which is less restrictive than (3.3), since

$$1 < \frac{2}{3} + \frac{2}{9/2} = \frac{10}{9},$$

but at the same time the space $L^3(0,T;L^{9/2}(\Omega))$ it is still between the class of existence and that of regularity, being $1 < \frac{2}{3} + \frac{3}{9/2} = \frac{4}{3} < \frac{3}{2}$.

The ranges obtained in Theorem 3.2.1 have, respectively, the following properties, which turn out easy to be compared with the condition (3.7).

We have in fact that:

(i) it holds $2 < \frac{2}{p} + \frac{3}{q} = 4 - \frac{3}{q} < \frac{7}{3}$, for $\frac{3}{2} < q < \frac{9}{5}$;

(ii) it holds $2 < \frac{2}{p} + \frac{3}{q} = 2 + \frac{3}{5q} < \frac{11}{5}$, for $\frac{9}{5} \le q < 3$;

(iii) it holds $2 < \frac{2}{p} + \frac{3}{q} = \frac{2q}{q+2} + \frac{3}{q} < \frac{11}{5}$, for $3 \le q < 6$.

Thus, we have that for $\frac{3}{2} < q < 6$ our conditions imply range of exponents which are not those of scaling invariance (3.7), hence not those implying full regularity of the solutions (and consequently also energy equality in a trivial way).

Remark 3.3.1. In cases (i) and (ii), i.e., when $\frac{3}{2} < q < 3$, the standard Sobolev embedding tells us that $u \in L^p(0,T;L^{q^*}(\Omega))$ where $q^* = \frac{3q}{3-q}$ and p is given as a function of q by Theorem 3.2.1. This means that by case (i) and case (ii) the exponent of integrability q^* in the space variable for the solution u can range from 3 to $+\infty$.

Recalling that $q^* = \frac{3q}{3-q}$, the ranges obtained in Theorem 3.2.1 have, respectively, the following properties:

(i) it holds $1 < \frac{2}{p} + \frac{2}{q^*} = \frac{2(5q-6)}{3q}$, for $\frac{12}{7} < q < \frac{9}{5}$;

(ii) it holds $1 < \frac{2}{p} + \frac{2}{q^*} = \frac{2(10q-3)}{15q}$, for $\frac{9}{5} \le q < 3$;

(iii) it holds $1 < \frac{2}{p} + \frac{2}{q^*} = \frac{2q}{(q+2)}$, for $q > 3$.

Thus, we showed that our range of exponents improves those in Shinbrot condition (3.3). We observe that Shinbrot condition for the space integrability of u (exponent ≥ 4) corresponds to $q > \frac{12}{7}$ in our classification.

Remark 3.3.2. The "best exponent," where best is measured in terms of giving the quantity $\frac{2}{p}+\frac{2}{q}$ as large as possible, turns out to be $q = 9/5$. This implies that by embedding $(9/5)^* = 9/2$ which gives as sufficient condition

$$u \in L^3(0,T; W^{1,9/5}(\Omega)) \subset L^3(0,T; L^{9/2}(\Omega)),$$

that is at the same level of scaling of [11, 17], even if the various conditions are not directly comparable each other.

We further remark that, in the range $q > \frac{12}{7}$, our result improves also the ranges obtained by Leslie and Shvydkoy in [29]. Indeed, they prove (see [29, Theorem 1.1]) the validity of the energy equality for $u \in L^r(0,T; L^s(\Omega))$ with $3 \le r \le s$ and

$$\frac{1}{r}+\frac{1}{s} \le \frac{1}{2},$$

while for $q > \frac{12}{7}$ we are assuming $\nabla u \in L^p(0,T; L^q(\Omega))$ and hence $u \in L^p(0,T; L^{q^*}(\Omega))$ with $p \le q^*$ and

$$\frac{1}{p}+\frac{1}{q^*} > \frac{1}{2}.$$

However, the authors in [29] studied also the case $s < 3$ corresponding in our case to $q < 3/2$ which is not covered here. Moreover, for the case $3 \le s < r$, they can prove the energy equality with exponents r and s satisfying $\frac{1}{r}+\frac{1}{s} < \frac{1}{2}$: this matches our interval $3/2 \le q < 12/7$, where $\frac{1}{p}+\frac{1}{q^*} < \frac{1}{2}$, thus showing that we are not improving [29] in this last range of exponents.

3.3.3 On the Shinbrot Condition (3.3) and Very-Weak Solutions

In the case of very-weak solutions, the sufficient condition for energy equality we find is the same as (3.3), hence showing that it has a universal role, since it applies to distributional solutions which can be outside of any Lebesgue space, hence not the common Leray–Hopf weak solutions.

We observe that the condition (3.3) breaks all the standard theory of scaling invariance, even if we consider different families of scaling transformation. In fact, beside the standard parabolic scaling (3.8), it is well-known that useful different approaches are those concerning invariance of the equation under the following space-time transformation, indexed by $\alpha \in \mathbb{R}$

$$u_{\lambda,\alpha}(t,x) := \lambda^{\alpha} u(\lambda^{\alpha+1}t, \lambda x) \qquad \text{and} \qquad p_{\lambda,\alpha}(t,x) := \lambda^{2\alpha} p(\lambda^{\alpha+1}t, \lambda x). \tag{3.9}$$

Note that the transformation in (3.8) corresponds to $\alpha = 1$. On varying α, with these transformations it is possible to extract some heuristic and useful information from the equations. In particular, they keep for all $\alpha \in \mathbb{R}$ the material derivative operator $\frac{D}{Dt} := \partial_t + u \cdot \nabla$ unchanged, while the viscosity changes from ν to $\lambda^{\alpha-1}\nu$. This explains the relevance of these transformation in the study of Euler equations or also in presence of very small viscosities. We point out that the case $\alpha = -1/3$ determines the so-called *Kolmogorov scaling*, which has connections with the conservation of energy input for stochastic statistically stationary solutions, stochastic equations, and Large Eddy Simulation for turbulence modeling. See [7, 8], Flandoli, Gubinelli, Hairer, and Romito [20], and Kupiainen [27]. If we try to consider the scaling of norms under the transformation (3.9) we obtain that

$$\|u_{\lambda,\alpha}\|_{L^r(0,T/\lambda^2;L^s(\Omega_\lambda))} = \lambda^{\alpha-\frac{3}{s}-\frac{\alpha+1}{r}}\|u\|_{L^r(0,T;L^s(\Omega))}$$

hence the norms cannot be invariant for the relevant values of α, since $\alpha - \frac{3}{s} - \frac{\alpha+1}{r}$ never vanishes for both $\alpha = 1$ and $\alpha = -1/3$. A possible explanation or justification of the condition (3.3) will come from a different analysis.

In particular, in condition (3.3) the space and time variables have the same "strength," as in hyperbolic equations, while in our studies the viscosity and hence the parabolic part play a fundamental role. To understand in which situation the hyperbolic nature of the convection plays a major role we have to consider the maximal regularity type results for the Stokes problem, which will be used to handle the adjoint problem. Coming back to Solonnikov [40] (see also Giga and Sohr [24] and Maremonti and Solonnikov [32] for the mixed estimates) it is well-known that for a wide class of domains the following result holds true.

Theorem 3.3.1. Let $\Omega \subset \mathbb{R}^3$ be smooth and bounded (or exterior or the full-space or even the half-space) and let $\mathcal{F} \in L^\alpha(0,T;L^\beta(\Omega))$, with $1 < \alpha,\, \beta < \infty$. Then, the boundary initial value problem associated with the linear Stokes system

$$\begin{aligned} \partial_t v - \Delta v + \nabla\pi = \mathcal{F} \qquad & \text{in } (0,T)\times\Omega, \\ \nabla\cdot v = 0 \qquad & \text{in } (0,T)\times\Omega, \\ v(t,x) = 0 \qquad & \text{on } (0,T)\times\partial\Omega, \\ v(0,x) = 0 \qquad & \text{in } \Omega \end{aligned} \tag{3.10}$$

has a unique solution (v,π) such that

$$\exists\, c>0 \qquad \|\partial_t v\|_{L^\alpha(L^\beta)} + \|\nabla^2 v\|_{L^\alpha(L^\beta)} + \|\nabla\pi\|_{L^\alpha(L^\beta)} \le c\,\|\mathcal{F}\|_{L^\alpha(L^\beta)}, \tag{3.11}$$

where the constant c depends on T, α, β, and on the domain.

Remark 3.3.3. In fact, the constant c in (3.11) can be taken independent of T if $\beta < 3/2$ (see, for instance, [32]), but this is not relevant in our context since we will be always working on the finite time interval $(0,T)$.

For our purposes, a very interesting application consists in using this result to find improved regularity for the solutions of a linear problem in which $\mathcal{F} = (u\cdot\nabla)\,v$, where u is an assigned function with a given space-time summability as, for instance, (3.3).

We have the following lemma.

Lemma 3.3.4. Let $\Omega \subset \mathbb{R}^3$ be smooth and bounded (or exterior or the full-space or even the half-space), and let $u \in L^r(0,T;L^s(\Omega))$ with $\frac{2}{r}+\frac{3}{s} \le \frac{7}{2}$, such that $\nabla \cdot u = 0$ in $\mathcal{D}'(\Omega)$, for a.e. $t \in [0,T]$. Then, the boundary initial value problem

$$\begin{aligned} \partial_t v - \Delta v + \nabla \pi &= -(u\cdot\nabla)\,v && \text{in } (0,T)\times\Omega,\\ \nabla\cdot v &= 0 && \text{in } (0,T)\times\Omega,\\ v(t,x) &= 0 && \text{on } (0,T)\times\partial\Omega,\\ v(0,x) &= 0 && \text{in } \Omega \end{aligned} \tag{3.12}$$

has a unique solution such that

$$v \in L^\infty(0,T;L^2(\Omega)) \cap L^2(0,T;H^1_0(\Omega)).$$

Moreover, if the solution v satisfies also

$$\nabla v \in L^\alpha(0,T;L^\beta(\Omega)), \quad \text{with } 1<\alpha,\beta<\infty, \text{ for } \frac{1}{r}+\frac{1}{\alpha}<1 \text{ and } \frac{1}{s}+\frac{1}{\beta}<1,$$

then the solution v itself satisfies also the following improved estimate

$$\exists c>0 \quad \|\partial_t v\|_{L^{\frac{r\alpha}{r+\alpha}}(L^{\frac{s\beta}{s+\beta}})} + \|\nabla^2 v\|_{L^{\frac{r\alpha}{r+\alpha}}(L^{\frac{s\beta}{s+\beta}})} + \|\nabla\pi\|_{L^{\frac{r\alpha}{r+\alpha}}(L^{\frac{s\beta}{s+\beta}})} \le c\,\|u\|_{L^r(L^s)}\|\nabla v\|_{L^\alpha(L^\beta)},$$

for some non-negative constant $c = c(T,r,s,\alpha,\beta,\Omega)$.

The existence part in the energy space comes from usual smoothing of u, Faedo-Galerkin approximation and eventually invading domains for unbounded domains (see [23, 31, 40]), while the further regularity follows by applying Theorem 3.3.1

In order to show the requested regularity which will make the calculations justified in the proof of Theorem 3.2.3 we need to apply the above result several (but a finite number of) times, making possible a sort of bootstrap. This is the main difference with the proof of the result for $r=s=4$ from [23], where a single step was enough.

In our case we need to apply an extremely sharp interpolation result, which has been kindly provided to us by Prof. Amann [2] in the following explicit form.

Lemma 3.3.5. Let $\phi \in W^{1,p}(0,T;L^q(\Omega)) \cap L^p(0,T;W^{2,q}\cap W^{1,q}_0(\Omega))$, with $\phi(0)=0$, for $1<p<2$, $1<q<\infty$. Then, it follows that

$$\phi \in L^{p_1}(0,T;W^{1,q}_0(\Omega)) \qquad \text{for all } p_1 \le p_*, \text{ where } \frac{1}{p_*} := \frac{1}{p}-\frac{1}{2}.$$

This Lemma could be deduced also from a more general result known as the mixed derivative theorem due to Sobolevskii [38]. We refer, for instance, to Lemma 9.7 in [15] where the particular choice of operators therein $A=-\Delta$ and $B=\partial_t$ would allow to recover Lemma 3.3.5.

Remark 3.3.6. We observe that in the embedding of Lemma 3.3.5 we have also compactness for any given $p_1 < p_*$ and this is a classical, e.g., see Simon [37, Corollary. 8]). We note that the case $r = s = 4$ can be treated directly, without Lemma 3.3.5 and also we observe that if instead of the sharp continuous embedding one uses the classical compact one (not valid in the endpoint case) the same result of Theorem 3.2.3 can be obtained, under the more restrictive condition $u \in L^r(0,T;L^s(\Omega))$ with $\frac{2}{r} + \frac{2}{s} < 1$, for $s > 4$.

The interpolation type inequality from Lemma 3.3.5 allows us to infer the following result.

Proposition 3.3.7. Let r, s, α, β as in Lemma 3.3.4 and if $u \in L^r(0,T;L^s(\Omega))$ is given, with $\frac{1}{r} + \frac{1}{s} = \gamma \leq \frac{1}{2}$. If the unique solution of (3.12) satisfies $\nabla v \in L^\alpha(0,T;L^\beta(\Omega))$, then in addition

$$\nabla v \in L^{\alpha_1}(0,T;L^{\frac{\beta s}{\beta+s}}(\Omega)) \qquad \forall \alpha_1 \leq \left(\frac{\alpha\, r}{\alpha + r}\right)_* . \tag{3.13}$$

We observe that since by definition $1/\left(\frac{\alpha\, r}{\alpha+r}\right)_* = \frac{\alpha+r}{\alpha\, r} - \frac{1}{2}$, then the following equalities hold true

$$\frac{1}{\left(\frac{\alpha\, r}{\alpha+r}\right)_*} + \frac{1}{\frac{\beta\, s}{\beta+s}} = \frac{\alpha + r}{\alpha\, r} - \frac{1}{2} + \frac{\beta + s}{\beta\, s} = \frac{1}{r} + \frac{1}{s} - \frac{1}{2} + \frac{1}{\alpha} + \frac{1}{\beta} = \left[\gamma - \frac{1}{2}\right] + \frac{1}{\alpha} + \frac{1}{\beta},$$

hence the value $\gamma = \frac{1}{2}$ is exactly the one which keeps the mixed space-time regularity unchanged. Since we will need this to iterate the procedure to infer further regularity for the solution of the linear problem, we give now the precise regularity which is obtained by using repeatedly the same argument. To this aim let $(\alpha_1, \beta_1) = (2, 2)$ (this because the starting point is the energy space, that is $\nabla v \in L^2((0,T) \times \Omega)$), we define then by recurrence

$$\alpha_{n+1} := \left(\frac{\alpha_n\, r}{\alpha_n + r}\right)_* \qquad \text{and} \qquad \beta_{n+1} := \frac{\beta_n\, s}{\beta_n + s}.$$

We observe that

$$\frac{1}{\alpha_{n+1}} + \frac{1}{\beta_{n+1}} = \left[\gamma - \frac{1}{2}\right] + \frac{1}{\alpha_n} + \frac{1}{\beta_n} \qquad \forall\, n \in \mathbb{N}, \tag{3.14}$$

and taking into account that the iteration stops if the limiting values 1 or $+\infty$ for either α_{n+1} or β_{n+1} are reached.

Remark 3.3.8. By direct computations using the recurrence definition, it follows that $\beta_{n+1} \leq \beta_n$, and consequently $\alpha_{n+1} \geq \alpha_n$. Hence, the use of the maximal regularity applied to the linear problem decreases the available regularity in the space variables, but increases the one of the time variable.

Concerning the linear Stokes problem we have the following result

Proposition 3.3.9. Let $u \in L^r(0,T;L^s(\Omega))$ be given with $\frac{1}{r}+\frac{1}{s}=\gamma\leq\frac{1}{2}$. Then, the unique solution v of (3.12) satisfies

$$\nabla v \in L^{\widetilde{\alpha}_n}(0,T;L^{\beta_n}(\Omega)), \tag{3.15}$$

for all $\widetilde{\alpha}_n \leq \alpha_n$, where the couple (α_n,β_n) is defined as in (3.14). Hence, by Hölder inequality we have also

$$(u\cdot\nabla)\,v,\ \nabla\pi \in L^{\frac{\widetilde{\alpha}_n r}{\widetilde{\alpha}_n+r}}(0,T;L^{\frac{\beta_n s}{\beta_n+s}}(\Omega)), \tag{3.16}$$

and

$$\frac{1}{\frac{\widetilde{\alpha}_n r}{\widetilde{\alpha}_n+r}}+\frac{1}{\frac{\beta_n s}{\beta_n+s}}\leq\gamma+1.$$

Observe that, while in the time variable we can consider by Hölder inequality also exponents smaller than α_n, in the space variables, this is not possible when the domain Ω is with infinite Lebesgue measure.

In terms of scaling the result of Proposition 3.3.9 can be viewed as a modulated and controlled exchange of regularity between space and time (which resembles some of the estimates obtainable with Fourier analysis and has connections also with the early estimates for the gradient of the vorticity in $L^{4/3+\varepsilon}((0,T)\times\Omega)$ from [13]). The above result shows that the standard theory of linear parabolic equations can be used to obtain a full scale of results with similar regularity drain from some variables towards others. This will be enough for our purposes of applications to very-weak solutions. The complete proof can be found in [6].

References

1. H. Amann. On the strong solvability of the Navier-Stokes equations. *J. Math. Fluid Mech.*, 2(1):16–98, 2000.

2. H. Amann. *Linear and quasilinear parabolic problems. Vol. II.* Monographs in Mathematics. Birkhäuser Boston, Inc., Boston, MA, 2019, to appear. Functions spaces and linear differential operators.

3. H. Beirão da Veiga. A new regularity class for the Navier-Stokes equations in $\mathbf{R}^n$. *Chinese Ann. Math. Ser. B*, 16(4):407–412, 1995. A Chinese summary appears in Chinese Ann. Math. Ser. A **16** (1995), no. 6, 797.

4. H. Beirão da Veiga and J. Yang. On the energy equality for solutions to Newtonian and non-Newtonian fluids. *Nonlinear Anal*, **185** (2019), 388–402.

5. L. C. Berselli. On a regularity criterion for the solutions to the 3D Navier-Stokes equations. *Differential Integral Equations*, 15(9):1129–1137, 2002.

6. L. C. Berselli and E. Chiodaroli On the energy equality for the 3D Navier-Stokes equations. *Nonlinear Anal*, *arXiv:1807.02667v3*, 2019.

7. L. C. Berselli and F. Flandoli. On a stochastic approach to eddy viscosity models for turbulent flows. In *Advances in mathematical fluid mechanics*, pages 55–81. Springer, Berlin, 2010.

8. L. C. Berselli, T. Iliescu, and W. J. Layton. *Mathematics of Large Eddy Simulation of turbulent flows*. Scientific Computation. Springer-Verlag, Berlin, 2006.

9. T. Buckmaster, C. De Lellis, L. Székelyhidi, Jr., and Vicol V. Onsager's conjecture for admissible weak solutions. *Comm. Pure Appl. Math.*, 72(2):229–274, 2019.

10. A. Cheskidov, P. Constantin, S. Friedlander, and R. Shvydkoy. Energy conservation and Onsager's conjecture for the Euler equations. *Nonlinearity*, 21(6):1233–1252, 2008.

11. A. Cheskidov, S. Friedlander, and R. Shvydkoy. On the energy equality for weak solutions of the 3D Navier-Stokes equations. In *Contributions to current challenges in mathematical fluid mechanics*, Adv. Math. Fluid Mech., pages 171–175. Birkhäuser, Basel, 2010.

12. A. Cheskidov and X. Luo. Energy equality for the Navier-Stokes equations in weak-in-time Onsager spaces. *Nonlinearity*, Technical Report 1802.05785, ArXiv, 2018.

13. P. Constantin. Navier-Stokes equations and area of interfaces. *Comm. Math. Phys.*, 129(2):241–266, 1990.

14. P. Constantin, W. E, and E.S. Titi. Onsager's conjecture on the energy conservation for solutions of Euler's equation. *Comm. Math. Phys.*, 165(1):207–209, 1994.

15. J. Escher, J. Prüss, and G. Simonett. Analytic solutions for a Stefan problem with Gibbs-Thomson correction. *J. Reine Angew. Math.*, 563:1–52, 2003.

16. E. B. Fabes, B. F. Jones, and N. M. Rivière. The initial value problem for the Navier-Stokes equations with data in L^p. *Arch. Rational Mech. Anal.*, 45:222–240, 1972.

17. R. Farwig. On regularity of weak solutions to the instationary Navier-Stokes system: a review on recent results. *Ann. Univ. Ferrara Sez. VII Sci. Mat.*, 60(1):91–122, 2014.

18. R. Farwig, H. Kozono, and H. Sohr. Very weak solutions of the Navier-Stokes equations in exterior domains with nonhomogeneous data. *J. Math. Soc. Japan*, 59(1):127–150, 2007.

19. R. Farwig and Y. Taniuchi. On the energy equality of Navier-Stokes equations in general unbounded domains. *Arch. Math. (Basel)*, 95(5):447–456, 2010.

20. F. Flandoli, M. Gubinelli, M. Hairer, and M. Romito. Rigorous remarks about scaling laws in turbulent fluids. *Comm. Math. Phys.*, 278(1):1–29, 2008.

21. C. Foias. Une remarque sur l'unicité des solutions des équations de Navier-Stokes en dimension *n*. *Bull. Soc. Math. France*, 89:1–8, 1961.

22. G. P. Galdi. On the relation between very weak and Leray-Hopf solutions to Navier–Stokes equations. *Proc. Amer. Math. Soc.*, Technical Report arXiv:1809.03991, ArXiv, 2018.

23. G. P. Galdi. On the energy equality for distributional solutions to Navier-Stokes equations. *Proc. Amer. Math. Soc.*, 147(2):785–792, 2019.

24. Y. Giga and H. Sohr. Abstract L^p estimates for the Cauchy problem with applications to the Navier-Stokes equations in exterior domains. *J. Funct. Anal.*, 102(1):72–94, 1991.

25. E. Hopf. Über die Anfangswertaufgabe für die hydrodynamischen Grundgleichungen. *Math. Nachr.*, 4:213–231, 1951.

26. P. Isett. A proof of Onsager's conjecture. *Ann. of Math. (2)*, 188(3):871–963, 2018.

27. A. Kupiainen. Statistical theories of turbulence. In: Advances in Mathematical Sciences and Applications, Gakkotosho, Tokyo, 2003.

28. J. Leray. Sur le mouvement d'un liquide visqueux emplissant l'espace. *Acta Math.*, 63(1):193–248, 1934.

29. T. M. Leslie and R. Shvydkoy. Conditions implying energy equality for weak solutions of the Navier-Stokes equations. *SIAM J. Math. Anal.*, 50(1):870–890, 2018.

30. J.-L. Lions. Sur la régularité et l'unicité des solutions turbulentes des équations de Navier Stokes. *Rend. Sem. Mat. Univ. Padova*, 30:16–23, 1960.

31. P. Maremonti. A note on Prodi-Serrin conditions for the regularity of a weak solution to the Navier-Stokes equations. *J. Math. Fluid Mech.*, 20(2):379–392, 2018.

32. P. Maremonti and V. A. Solonnikov. On nonstationary Stokes problem in exterior domains. *Ann. Scuola Norm. Sup. Pisa Cl. Sci. (4)*, 24(3):395–449, 1997.

33. L. Onsager. Statistical hydrodynamics. *Nuovo Cimento (9)*, 6(Supplemento, 2 (Convegno Internazionale di Meccanica Statistica)):279–287, 1949.

34. G. Prodi. Un teorema di unicità per le equazioni di Navier-Stokes. *Ann. Mat. Pura Appl. (4)*, 48:173–182, 1959.

35. J. Serrin. The initial value problem for the Navier-Stokes equations. In *Nonlinear Problems (Proc. Sympos., Madison, Wis.*, pages 69–98. Univ. of Wisconsin Press, Madison, Wis., 1963.

36. M. Shinbrot. The energy equation for the Navier-Stokes system. *SIAM J. Math. Anal.*, 5:948–954, 1974.

37. J. Simon. Compact sets in the space $L^p(0,T;B)$. *Ann. Mat. Pura Appl. (4)*, 146:65–96, 1987.

38. P. E. Sobolevskiĭ. Fractional powers of coercively positive sums of operators. *Dokl. Akad. Nauk SSSR*, 225(6):1271–1274, 1975.

39. H. Sohr. Zur Regularitätstheorie der instationären Gleichungen von Navier-Stokes. *Math. Z.*, 184(3):359–375, 1983.

40. V. A. Solonnikov. Estimates of the solution of a certain initial-boundary value problem for a linear nonstationary system of Navier-Stokes equations. *Zap. Naučn. Sem. Leningrad. Otdel Mat. Inst. Steklov. (LOMI)*, 59:178–254, 257, 1976. Boundary value problems of mathematical physics and related questions in the theory of functions, 9.

Chapter 4

Existence, Uniqueness, and Asymptotic Behavior of Regular Time-Periodic Viscous Flow Around a Moving Body

Giovanni Paolo Galdi (✉)
University of Pittsburgh, Swanson School of Engineering, Pittsburgh, PA, USA
e-mail: galdi@pitt.edu

We show existence and uniqueness of regular time-periodic solutions to the Navier–Stokes problem in the exterior of a rigid body, $\mathcal{B}$, that moves by arbitrary (sufficiently smooth) time-periodic translational motion of the same period, provided the size of the data is suitably restricted. Moreover, we characterize the spatial asymptotic behavior of such solutions and prove, in particular, that if $\mathcal{B}$ has a nonzero net motion identified by a constant velocity $\overline{\boldsymbol{\xi}}$ (say), then the solution exhibits a wake-like behavior in the direction $-\overline{\boldsymbol{\xi}}$ entirely analogous to that of a steady-state flow around a body that moves with velocity $\overline{\boldsymbol{\xi}}$.

T. Bodnár et al. (eds.), *Waves in Flows*, Advances in Mathematical Fluid Mechanics, https://doi.org/10.1007/978-3-030-68144-9_4

4.1 Introduction

Rigorous mathematical analysis of time-periodic flow of a Navier–Stokes liquid $\mathcal{L}$, around a *moving* rigid body, $\mathcal{B}$, is a relatively recent area of research.[1] In fact, the first contribution, due to A. L. Silvestre and the present author, can be found in [5]. In that paper the authors considered the general case where $\mathcal{B}$ moves by arbitrary motion characterized by (sufficiently smooth) time-periodic translational velocity $\boldsymbol{\xi} = \boldsymbol{\xi}(t)$, and angular velocity $\boldsymbol{\omega} = \boldsymbol{\omega}(t)$. In particular, they showed the existence of corresponding solutions to the associated Navier–Stokes problem in a "weak" class (a la Leray–Hopf) for data of arbitrary size, and in a "strong" class (a la Ladyzhenskaya) if the size of the data is appropriately restricted. However, the important problem of uniqueness of these solutions was left open.

The question was successively reconsidered and thoroughly investigated by a number of authors who, by entirely different methods, were able to prove existence and uniqueness of time-periodic solutions of period T (from now on, referred to as "T-periodic" solutions) in appropriate function classes, under the assumption that both characteristic vectors $\boldsymbol{\xi}$ and $\boldsymbol{\omega}$ are constant [3, 4, 7, 9–11, 16], and a T-periodic body force is acting on $\mathcal{L}$.

Very recently, in [2] we began to investigate the above properties in the general situation when $\boldsymbol{\xi}$ is not constant while assuming $\boldsymbol{\omega} \equiv \mathbf{0}$. Converted in mathematical terms, this amounts to finding T-periodic solutions $(\boldsymbol{u}, p)$ to the following system of equations:

$$\left.\begin{array}{l}\partial_t \boldsymbol{u} - \boldsymbol{\xi}(t) \cdot \nabla \boldsymbol{u} + \boldsymbol{u} \cdot \nabla \boldsymbol{u} = \Delta \boldsymbol{u} - \nabla p + \boldsymbol{b} \\ \operatorname{div} \boldsymbol{u} = 0\end{array}\right\} \text{ in } \Omega \times (-\infty, \infty)$$
$$\boldsymbol{u}(x,t) = \boldsymbol{\xi}(t)\,, \quad (x,t) \in \partial\Omega \times (-\infty, \infty)\,, \tag{4.1.1}$$

where $\boldsymbol{u}$ and p are velocity and pressure fields of $\mathcal{L}$, and Ω is the exterior of a connected compact region of $\mathbb{R}^3$ (the body $\mathcal{B}$). Moreover, for the sake of generality, we include also a (prescribed) T-periodic body force $\boldsymbol{b} = \boldsymbol{b}(x,t)$ acting on $\mathcal{L}$. In [2] we analyzed the case when $\boldsymbol{\xi}$ has zero average over a period, namely,

$$\overline{\boldsymbol{\xi}} := \frac{1}{T} \int_0^T \boldsymbol{\xi}(t)\, \mathrm{d}t = \mathbf{0}, \tag{4.1.2}$$

which implies that $\mathcal{B}$ has zero net motion. This happens, for example, if $\mathcal{B}$ oscillates between two fixed configurations. We then showed, in particular, existence, uniqueness, and regularity of such a flow, on condition that $\boldsymbol{\xi}$ is, in suitable norm, below a certain constant depending only on Ω and T. Furthermore, we proved that $\boldsymbol{u}(x,t)$ decays like $|x|^{-1}$, uniformly in time $t \in [0,T]$, where $|x|$ denotes the distance of a generic point in Ω from the origin located in $\mathcal{B}$. Notice that this

[1] If the body is fixed, we refer the reader to [6, 12–15, 18].

behavior is the same as that of a steady-state flow around an immovable body [1, Section X.9].

Objective of the present work is to continue and, to an extent, complete the research initiated in [2]. Specifically, we shall investigate the same problem as in [2], but relaxing the assumption (4.1.2), thus allowing $\mathcal{B}$ to have a nonzero net motion over a period of time. Also for this more general problem, we are able to show existence and uniqueness of regular solutions if the data are suitably restricted. Concerning the asymptotic spatial behavior, we demonstrate the following. Without loss of generality, take $\overline{\boldsymbol{\xi}} = \lambda\, \boldsymbol{e}_1$, with $\boldsymbol{e}_1$ unit vector in the direction x_1 and $\lambda \geq 0$. Then $\boldsymbol{u}(x,t)$ decays like $|x|^{-1}[1+\lambda\,(|x|+x_1)]^{-1}$, uniformly in $t \in [0,T]$. As a consequence, if $\lambda = 0$, we find a result in agreement with that obtained in [2]. However, if $\lambda > 0$, the velocity field shows a "wake" behavior in the direction opposite to $\overline{\boldsymbol{\xi}}$, entirely analogous to that of a steady-state flow around a body that moves with constant velocity $\overline{\boldsymbol{\xi}}$ [1, Section X.8].

The method we use here is a generalization of that employed in [2] and relies upon the proof of existence, uniqueness, and corresponding estimates of solutions to the linear counterpart of problem (4.1.1) in a specific function class. Members of this class are regular in a well-defined sense, on the one hand, and, on the other hand, they decay at large spatial distances uniformly in time in a suitable fashion, provided the data decay appropriately as well; see Proposition 4.3.1. With this result in hand, it is then quite straightforward to apply the contraction mapping theorem and prove analogous results for the full nonlinear problem (4.1.1) on condition that $\boldsymbol{\xi}$ and $\boldsymbol{b}$ are below a constant depending on Ω and T; see Theorem 4.4.1.

The plan of the paper is as follows. After recalling some preliminary lemmas in Sect. 4.2, in the following Sect. 4.3 we prove the well-posedness results mentioned above for the linear problem obtained from (4.1.1) by neglecting the nonlinear term $\boldsymbol{u}\cdot\nabla\boldsymbol{u}$ and replacing $\boldsymbol{b}$ with a function $\boldsymbol{f}$ satisfying suitable regularity and spatial decay conditions. This result is achieved in two steps. In the first one, we construct unique regular solutions by combining Galerkin's method with the classical "invading domains" procedure; see Lemma 4.3.1. This finding requires that $\boldsymbol{\xi}$ and $\boldsymbol{f}$ possess a certain degree of smoothness. If, in addition, $\boldsymbol{f}$ decays at large distances and uniformly in time at a suitable rate, we then show that the above solutions must decay as well; see Lemma 4.3.2. The two lemmas are then combined in Proposition 4.3.1, to provide the desired well-posedness result. In the final Section 4.4, we employ Proposition 4.3.1 in combination with a classical fixed-point argument to extend the results of that proposition to the full nonlinear problem (4.1.1), under suitable restrictions on the magnitude of $\boldsymbol{\xi}$ and $\boldsymbol{b}$ in suitable norms; see Theorem 4.4.1.

4.2 Preliminaries

We begin to recall some notation. Throughout, Ω denotes the complement of the closure of a bounded domain $\Omega_0 \subset \mathbb{R}^3$, which we assume of class C^2, and take the origin of the coordinate system in the interior of Ω_0. For $R \geq R_* := 2\mathrm{diam}\,(\Omega_0)$, we set $\Omega_R = \Omega \cap \{|x| < R\}$, $\Omega^R = \Omega \cap \{|x| > R\}$. For a given domain $A \subseteq \mathbb{R}^3$, by $L^q(A)$, $1 \leq q \leq \infty$, $W^{m,q}(A)$, $W_0^{m,q}(A)$, $m \geq 0$, ($W^{0,q} \equiv W_0^{0,q} \equiv L^q$), we denote usual Lebesgue and Sobolev classes, with corresponding norms $\|.\|_{q,A}$ and $\|.\|_{m,q,A}$.[2] The letter P stands for the (Helmholtz) projector from $L^2(A)$ onto its subspace constituted by solenoidal (vector) function with vanishing normal component, in distributional sense, at ∂A. We also set $\int_A u \cdot v = \langle u, v\rangle_A$. $D^{m,2}(A)$ is the space of (equivalence classes of) functions u such that $\sum_{|k|=m} \|D^k u\|_{2,A} < \infty$. Obviously, the latter defines a seminorm in $D^{m,2}(A)$. Also, by $D_0^{1,2}(A)$ we denote the completion of $C_0^\infty(A)$ in the norm $\|\nabla(\cdot)\|_2$. In the above notation, the subscript "A" will be omitted, unless confusion arises. A function $u : A \times \mathbb{R} \mapsto \mathbb{R}^3$ is *T-periodic*, $T > 0$, if $u(\cdot, t+T) = u(\cdot\, t)$, for a.a. $t \in \mathbb{R}$, and we set $\overline{u} := \frac{1}{T}\int_0^T u(t)\mathrm{d}t$. Let B be a function space endowed with seminorm $\|\cdot\|_B$, $r = [1, \infty]$, and $T > 0$. $L^r(0, T; B)$ is the class of functions $u : (0, T) \to B$ such that

$$\|u\|_{L^r(B)} \equiv \begin{cases} \left(\int_0^T \|u(t)\|_B^r\right)^{\frac{1}{r}} < \infty, & \text{if } r \in [1, \infty)\,; \\ \operatorname*{ess\,sup}_{t\in[0,T]} \|u(t)\|_B < \infty, & \text{if } r = \infty. \end{cases}$$

Likewise, we put

$$W^{m,r}(0, T; B) = \Big\{u \in L^r(0, T; B) : \partial_t^k u \in L^r(0, T; B),\ k = 1, \dots, m\Big\}.$$

Unless confusion arises, we shall simply write $L^r(B)$ for $L^r(0, T; B)$, etc. Finally, if $A := \Omega, \mathbb{R}^3$, $m \geq 1$, and $\lambda \geq 0$, we set

$$\begin{aligned} [\![f]\!]_{m,\lambda,A} &:= \sup_{x\in A} |(1 + |x|)^m (1 + 2\lambda\, s(x))^m f(x)|\,, \\ [\![f]\!]_{\infty,m,\lambda,A} &:= \sup_{(x,t)\in A\times(0,\infty)} |(1 + |x|)^m (1 + 2\lambda\, s(x))^m f(x, t)|\,, \end{aligned}$$

where $s(x) = |x|+x_1$, $x \in \mathbb{R}^3$, and the subscript A will be omitted, unless necessary.

We next collect some preliminary results whose proof is given elsewhere. We begin with the following one, a special case of [1, Lemma II.6.4].

Lemma 4.2.1. *There exists a function $\psi_R \in C_0^\infty(\mathbb{R}^n)$ defined for all $R > 0$ such that $0 \leq \psi_R(x) \leq 1$, $x \in \mathbb{R}^n$, and satisfying the following properties:*

$$\lim_{R\to\infty} \psi_R(x) = 1\,, \ \textit{uniformly pointwise}\,; \quad \left\|\frac{\partial \psi_R}{\partial x_1}\right\|_{\frac{3}{2}} \leq C_1\,,$$

where C_1 is independent of R. Moreover, the support of $\partial\psi_R/\partial x_j$, $j = 1, \dots, n$, is contained in $\Omega^{\frac{R}{\sqrt{2}}}$ and

$$\|u\,|\nabla\psi_R|\,\|_2 \leq C_2\, \|\nabla u\|_{2,\Omega^{\frac{R}{\sqrt{2}}}}\,, \quad \textit{for all } u \in D_0^{1,2}(\Omega)\,,$$

where C_2 is independent of R.

[2] We shall use the same font style to denote scalar, vector, and tensor function spaces.

The following result can be found in [1, Exercise III.3.7].

Lemma 4.2.2. *Let $\mathcal{A}$ be a bounded Lipschitz domain in $\mathbb{R}^3$, and let $f \in W_0^{1,2}(\mathcal{A})$ with $\int_{\mathcal{A}} f = 0$. Then the problem*

$$\operatorname{div} \boldsymbol{z} = f \ \text{ in } \mathcal{A}\,, \ \ \boldsymbol{z} \in W_0^{2,2}(\mathcal{A})\,, \ \ \|\boldsymbol{z}\|_{2,2} \le C_0 \, \|f\|_{1,2}\,, \tag{4.2.3}$$

for some $C_0 = C_0(\mathcal{A}) > 0$, has at least one solution. Moreover, if $f = f(t)$ with $\partial_t f \in L^\infty(L^2(\mathcal{A}))$, then we have also $\partial_t \boldsymbol{z} \in L^\infty(W_0^{1,2}(\mathcal{A}))$ and

$$\|\partial_t \boldsymbol{z}\|_{1,2} \le C_0 \, \|\partial_t f\|_2\,.$$

The next result is proved in [5, Lemma 2.2].

Lemma 4.2.3. *Let $\boldsymbol{\xi} \in W^{2,2}(0,T)$ be T-periodic. There exists a solenoidal, T-periodic function $\widetilde{\boldsymbol{u}} \in W^{1,2}(W^{m,q})$, $m \in \mathbb{N}$, $q \in [1,\infty]$, such that*

$$\begin{aligned}
&\widetilde{\boldsymbol{u}}(x,t) = \boldsymbol{\xi}(t)\,, \ \ (t,\boldsymbol{x}) \in [0,T] \times \partial\Omega\,,\\
&\widetilde{\boldsymbol{u}}(x,t) = 0\,, \ \ \text{for all } t \in [0,T], \text{ all } |\boldsymbol{x}| \ge \rho, \text{ and some } \rho > R_*\,,\\
&\|\widetilde{\boldsymbol{u}}\|_{W^{2,2}(W^{m,q})} \le C\, \|\boldsymbol{\xi}\|_{W^{2,2}(0,T)}\,,
\end{aligned}$$

where $C = C(\Omega, m, q)$.

We conclude by recalling the following lemma showing suitable existence and uniqueness properties for a linear Cauchy problem [1, Theorem VIII.4.4].

Lemma 4.2.4. *Let $\boldsymbol{\mathcal{G}}$ be a second-order tensor field in $\mathbb{R}^3 \times (0,\infty)$ such that*

$$[\![\boldsymbol{\mathcal{G}}(t)]\!]_{\infty,2,\lambda} + \operatorname*{ess\,sup}_{t\ge 0} \|\nabla \cdot \boldsymbol{\mathcal{G}}(t)\|_2 < \infty\,,$$

and let $\boldsymbol{h} \in L^{\infty,q}(\mathbb{R}^3 \times (0,\infty))$, $q \in (3,\infty)$, with spatial support contained in a ball of radius ρ, some $\rho > 0$, centered at the origin. Then, the problem

$$\left.\begin{aligned}
\frac{\partial \boldsymbol{w}}{\partial t} &= \Delta \boldsymbol{w} + \lambda \frac{\partial \boldsymbol{w}}{\partial x_1} - \nabla \phi + \nabla \cdot \boldsymbol{\mathcal{G}} + \boldsymbol{h}\\
\nabla \cdot \boldsymbol{w} &= 0
\end{aligned}\right\} \ \text{in } \mathbb{R}^3 \times (0,T) \qquad \boldsymbol{w}(x,0) = \boldsymbol{0} \tag{4.2.4}$$

has one and only one solution such that for all $T > 0$,

$$\boldsymbol{w} \in L^2(0,T;W^{2,2})\,, \ \partial_t \boldsymbol{w} \in L^2(0,T;L^2)\,; \ \ \nabla\phi \in L^2(0,T;L^2). \tag{4.2.5}$$

Moreover,

$$[\![\boldsymbol{w}(t)]\!]_{\infty,1,\lambda} < \infty\,,$$

and the following inequality holds:

$$[\![\boldsymbol{w}(t)]\!]_{\infty,1,\lambda} \le C \left([\![\boldsymbol{\mathcal{G}}(t)]\!]_{\infty,2,\lambda} + \operatorname*{ess\,sup}_{t\ge 0} \|\boldsymbol{h}(t)\|_q \right) \tag{4.2.6}$$

with $C = C(q,\rho,B)$, whenever $\lambda \in [0,B]$, for some $B > 0$.

4.3 On the Unique Solvability of the Linear Problem

The main objective of this section is to prove existence and uniqueness of T-periodic solutions, in appropriate function classes, to the following set of linear equations:

$$\left.\begin{aligned}&\partial_t \boldsymbol{u} - \boldsymbol{\xi}(t)\cdot\nabla\boldsymbol{u} = \Delta\boldsymbol{u} - \nabla p + \boldsymbol{f}\\ &\operatorname{div}\boldsymbol{u} = 0\end{aligned}\right\}\ \text{in } \Omega\times(0,T)$$
$$\boldsymbol{u}(x,t) = \boldsymbol{\xi}(t)\,,\quad (x,t)\in\partial\Omega\times[0,T]\,, \tag{4.3.1}$$

where $\boldsymbol{\xi} = \boldsymbol{\xi}(t)$ and $\boldsymbol{f} = \boldsymbol{f}(x,t)$ are suitably prescribed T-periodic functions. Without loss, we take

$$\overline{\boldsymbol{\xi}} = \lambda\,\boldsymbol{e}_1\,,\quad \lambda\geq 0\,,$$

where $\boldsymbol{e}_1$ is the unit vector along the x_1-axis.

We begin by showing the following result.

Lemma 4.3.1. *Let*

$$\boldsymbol{f} = \operatorname{div}\boldsymbol{\mathcal{F}} \in W^{1,2}(L^2(\Omega))\,,\ \boldsymbol{\mathcal{F}}\in L^2(L^2(\Omega))\,,$$

and $\boldsymbol{\xi}\in W^{2,2}(0,T)$ be prescribed T-periodic functions. Then, there exists one and only one T-periodic solution $(\boldsymbol{u},p)$ to (4.3.1) such that

$$\boldsymbol{u}\in L^\infty(L^6)\,,\ \partial_t\boldsymbol{u}\in L^\infty(W^{1,2})\cap L^2(D^{2,2})\,,\ \nabla\boldsymbol{u}\in L^\infty(W^{1,2})\,;\ \ p\in L^\infty(D^{1,2})\,. \tag{4.3.2}$$

Furthermore,

$$\begin{aligned}&\|\partial_t\boldsymbol{u}\|_{L^\infty(W^{1,2})\cap L^2(D^{2,2})} + \|\boldsymbol{u}\|_{L^\infty(L^6)}\\ &\quad+\|\nabla\boldsymbol{u}\|_{L^\infty(W^{1,2})} + \|\nabla p\|_{L^\infty(D^{1,2})} + \|p\|_{L^\infty(L^2(\Omega_R))}\\ &\quad\leq C\left(\|\boldsymbol{f}\|_{W^{1,2}(L^2)} + \|\boldsymbol{\mathcal{F}}\|_{L^2(L^2)} + \|\boldsymbol{\xi}\|_{W^{2,2}(0,T)}\right),\end{aligned} \tag{4.3.3}$$

where $C = C(\Omega,T,R,\xi_0)$, for any fixed ξ_0 such that $\|\boldsymbol{\xi}\|_{W^{2,2}(0,T)}\leq\xi_0$.

Proof. We follow the argument of [5, Sections 3 & 4] to show existence, by combining the classical Galerkin method with the "invading domains" procedure. We shall limit ourselves to prove the basic a priori estimates, referring the reader to that article for the (classical) procedure of how these estimates can be used to prove the stated existence result. Let $\boldsymbol{u} = \boldsymbol{v} + \widetilde{\boldsymbol{u}}$, with $\widetilde{\boldsymbol{u}}$ given in Lemma 4.2.3, and consider problem (4.3.1) along an increasing, unbounded sequence of (bounded) domains $\{\Omega_{R_k}\}$ with $\cup_{k\in\mathbb{N}}\Omega_{R_k} = \Omega$, that is,

$$\left.\begin{aligned}&\partial_t \boldsymbol{v}_k - \boldsymbol{\xi}(t)\cdot\nabla\boldsymbol{v}_k = \Delta\boldsymbol{v}_k - \nabla\widetilde{p}_k + \boldsymbol{f} + \boldsymbol{f}_c\\ &\operatorname{div}\boldsymbol{v}_k = 0\end{aligned}\right\}\ \text{in } \Omega_{R_k}\times(0,T)$$
$$\boldsymbol{v}_k(x,t) = \boldsymbol{0}\,,\quad (x,t)\in\partial\Omega_{R_k}\times[0,T]\,, \tag{4.3.4}$$

where

$$\boldsymbol{f}_c := \Delta\widetilde{\boldsymbol{u}} - \partial_t\widetilde{\boldsymbol{u}} + \boldsymbol{\xi}(t)\cdot\nabla\widetilde{\boldsymbol{u}}.$$

If we formally dot-multiply $(4.3.4)_1$ by $\boldsymbol{v}_k$ and integrate by parts over Ω_{R_k}, we get

$$\tfrac{1}{2}\frac{d}{dt}\|\boldsymbol{v}_k(t)\|_2^2 + \|\nabla\boldsymbol{v}_k(t)\|_2^2 = \langle \boldsymbol{f}+\boldsymbol{f}_c, \boldsymbol{v}_k\rangle \le c_0\left(\|\boldsymbol{\mathcal{F}}\|_2 + \|\boldsymbol{f}_c\|_{\frac{6}{5}}\right)\|\nabla\boldsymbol{v}_k\|_2\,, \quad (4.3.5)$$

where we have used the Sobolev inequality

$$\|\boldsymbol{z}\|_6 \le \gamma_0\,\|\nabla\boldsymbol{z}\|_2, \quad \boldsymbol{z}\in D_0^{1,2}(\mathbb{R}^3)\,, \quad (4.3.6)$$

with γ_0 numerical constant. Employing in (4.3.5) Cauchy inequality along with Poincarè inequality $\|\boldsymbol{v}_k\|_2 \le c_{R_k}\|\nabla\boldsymbol{v}_k\|_2$, we get, in particular,

$$\frac{d}{dt}\|\boldsymbol{v}_k(t)\|_2^2 + c_{1R_k}\|\boldsymbol{v}_k(t)\|_2^2 \le c_2\left(\|\boldsymbol{\mathcal{F}}\|_2^2 + \|\boldsymbol{f}_c\|_{\frac{6}{5}}^2\right).$$

Proceeding as in [5, Lemma 3.1], we may combine this inequality with Galerkin method to prove the existence of a T-periodic (distributional) solution $\boldsymbol{v}_k$ to (4.3.4) with $\boldsymbol{v}_k \in L^\infty(L^2(\Omega_{R_k}))\cap L^2(D_0^{1,2}(\Omega_{R_k}))$. In addition,

$$\|\nabla\boldsymbol{v}_k\|_{L^2(L^2)} \le c\left(\|\boldsymbol{\mathcal{F}}\|_{L^2(L^2)} + \|\boldsymbol{f}_c\|_{L^2(L^{\frac{6}{5}})}\right), \quad (4.3.7)$$

where the constant c is independent of R_k; see [5, Section 3] for details. We will next show uniform (in k) estimates for $\boldsymbol{v}_k$ in spaces of higher regularity. In this regard, we notice that by the mean value theorem, from (4.3.7) it follows that there is $t_0\in(0,T)$ such that

$$\|\nabla\boldsymbol{v}_k(t_0)\|_2^2 \le c_3\left(\|\boldsymbol{\mathcal{F}}\|_{L^2(L^2)}^2 + \|\boldsymbol{f}_c\|_{L^2(L^{\frac{6}{5}})}^2\right). \quad (4.3.8)$$

If we formally dot-multiply both sides of $(4.3.4)_1$ a first time by $P\Delta\boldsymbol{v}_k$ and a second time by $\partial_t\boldsymbol{v}_k$ and integrate by parts over Ω_{R_k}, we deduce

$$\begin{aligned}
&\tfrac{1}{2}\frac{d}{dt}\|\nabla\boldsymbol{v}_k(t)\|_2^2 + \|P\Delta\boldsymbol{v}_k(t)\|_2^2 = \langle\boldsymbol{\xi}\cdot\nabla\boldsymbol{v}_k, P\Delta\boldsymbol{v}_k\rangle + \langle\boldsymbol{f}+\boldsymbol{f}_c, P\Delta\boldsymbol{v}_k(t)\rangle\\
&\tfrac{1}{2}\frac{d}{dt}\|\nabla\boldsymbol{v}_k(t)\|_2^2 + \|\partial_t\boldsymbol{v}_k(t)\|_2^2 = \langle\boldsymbol{\xi}\cdot\nabla\boldsymbol{v}_k, \partial_t\boldsymbol{v}_k\rangle + \langle\boldsymbol{f}+\boldsymbol{f}_c, \partial_t\boldsymbol{v}_k(t)\rangle\,.
\end{aligned} \quad (4.3.9)$$

Therefore, summing side-by-side the two equations in (4.3.9) and employing Cauchy-Schwarz inequality allow us to infer

$$\frac{d}{dt}\|\nabla\boldsymbol{v}_k(t)\|_2^2 + c_4((\|\partial_t\boldsymbol{v}_k(t)\|_2^2 + \|P\Delta\boldsymbol{v}_k(t)\|_2^2) \le c_4(\|\boldsymbol{f}\|_2^2 + \|\boldsymbol{f}_c\|_2^2 + \|\nabla\boldsymbol{v}_k(t)\|_2^2)\,, \quad (4.3.10)$$

with $c_4 = c_4(\xi_0)$. We now recall the inequality

$$\|D^2\boldsymbol{z}\|_{2,\Omega_R} \le c_\Omega\left(\|P\Delta\boldsymbol{z}\|_{2,\Omega_R} + \|\nabla\boldsymbol{z}\|_{2,\Omega_R}\right), \quad \boldsymbol{z}\in D^{1,2}(\Omega_R)\cap D^{2,2}(\Omega_R)\,, \quad (4.3.11)$$

with c_Ω depending only on the regularity of Ω but *not* on R [8, Lemma 1]. Thus, integrating both sides of (4.3.10) over $[t_0, t]$, using the T-periodicity property along with (4.3.8), (4.3.11), and Lemma 4.2.3 we show that $\boldsymbol{v}_k \in W^{1,2}(L^2(\Omega_{R_k})) \cap L^\infty(D_0^{1,2}(\Omega_{R_k})) \cap L^2(D^{2,2}(\Omega_{R_k}))$ and, in addition, $\boldsymbol{v}_k$ satisfies the uniform bound [5, Lemma 4.1]

$$\begin{aligned}\|\boldsymbol{v}_k\|_{L^\infty(L^6)} + \|\nabla \boldsymbol{v}_k\|_{L^\infty(L^2)} + \quad & \|\partial_t \boldsymbol{v}_k\|_{L^2(L^2)} + \|D^2 \boldsymbol{v}_k\|_{L^2(L^2)} \\ & \le c\left(\|\boldsymbol{f}\|_{L^2(L^2)} + \|\boldsymbol{\mathcal{F}}\|_{L^2(L^2)} + \|\boldsymbol{f}_c\|_{L^2(L^{\frac{6}{5}})}\right) \\ & \le C\left(\|\boldsymbol{f}\|_{L^2(L^2)} + \|\boldsymbol{\mathcal{F}}\|_{L^2(L^2)} + \|\boldsymbol{\xi}\|_{W^{2,2}(0,T)}\right),\end{aligned} \tag{4.3.12}$$

with C independent of R_k. Next, we take the time derivative of both sides of $(4.3.4)_1$, and dot-multiply both sides of the resulting equation a first time by $\partial_t \boldsymbol{v}_k$ and a second time by $P\Delta\partial_t \boldsymbol{v}_k$ and then integrate over Ω_{R_k}. We then obtain

$$\tfrac{1}{2}\frac{d}{dt}\|\partial_t \boldsymbol{v}_k(t)\|_2^2 + \|\nabla\partial_t \boldsymbol{v}_k(t)\|_2^2 = \langle \boldsymbol{\xi}' \cdot \nabla \boldsymbol{v}_k, \partial_t \boldsymbol{v}_k\rangle + \langle \partial_t \boldsymbol{f} + \partial_t \boldsymbol{f}_c, \partial_t \boldsymbol{v}_k(t)\rangle\,, \tag{4.3.13}$$

and

$$\begin{aligned}\tfrac{1}{2}\frac{d}{dt}\|\nabla\partial_t \boldsymbol{v}_k(t)\|_2^2 + \|P\,\Delta\partial_t \boldsymbol{v}_k(t)\|_2^2 & \\ = \langle \boldsymbol{\xi}' \cdot \nabla \boldsymbol{v}_k, P\Delta\partial_t \boldsymbol{v}_k(t)\rangle & + \langle \partial_t \boldsymbol{f} + \partial_t \boldsymbol{f}_c, P\Delta\partial_t \boldsymbol{v}_k(t)\rangle\,.\end{aligned} \tag{4.3.14}$$

From (4.3.12) and the mean value theorem, we find that there is $t_1 \in (0, T)$ such that

$$\|\partial_t \boldsymbol{v}_k(t_1)\|_2 \le c\left(\|\boldsymbol{f}\|_{L^2(L^2)} + \|\boldsymbol{\mathcal{F}}\|_{L^2(L^2)} + \|\boldsymbol{\xi}\|_{W^{2,2}(0,T)}\right), \tag{4.3.15}$$

and so, integrating (4.3.13) over $[t_1, t]$ and using Cauchy-Schwarz inequality, (4.3.15), (4.3.12), and the T-periodicity of $\boldsymbol{v}_k$, we arrive at

$$\|\partial_t \boldsymbol{v}_k\|_{L^\infty(L^2)} + \|\nabla\partial_t \boldsymbol{v}_k\|_{L^2(L^2)} \le C\left(\|\boldsymbol{f}\|_{W^{1,2}(L^2)} + \|\boldsymbol{\mathcal{F}}\|_{L^2(L^2)} + \|\boldsymbol{\xi}\|_{W^{2,2}(0,T)}\right). \tag{4.3.16}$$

By a similar token, from (4.3.14), (4.3.16), and (4.3.11), we get

$$\begin{aligned}\|\nabla\partial_t \boldsymbol{v}_k\|_{L^\infty(L^2)} + \|\,D^2\partial_t \boldsymbol{v}_k\|_{L^2(L^2)} & \\ \le C\left(\|\boldsymbol{f}\|_{W^{1,2}(L^2)} + \|\boldsymbol{\mathcal{F}}\|_{L^2(L^2)} + \|\boldsymbol{\xi}\|_{W^{2,2}(0,T)}\right). &\end{aligned} \tag{4.3.17}$$

Therefore, combining (4.3.12), (4.3.16), and (4.3.17), we infer

$$\begin{aligned}\|\partial_t \boldsymbol{v}_k\|_{L^\infty(W^{1,2})} + \|\boldsymbol{v}_k\|_{L^\infty(L^6)} + \|\nabla \boldsymbol{v}_k\|_{L^\infty(L^2)} + \|D^2 \boldsymbol{v}_k\|_{W^{1,2}(L^2)} & \\ \le C\left(\|\boldsymbol{f}\|_{W^{1,2}(L^2)} + \|\boldsymbol{\mathcal{F}}\|_{L^2(L^2)} + \|\boldsymbol{\xi}\|_{W^{2,2}(0,T)}\right), &\end{aligned} \tag{4.3.18}$$

where C is independent of k. Finally, setting $\boldsymbol{F}_k := \Delta \boldsymbol{v}_k + \boldsymbol{f} + \boldsymbol{f}_c$, from $(4.3.4)_1$ we get, formally, that $\widetilde{p}_k$ obeys for a.a. $t \in [0,T]$ the following Neumann problem:[3]

$$\Delta \widetilde{p}_k = \operatorname{div} \boldsymbol{F}_k \ \text{ in } \Omega_{R_k}, \quad \partial \widetilde{p}_k / \partial \boldsymbol{n}|_{\partial \Omega_{R_k}} = \boldsymbol{F}_k \cdot \boldsymbol{n}. \tag{4.3.19}$$

Therefore, multiplying both sides of the first equation by $\widetilde{p}_k$ and integrating by parts over Ω_{R_k}, we easily establish that the pressure field p_k associated with $\boldsymbol{v}_k$ satisfies the estimate [5, Lemma 4.3]

$$\|\nabla \widetilde{p}_k\|_2 \le c \left(\|D^2 \boldsymbol{v}_k\|_2 + \|\boldsymbol{f}\|_2 + \|\boldsymbol{f}_c\|_2 \right) \tag{4.3.20}$$

with c independent of k. We may now let $R_k \to \infty$ and use the uniform estimate (4.3.18) and Lemma 4.2.3, to show the existence of a pair $(\boldsymbol{u} := \boldsymbol{v} + \widetilde{\boldsymbol{u}}, \widetilde{p})$, with $\boldsymbol{u}$ T-periodic, in the class

$$\partial_t \boldsymbol{u} \in L^\infty(W^{1,2}) \cap L^2(D^{2,2}), \boldsymbol{u} \in L^\infty(L^6), \nabla \boldsymbol{u} \in L^\infty(L^2) \cap L^2(D^{1,2}), \widetilde{p} \in L^2(D^{1,2}), \tag{4.3.21}$$

such that

$$\begin{aligned} \|\partial_t \boldsymbol{u}\|_{L^\infty(W^{1,2})} + \|\boldsymbol{u}\|_{L^\infty(L^6)} + \|\nabla \boldsymbol{u}\|_{L^\infty(L^2)} + \|D^2 \boldsymbol{u}\|_{W^{1,2}(L^2)} + \|\nabla \widetilde{p}\|_{L^2(L^2)} \\ \le C \left(\|\boldsymbol{f}\|_{W^{1,2}(L^2)} + \|\boldsymbol{\mathcal{F}}\|_{L^2(L^2)} + \|\boldsymbol{\xi}\|_{W^{2,2}(0,T)} \right) \end{aligned} \tag{4.3.22}$$

and which, in addition, solves the original problem (4.3.1). The proof of this convergence property is entirely analogous to that given in [5, Lemma 3.4 and Section 4], to which we refer for the missing details. Finally, the T-periodicity property of the pressure field is proved exactly as in [2, Lemma 2], and its proof will be omitted. In order to complete the existence part of the lemma, we recall some classical properties of solutions to the Stokes problem:

$$\left.\begin{array}{l} \Delta \mathsf{w} = \nabla \mathsf{p} + \mathsf{F} \\ \operatorname{div} \mathsf{w} = 0 \end{array}\right\} \text{ in } \Omega \qquad \mathsf{w}(x) = \mathsf{w}_\star, \quad x \in \partial\Omega. \tag{4.3.23}$$

In particular, we know that any distributional solution to (4.3.23) satisfies the following estimate [1, Lemma V.4.3]:

$$\|D^2 \mathsf{w}\|_2 + \|\nabla \mathsf{p}\|_2 \le C \left(\|\mathsf{F}\|_2 + \|\mathsf{w}_\star\|_{3/2,2,\partial\Omega} + \|\mathsf{w}\|_{2,\Omega_R} + \|\mathsf{p}\|_{2,\Omega_R} \right), \tag{4.3.24}$$

with $C = C(\Omega, R)$. Let $h \in L^2(\Omega_R)$ with $\int_{\Omega_R} h = 0$, and let $\boldsymbol{\varphi} \in W_0^{1,2}(\Omega_R)$ be a solution to the problem $\operatorname{div} \boldsymbol{\varphi} = h$ in Ω_R, satisfying $\|\boldsymbol{\varphi}\|_{1,2} \le c_R \|h\|_2$. The existence of such a $\boldsymbol{\varphi}$ is well-known [1, Theorem III.3.1]. Dot-multiplying both sides of $(4.3.23)_1$ by $\boldsymbol{\varphi}$ and integrating by parts over Ω_R, we get

$$\langle \mathsf{F}, \boldsymbol{\varphi} \rangle + \langle \nabla \mathsf{w}, \nabla \boldsymbol{\varphi} \rangle = \langle \mathsf{p}, \operatorname{div} \boldsymbol{\varphi} \rangle = \langle \mathsf{p}, h \rangle .$$

[3] Note that $\boldsymbol{\xi}(t) \cdot \nabla \boldsymbol{v}_k \cdot \boldsymbol{n}|_{\partial \Omega_{R_k}} = 0$.

From this relation, the properties of φ, and the arbitrariness of h, we deduce that p, modified by a possible addition of a (T-periodic) function of time, must obey the following inequality:

$$\|\mathsf{p}\|_{2,\Omega_R} \leq c_R \left(\|\mathbf{F}\|_{2,\Omega_R} + \|\nabla \mathbf{w}\|_{2,\Omega_R}\right).$$

As a result, (4.3.24) furnishes

$$\|D^2\mathbf{w}\|_2 + \|\nabla \mathsf{p}\|_2 + \|\mathsf{p}\|_{2,\Omega_R} \leq C\left(\|\mathbf{F}\|_2 + \|\mathbf{w}_\star\|_{3/2,2,\partial\Omega} + \|\mathbf{w}\|_{1,2,\Omega_R}\right). \tag{4.3.25}$$

We next observe that, for each $t \in [0,T]$, (4.3.1) can be put in the form (4.3.23) with

$$\mathbf{w} \equiv \boldsymbol{u}\,, \quad \mathsf{p} \equiv p\,, \quad \mathbf{F} \equiv \partial_t \boldsymbol{u} + \boldsymbol{\xi}\cdot\nabla\boldsymbol{u} - \boldsymbol{f}\,, \quad \mathbf{w}_\star \equiv \boldsymbol{\xi}\,,$$

so that (4.3.25) leads to

$$\|D^2\boldsymbol{u}(t)\|_2 + \|\nabla p(t)\|_2 + \|p(t)\|_{2,\Omega_R} \leq C_1\left(\|\boldsymbol{f}(t)\|_2 + |\boldsymbol{\xi}(t)| + \|\partial_t\boldsymbol{u}(t)\|_2 + \|\nabla\boldsymbol{u}(t)\|_2 + \|\boldsymbol{u}(t)\|_{2,\Omega_R}\right), \tag{4.3.26}$$

with $C_1 = C_1(\Omega, R, \xi_0)$. If we combine (4.3.26) and use (4.3.22) we then show

$$\|D^2\boldsymbol{u}\|_{L^\infty(L^2)} + \|\nabla p\|_{L^\infty(L^2)} + \|p(t)\|_{L^\infty(L^2(\Omega_R))} \leq C\left(\|\boldsymbol{f}\|_{W^{1,2}(L^2)} + \|\boldsymbol{\mathcal{F}}\|_{L^2(L^2)} + \|\boldsymbol{\xi}\|_{W^{3,2}(0,T)}\right). \tag{4.3.27}$$

In view of (4.3.22) and (4.3.27), the proof of the existence property is thus completed. We shall now prove uniqueness, namely, that $\boldsymbol{u} \equiv \nabla p \equiv \mathbf{0}$ is the only T-periodic solution in the class (4.3.2) to the following system:

$$\left.\begin{array}{l}\partial_t\boldsymbol{u} - \boldsymbol{\xi}(t)\cdot\nabla\boldsymbol{u} = \Delta\boldsymbol{u} - \nabla p \\ \operatorname{div}\boldsymbol{u} = 0\end{array}\right\} \text{ in } \Omega\times(0,T) \\ \boldsymbol{u}(x,t) = \mathbf{0}\,, \quad (x,t) \in \partial\Omega\times[0,T]\,. \tag{4.3.28}$$

To this end, we write

$$\boldsymbol{u} = (\boldsymbol{u} - \overline{\boldsymbol{u}}) + \overline{\boldsymbol{u}} := \boldsymbol{w} + \overline{\boldsymbol{u}}\,, \quad \boldsymbol{\xi} = (\boldsymbol{\xi} - \overline{\boldsymbol{\xi}}) + \overline{\boldsymbol{\xi}}) := \boldsymbol{\chi} + \overline{\boldsymbol{\xi}}. \tag{4.3.29}$$

Since $\overline{\boldsymbol{w}} = 0$, by Poincaré inequality, Fubini's theorem, and (4.3.2), we deduce $\boldsymbol{w} \in L^2(L^2)$, so that, in particular,

$$\boldsymbol{w} \in W^{1,2}(L^2) \cap L^2(W^{2,2})\,. \tag{4.3.30}$$

From classical embedding theorems (e.g., [17, Theorem 2.1]) and (4.3.30), we deduce

$$\boldsymbol{w} \in L^\infty(L^2)\,. \tag{4.3.31}$$

Furthermore, from (4.3.28) it follows that p obeys the following Neumann problem for a.a. $t \in [0,T]$:

$$\Delta p = 0 \ \text{ in } \Omega\,, \quad \frac{\partial p}{\partial \boldsymbol{n}} = -\operatorname{curl}(\psi\operatorname{curl}\boldsymbol{u})\cdot\boldsymbol{n} \ \text{ at } \partial\Omega, \tag{4.3.32}$$

where ψ is a smooth function of bounded support that is 1 in a neighborhood of $\partial\Omega$, and we used the identity $\Delta\boldsymbol{u} = -\mathrm{curl}\,\mathrm{curl}\,\boldsymbol{u}$. Employing well-known results on the Neumann problem [1, Theorem III.3.2] and the fact that $\boldsymbol{u}$ is in the class (4.3.2), we get

$$\|\nabla p\|_{L^2(L^q)} \le c\,\|\mathrm{curl}\,\mathrm{curl}\,\boldsymbol{u}\|_{L^2(L^q(K))} + \|\mathrm{curl}\,\boldsymbol{u}\|_{L^2(L^q(K))}\,, \quad \text{all } q \in (1,2]\,,$$

with $K = \mathrm{supp}\,(\psi)$. From this and Sobolev inequality, we may then modify p by adding to it a suitable T-periodic function of time, in such a way that the redefined pressure field, that we continue to denote by p, satisfies

$$p \in L^2(L^r)\,, \quad \text{all } r \in (3/2, 6]\,. \tag{4.3.33}$$

Let $\psi_R = \psi_R(x)$ be the function defined in Lemma 4.2.1. We dot-multiply both sides of $(4.3.28)_1$ by $\psi_R\boldsymbol{u}$ and integrate by parts over $\Omega \times (0,T)$. Noticing that $\boldsymbol{u} \in L^2(L^2(\Omega_\rho))$, all $\rho \ge R_*$, and using T-periodicity, we thus show

$$\begin{aligned}\int_0^T\int_\Omega \psi_R\,|\nabla\boldsymbol{u}|^2 &= -\tfrac{1}{2}\int_0^T\int_{\Omega_{\frac{R}{\sqrt{2}}}} \nabla\psi_R\cdot\boldsymbol{\xi}(t)|\boldsymbol{u}|^2 + \int_0^T\int_{\Omega_{\frac{R}{\sqrt{2}}}} p\,\nabla\psi_R\cdot\boldsymbol{u}\\ &:= -\tfrac{1}{2}I_{1R} + I_{2R}\,.\end{aligned} \tag{4.3.34}$$

From Schwarz inequality, the properties of ψ_R and (4.3.2), we get

$$|I_{2R}| \le C_1 \sup_{t\in[0,T]} \|\nabla\boldsymbol{u}(t)\|_2 \int_0^T \|p(t)\|_{2,\Omega_{\frac{R}{\sqrt{2}}}}\,,$$

which, by (4.3.33), entails

$$\lim_{R\to\infty} |I_{2R}| = 0\,. \tag{4.3.35}$$

Next, recalling that $\overline{\boldsymbol{\xi}} = \lambda\,\boldsymbol{e}_1$, we may employ (4.3.29) and Fubini's theorem to show

$$\begin{aligned}I_{1R} &= \int_0^T\int_{\Omega_{\frac{R}{\sqrt{2}}}} \left[\lambda\,\frac{\partial\psi_R}{\partial x_1}\,|\overline{\boldsymbol{u}}|^2 + \nabla\psi_R\cdot\boldsymbol{\xi}(t)(|\boldsymbol{w}|^2 + 2\overline{\boldsymbol{u}}\cdot\boldsymbol{w})\right]\\ &:= I_{1R}^1 + I_{1R}^2\,,\end{aligned}$$

where we have used the fact that $\overline{\boldsymbol{\chi}} = \boldsymbol{0}$. By Hölder inequality and the summability properties of $\partial\psi_R/\partial x_1$, we show

$$|I_{1R}^1| \le c\,\|\boldsymbol{\xi}\|_{W^{1,2}(0,T)} \int_0^T \|\overline{\boldsymbol{u}}\|^2_{6,\Omega_{\frac{R}{\sqrt{2}}}}\,,$$

which, in view of (4.3.2), implies

$$\lim_{R\to\infty} |I_{1R}^1| = 0\,. \tag{4.3.36}$$

Finally, by using one more time Schwarz inequality and the properties of ψ_R, we infer

$$|I_{2R}^2| \le 2\|\boldsymbol{\xi}\|_{W^{1,2}(0,T)}\||\boldsymbol{u}|\,\nabla\psi_R\|_{2,\Omega_{\frac{R}{\sqrt{2}}}}\|\boldsymbol{w}\|_{L^\infty(L^2)} \le c\,\|\boldsymbol{\xi}\|_{W^{1,2}(0,T)}\|\nabla\boldsymbol{u}\|_{2,\Omega_{\frac{R}{\sqrt{2}}}}\|\boldsymbol{w}\|_{L^\infty(L^2)}\,,$$

and so from the latter, (4.3.31), and (4.3.2), we deduce

$$\lim_{R\to\infty}|I_{2R}^2| = 0\,. \tag{4.3.37}$$

Uniqueness then follows by letting $R\to\infty$ in (4.3.34) and using (4.3.35)–(4.3.37). The lemma is completely proved.

□

The following result provides, under further assumptions on $\boldsymbol{f}$, the spatial asymptotic behavior of solutions determined in the previous lemma.

Lemma 4.3.2. *Let* $(\boldsymbol{u},p)$ *be the solution to* (4.3.1) *constructed in* Lemma 4.3.1. *Then, if, in addition,* $[\![\boldsymbol{\mathcal{F}}]\!]_{\infty,2,\lambda}<\infty$, *it follows that* $[\![\boldsymbol{u}]\!]_{\infty,1,\lambda}<\infty$, *and, moreover,*

$$[\![\boldsymbol{u}]\!]_{\infty,1,\lambda} \le C\left([\![\boldsymbol{\mathcal{F}}]\!]_{\infty,2,\lambda} + \|\boldsymbol{f}\|_{W^{1,2}(L^2)} + \|\boldsymbol{\xi}\|_{W^{2,2}(0,T)}\right),$$

where $C=C(\Omega,T,\xi_0)$, *whenever* $\|\boldsymbol{\xi}\|_{W^{2,2}(0,T)}\in[0,\xi_0]$, *for some* $\xi_0>0$.

Proof. Let ψ be the "cut-off" function introduced in (4.3.32), and let $\boldsymbol{z}$ be a solution to problem (4.2.3) with $f\equiv -\nabla\psi\cdot\boldsymbol{u}$. Since $\int_K f = 0$, where $K=\operatorname{supp}(f)$, Lemma 4.2.2 guarantees the existence of such a $\boldsymbol{z}$. Thus, setting

$$\boldsymbol{w} := \psi\,\boldsymbol{u} + \boldsymbol{z}\,,\quad \mathsf{p} := \psi\,p\,,\quad \boldsymbol{\mathcal{H}} = \psi\boldsymbol{\mathcal{F}}$$

from (4.3.1) we deduce that $(\boldsymbol{w},\mathsf{p})$ is a T-periodic solution to the following problem:

$$\left.\begin{array}{l}\partial_t\boldsymbol{w} - \boldsymbol{\xi}(t)\cdot\nabla\boldsymbol{w} = \Delta\boldsymbol{w} - \nabla\mathsf{p} + \operatorname{div}\boldsymbol{\mathcal{H}} + \boldsymbol{g}\\ \operatorname{div}\boldsymbol{w} = 0\end{array}\right\}\ \text{in } \mathbb{R}^3\times(0,T)\,, \tag{4.3.38}$$

where

$$\boldsymbol{g} := -\partial_t\boldsymbol{z} + \boldsymbol{\xi}(t)\cdot\nabla\boldsymbol{z} + \Delta\boldsymbol{z} - 2\nabla\psi\cdot\nabla\boldsymbol{u} + p\,\nabla\psi - \boldsymbol{\xi}(t)\cdot\nabla\psi\,\boldsymbol{u}\,.$$

If we extend $\boldsymbol{z}$ to 0 outside its support, we infer that $\boldsymbol{g}$ is of bounded support. Also with the help of Lemma 4.2.2 and Lemma 4.3.1, we easily deduce

$$\begin{array}{l}\sup\limits_{t\ge 0}\|\boldsymbol{g}(t)\|_2 \le c\,(\|\boldsymbol{f}\|_{W^{1,2}(L^2)} + [\![\boldsymbol{\mathcal{F}}]\!]_{\infty,2,\lambda} + \|\boldsymbol{\xi}\|_{W^{2,2}(0,T)})\,,\\ \operatorname{div}\boldsymbol{\mathcal{H}}(t)\in L^\infty(L^2)\,,\end{array} \tag{4.3.39}$$

where we have used the obvious inequality $\|\boldsymbol{\mathcal{F}}\|_{L^2(L^2)}\le c\,[\![\boldsymbol{\mathcal{F}}]\!]_{\infty,2,\lambda}$. We now introduce the new variable $\boldsymbol{y}$ defined by

$$\boldsymbol{y} = \boldsymbol{x} + \boldsymbol{x}_0(t), \tag{4.3.40}$$

where

$$\boldsymbol{x}_0(t) := \int_0^t (\boldsymbol{\xi}(s) - \overline{\boldsymbol{\xi}})\,\mathrm{d}s\,. \tag{4.3.41}$$

Since $\overline{(\boldsymbol{\xi}(t) - \overline{\boldsymbol{\xi}})} = 0$, one can show

$$\sup_{t\geq 0} |\boldsymbol{x}_0(t)| \leq M, \tag{4.3.42}$$

where

$$M := C\,\frac{T^{\frac{1}{2}}}{\pi}\Big(\int_0^T |\boldsymbol{\xi}(t) - \overline{\boldsymbol{\xi}}|^2\Big)^{\frac{1}{2}}$$

and C is a numerical constant; see [2]. Thus, in particular,

$$|\boldsymbol{x}| - M \leq |\boldsymbol{y}| \leq |\boldsymbol{x}| + M\,, \tag{4.3.43}$$

Setting

$$\begin{aligned} &\boldsymbol{v}(\boldsymbol{y},t) = \boldsymbol{w}(\boldsymbol{y} + \boldsymbol{x}_0(t), t), \quad \mathsf{P}(\boldsymbol{y},t) = \mathsf{p}(\boldsymbol{y} + \boldsymbol{x}_0(t), t),\\ &\boldsymbol{\mathcal{G}}(\boldsymbol{y},t) = \boldsymbol{\mathcal{H}}(\boldsymbol{y} + \boldsymbol{x}_0(t), t)\,, \quad \boldsymbol{h} = \boldsymbol{g}(\boldsymbol{y} + \boldsymbol{x}_0(t), t) \end{aligned} \tag{4.3.44}$$

from (4.3.38) we easily deduce that $(\boldsymbol{v}, \mathsf{P})$ solves the following Cauchy problem:

$$\left.\begin{aligned} &\partial_t \boldsymbol{v} - \lambda\,\partial_1 \boldsymbol{v} = \Delta \boldsymbol{v} - \nabla \mathsf{P} + \operatorname{div} \boldsymbol{\mathcal{G}} + \boldsymbol{h}\\ &\operatorname{div} \boldsymbol{v} = 0 \end{aligned}\right\} \text{ in } \mathbb{R}^3 \times (0,\infty)\,, \qquad \boldsymbol{v}(x,0) = \boldsymbol{w}(x,0)\,. \tag{4.3.45}$$

We look for a solution to (4.3.45) of the form $(\boldsymbol{v}_1 + \boldsymbol{v}_2, \mathsf{P}_1 + \mathsf{P}_2)$ where

$$\left.\begin{aligned} &\partial_t \boldsymbol{v}_1 - \lambda\,\partial_1 \boldsymbol{v}_1 = \Delta \boldsymbol{v}_1 - \nabla \mathsf{P}_1 + \operatorname{div} \boldsymbol{\mathcal{G}} + \boldsymbol{h}\\ &\operatorname{div} \boldsymbol{v}_1 = 0 \end{aligned}\right\} \text{ in } \mathbb{R}^3 \times (0,\infty)\,, \qquad \boldsymbol{v}_1(x,0) = \boldsymbol{0}\,, \tag{4.3.46}$$

and

$$\left.\begin{aligned} &\partial_t \boldsymbol{v}_2 - \lambda\,\partial_1 \boldsymbol{v}_2 = \Delta \boldsymbol{v}_2 - \nabla \mathsf{P}_2\\ &\operatorname{div} \boldsymbol{v}_2 = 0 \end{aligned}\right\} \text{ in } \mathbb{R}^3 \times (0,\infty)\,, \qquad \boldsymbol{v}_2(x,0) = \boldsymbol{w}(x,0)\,. \tag{4.3.47}$$

From (4.3.39) and (4.3.44), we readily deduce

$$\begin{aligned} &\sup_{t\geq 0} \|\boldsymbol{h}(t)\|_2 \leq c\,(\|\boldsymbol{f}\|_{W^{1,2}(L^2)} + [\![\boldsymbol{\mathcal{F}}]\!]_{\infty,2,\lambda} + \|\boldsymbol{\xi}\|_{W^{2,2}(0,T)})\,,\\ &\operatorname{div} \boldsymbol{\mathcal{G}}(t) \in L^\infty(L^2)\,. \end{aligned} \tag{4.3.48}$$

Furthermore, by (4.3.42) and (4.3.43), it follows that

$$\begin{aligned} (1 + |x|)(1 + 2\lambda\, s(x)) &\leq (1 + |y| + M)\big(1 + 2\lambda\, s(y) + 2\lambda\,(M + x_{01}(t))\big)\\ &\leq C\,(1 + |y|)\,\big(1 + 2\lambda\, s(y)\big)\,, \end{aligned} \tag{4.3.49}$$

where here and in the rest of the proof C denotes a constant depending, at most, on Ω, T, and ξ_0. Likewise,

$$(1+|y|)(1+2\lambda\, s(y)) \le C\,(1+|x|)\big(1+2\lambda\, s(x)\big)\,. \tag{4.3.50}$$

By (4.3.44) and the assumption on $\boldsymbol{\mathcal{F}}$, (4.3.50) implies, in particular, $[\![\boldsymbol{\mathcal{G}}]\!]_{\infty,2,\lambda} < \infty$ and that

$$[\![\boldsymbol{\mathcal{G}}]\!]_{\infty,2,\lambda} \le C\,[\![\boldsymbol{\mathcal{F}}]\!]_{\infty,2,\lambda} \tag{4.3.51}$$

Thus, combining (4.3.48), (4.3.51), and the assumption on $\boldsymbol{\mathcal{F}}$ with Lemma 4.2.4, we conclude that the Cauchy problem (4.3.46) has one (and only one) solution $(\boldsymbol{v}_1, \mathsf{P}_1)$ in the class (4.2.5) for all $T > 0$. Further, we have $[\![\boldsymbol{v}_1]\!]_{\infty,1,\lambda} < \infty$ with

$$[\![\boldsymbol{v}_1]\!]_{\infty,1,\lambda} \le C\,\big(\|\boldsymbol{f}\|_{W^{1,2}(L^2)} + [\![\boldsymbol{\mathcal{F}}]\!]_{\infty,2,\lambda} + \|\boldsymbol{\xi}\|_{W^{2,2}(0,T)}\big). \tag{4.3.52}$$

Concerning (4.3.47), a solution is given by

$$v_{2i}(y,t) = \int_{\mathbb{R}^3} \Gamma_{i\ell}(y-z,s;\lambda)\, w_\ell(z,0)\,\mathrm{d}z\,, \tag{4.3.53}$$

where $\boldsymbol{\Gamma}$ is the (time-dependent) Oseen fundamental tensor solution to $(4.3.47)_{1,2}$ [1, Theorem VIII.4.3]. Since $\boldsymbol{w}(x,0) \in L^6(\mathbb{R}^3)$, it follows that [1, Theorem VIII.4.3]

$$\begin{aligned}
&\boldsymbol{v}_2, \partial_t \boldsymbol{v}_2\, D^2\boldsymbol{v}_2 \in L^r([\varepsilon,\tau]\times\mathbb{R}^3)\,, \quad \text{all } \varepsilon\in(0,\tau),\ \tau>0, \text{ and } r\in[6,\infty]\,,\\
&\|\boldsymbol{v}_2(t)\|_\infty \le C_1\, t^{-\frac14}\,\|\boldsymbol{w}(0)\|_6\,, \qquad \sup_{t\in(0,\infty)} \|\boldsymbol{v}_2(t)\|_6 \le C_1\,\|\boldsymbol{w}(0)\|_6\,.
\end{aligned} \tag{4.3.54}$$

In view of the regularity properties of $\boldsymbol{u}$ (and hence of $\boldsymbol{w}$) and those in (4.2.5) and (4.3.54) for $\boldsymbol{v}_i$, $i = 1, 2$, respectively, we may use the results proved in [1, Lemma VIII.4.2] to guarantee $\boldsymbol{w} = \boldsymbol{v}_1 + \boldsymbol{v}_2$. As a consequence, due to the T-periodicity of $\boldsymbol{w}$ and $(4.3.44)_1$, for arbitrary positive integer n and $t \in [0, T]$, we obtain

$$\begin{aligned}
|\boldsymbol{w}(x,t)|(1+|x|)(1+2\lambda\, s(x)) &= |\boldsymbol{v}(y,t+nT)|(1+|x|)(1+2\lambda\, s(x))\\
&\le \big(|\boldsymbol{v}_1(y,t+nT)| + |\boldsymbol{v}_2(y,t+nT)|\big)(1+|x|)(1+2\lambda\, s(x)).
\end{aligned} \tag{4.3.55}$$

Employing (4.3.49), (4.3.52), and $(4.3.54)_2$ in this inequality, we get

$$\begin{aligned}
&|\boldsymbol{w}(x,t)|(1+|x|)(1+2\lambda\, s(x))\\
&\quad \le C\,\Big[(1+|x|)(1+2\lambda\, s(x))(t+nT)^{-\frac14}\|\boldsymbol{w}(0)\|_6 + \|\boldsymbol{f}\|_{W^{1,2}(L^2)} + [\![\boldsymbol{\mathcal{F}}]\!]_{\infty,2,\lambda} + \|\boldsymbol{\xi}\|_{W^{2,2}(0,T)}\Big]
\end{aligned}$$

so that, by letting $n \to \infty$ and recalling that, uniformly in $t \ge 0$, $\boldsymbol{u}(x,t) \equiv \boldsymbol{w}(x,t)$ for $|x|$ sufficiently large ($> \overline{R}$, say), we deduce

$$[\![\boldsymbol{u}]\!]_{\infty,1,\lambda,\Omega^{\overline{R}}} \le C\,\big(\|\boldsymbol{f}\|_{W^{1,2}(L^2)} + [\![\boldsymbol{\mathcal{F}}]\!]_{\infty,2,\lambda} + \|\boldsymbol{\xi}\|_{W^{2,2}(0,T)}\big)\,. \tag{4.3.56}$$

Since by classical embedding theorems, we have

$$\|\boldsymbol{u}\|_{L^\infty(L^\infty)} \le C\,\big(\|\boldsymbol{u}\|_{L^\infty(L^6)} + \|D^2\boldsymbol{u}\|_{L^\infty(L^2)}\big)\,, \tag{4.3.57}$$

the desired result then follows from (4.3.56), (4.3.57), and (4.3.3).

□

The findings of Lemma 4.3.1 and Lemma 4.3.2 can be combined to arrive at the following one that represents the main achievement of this section.

Proposition 4.3.1. *Let*

$$\boldsymbol{f} = \operatorname{div} \boldsymbol{\mathcal{F}} \in W^{1,2}(L^2)\,, \quad [\![\boldsymbol{\mathcal{F}}]\!]_{\infty,2,\lambda} < \infty\,, \quad \boldsymbol{\xi} \in W^{2,2}(0,T)$$

be prescribed T-periodic functions with $\overline{\boldsymbol{\xi}} = \lambda\, \boldsymbol{e}_1$, $\lambda \geq 0$. Then, there exists one and only one T-periodic solution $(\boldsymbol{u}, p)$ to (4.3.1) such that

$$[\![\boldsymbol{u}]\!]_{\infty,1,\lambda} < \infty\,, \;\; \partial_t \boldsymbol{u} \in L^\infty(W^{1,2}) \cap L^2(D^{2,2})\,, \;\; \nabla \boldsymbol{u} \in L^\infty(W^{1,2})\,; \;\; p \in L^\infty(D^{1,2})\,.$$

Furthermore,

$$\begin{aligned} &[\![\boldsymbol{u}]\!]_{\infty,1,\lambda} + \|\partial_t \boldsymbol{u}\|_{L^\infty(W^{1,2})\cap L^2(D^{2,2})} + \|\nabla \boldsymbol{u}\|_{L^\infty(W^{1,2})} + \|\nabla p\|_{L^\infty(D^{1,2})} + \|p\|_{L^\infty(L^2(\Omega_R))} \\ &\qquad \leq C\left(\|\boldsymbol{f}\|_{W^{1,2}(L^2)} + [\![\boldsymbol{\mathcal{F}}]\!]_{\infty,2,\lambda} + \|\boldsymbol{\xi}\|_{W^{2,2}(0,T)}\right), \end{aligned} \tag{4.3.58}$$

where $C = C(\Omega, T, R, \xi_0)$, for any fixed ξ_0 such that $\|\boldsymbol{\xi}\|_{W^{2,2}(0,T)} \leq \xi_0$.

4.4 On the Unique Solvability of the Nonlinear Problem

The main objective of this section is to study the properties of T-periodic solutions to the full nonlinear problem (4.1.1). This will be achieved by combining the results proved in Proposition 4.3.1 with a classical contraction mapping argument. To this end, we introduce the Banach space

$$\begin{aligned} \mathcal{S} := \big\{ &T\text{-periodic } \boldsymbol{u} : \Omega \times [0,T] \mapsto \mathbb{R}^3 \mid \\ &[\![\boldsymbol{u}]\!]_{\infty,1,\lambda} < \infty\,, \; \partial_t \boldsymbol{u} \in L^\infty(W^{1,2}) \cap L^2(D^{2,2})\,, \; \nabla \boldsymbol{u} \in L^\infty(W^{1,2})\,; \; \operatorname{div} \boldsymbol{u} = 0 \big\}\,, \end{aligned}$$

endowed with the norm

$$\|\boldsymbol{u}\|_{\mathcal{S}} := [\![\boldsymbol{u}]\!]_{\infty,1,\lambda} + \|\partial_t \boldsymbol{u}\|_{L^\infty(W^{1,2})\cap L^2(D^{2,2})} + \|\nabla \boldsymbol{u}\|_{L^\infty(W^{1,2})}. \tag{4.4.1}$$

Lemma 4.4.1. *Let* $\mathbf{u}, \mathbf{w} \in \mathcal{S}$. *Then* $\mathbf{u} \cdot \nabla \mathbf{w} \in W^{1,2}(L^2)$ *and*

$$\|\mathbf{u} \cdot \nabla \mathbf{w}\|_{W^{1,2}(L^2)} \leq c\, \|\mathbf{u}\|_{\mathcal{S}} \|\mathbf{w}\|_{\mathcal{S}}\,.$$

Proof. Clearly,

$$\|\mathbf{u} \cdot \nabla \mathbf{w}\|_{L^2(L^2)} \leq [\![\mathbf{u}]\!]_{\infty,1,\lambda}\, \|\nabla \mathbf{w}\|_{L^\infty(L^2)} \leq \|\mathbf{u}\|_{\mathcal{S}} \|\mathbf{w}\|_{\mathcal{S}}\,.$$

Moreover, by using the embedding $L^4 \subset W^{1,2}$ along with Schwarz inequality, we get

$$\begin{aligned}\|\partial_t \mathbf{u}\cdot\nabla\mathbf{w}\|_{L^2(L^2)}+\|\mathbf{u}\cdot\nabla\partial_t\mathbf{w}\|_{L^2(L^2)} &\le \|\partial_t\mathbf{u}\|_{L^4(L^4)}\|\nabla\mathbf{w}\|_{L^4(L^4)}+[\![\mathbf{u}]\!]_{\infty,1,\lambda}\,\|\nabla\partial_t\mathbf{w}\|_{L^2(L^2)}\\ &\le c\left(\|\partial_t\mathbf{u}\|_{L^\infty(W^{1,2})}\|\nabla\mathbf{w}\|_{L^\infty(W^{1,2})}\right.\\ &\qquad\left.+[\![\mathbf{u}]\!]_{\infty,1,\lambda}\|\partial_t\mathbf{w}\|_{L^\infty(W^{1,2})}\right)\\ &\le c\,\|\mathbf{u}\|_{\mathcal{S}}\,\|\mathbf{w}\|_{\mathcal{S}}\,.\end{aligned}$$

The proof of the lemma is completed.

□

We are now in a position to prove the main result of this paper.

Theorem 4.4.1. *Let*

$$\boldsymbol{b}=\operatorname{div}\boldsymbol{\mathcal{B}}\in W^{1,2}(L^2)\,,\quad [\![\boldsymbol{\mathcal{B}}]\!]_{\infty,2,\lambda}<\infty\,,\quad \boldsymbol{\xi}\in W^{2,2}(0,T)$$

be prescribed T-periodic functions with $\overline{\boldsymbol{\xi}}=\lambda\,\boldsymbol{e}_1$, $\lambda\ge 0$. Then, there exists $\varepsilon_0=\varepsilon_0(\Omega,T)>0$ such that if

$$\mathsf{D}:=\|\boldsymbol{b}\|_{W^{1,2}(L^2)}+[\![\boldsymbol{\mathcal{B}}]\!]_{\infty,2,\lambda}+\|\boldsymbol{\xi}\|_{W^{2,2}(0,T)}<\varepsilon_0\,,$$

problem (4.1.1) *has one and only one T-periodic solution $(\boldsymbol{u},p)\in\mathcal{S}\times L^\infty(D^{1,2})$. Moreover, $\|\boldsymbol{u}\|_{\mathcal{S}}\le c\,\mathsf{D}$, for some $c=c(\Omega,T)$.*

Proof. We employ the contraction mapping theorem. To this end, define the map

$$M:\mathbf{u}\in\mathcal{S}\mapsto\boldsymbol{u}\in\mathcal{S}\,,$$

with $\boldsymbol{u}$ solving the linear problem

$$\left.\begin{array}{l}\partial_t\boldsymbol{u}-\boldsymbol{\xi}(t)\cdot\nabla\boldsymbol{u}=\Delta\boldsymbol{u}-\nabla p+\mathbf{u}\cdot\nabla\mathbf{u}+\boldsymbol{b}\\ \operatorname{div}\boldsymbol{u}=0\end{array}\right\}\ \text{in }\Omega\times(0,T)$$
$$\boldsymbol{u}(x,t)=\boldsymbol{\xi}(t)\,,\quad (x,t)\in\partial\Omega\times[0,T]\,. \tag{4.4.2}$$

Set

$$\mathbf{f}:=\mathbf{u}\cdot\nabla\mathbf{u}=\operatorname{div}(\mathbf{u}\otimes\mathbf{u}):=\operatorname{div}\mathbf{F}\,, \tag{4.4.3}$$

where we used the condition $\operatorname{div}\mathbf{u}=0$. In virtue of Lemma 4.4.1, by assumption, and by the obvious inequality

$$[\![\mathbf{F}]\!]_{\infty,2,\lambda}\le c_1[\![\mathbf{u}]\!]^2_{\infty,1,\lambda}\,,\quad \mathbf{u}\in\mathcal{S}\,,$$

we infer that $\mathbf{F}$, $\boldsymbol{b}$, and $\boldsymbol{\xi}$ satisfy the assumption of Proposition 4.3.1. Therefore, by that proposition we conclude that the map M is well-defined and, in particular, that

$$\|\boldsymbol{u}\|_{\mathcal{S}}\le c_2\left(\|\mathbf{u}\|^2_{\mathcal{S}}+\mathsf{D}\right)\,, \tag{4.4.4}$$

with $c_2=c_2(\Omega,T,\xi_0)$. If we now take

$$\|\mathbf{u}\|_{\mathcal{S}}<\delta\,,\quad \delta:=4c_2\mathsf{D}\,,\quad \mathsf{D}<\frac{1}{16c_2^2}\,, \tag{4.4.5}$$

from (4.4.4) we deduce $\|\boldsymbol{u}\|_{\mathcal{S}} < \frac{1}{2}\delta$. Let $\mathbf{u}_i \in \mathcal{S}$ $i = 1, 2$, and set

$$\mathbf{u} := \mathbf{u}_1 - \mathbf{u}_2\,, \quad \boldsymbol{u} := M(\mathbf{u}_1) - M(\mathbf{u}_2)\,.$$

From (4.4.2), we then show

$$\left.\begin{array}{l}\partial_t \boldsymbol{u} - \boldsymbol{\xi}(t)\cdot\nabla\boldsymbol{u} = \Delta\boldsymbol{u} - \nabla p + \mathbf{u}_1\cdot\nabla\mathbf{u} + \mathbf{u}\cdot\nabla\mathbf{u}_2 \\ \operatorname{div}\boldsymbol{u} = 0\end{array}\right\} \text{ in } \Omega\times(0,T) \\ \boldsymbol{u}(x,t) = \mathbf{0}\,, \quad (x,t)\in\partial\Omega\times[0,T]\,. \tag{4.4.6}$$

Proceeding as in the proof of (4.4.4), we can show

$$\|\boldsymbol{u}\|_{\mathcal{S}} \le c_2\,(\|\mathbf{u}_1\|_{\mathcal{X}} + \|\mathbf{u}_2\|_{\mathcal{S}})\,\|\mathbf{u}\|_{\mathcal{S}}\,.$$

As a result, if $\|\mathbf{u}_i\|_{\mathcal{S}} < \delta$, $i = 1, 2$, from the previous inequality we infer

$$\|\boldsymbol{u}\|_{\mathcal{S}} < 2c_2\delta\|\mathbf{u}\|_{\mathcal{X}}\,,$$

and since by (4.4.5) $2c_2\delta < 1/2$, we may conclude that M is a contraction, which, along with (4.4.5), completes the proof of the theorem.

□

References

1. Galdi, G.P., *An introduction to the mathematical theory of the Navier-Stokes equations. Steady-state problems*, Second edition. Springer Monographs in Mathematics, Springer, New York (2011)

2. Galdi, G.P., Viscous flow past a body translating by time-periodic motion with zero average, *Arch. Rational Mech. Anal.*, **237**, 1237–1269 (2020)

3. Galdi, G.P., and Kyed, M., Time periodic solutions to the Navier-Stokes equations, in *Handbook of Mathematical Analysis in Mechanics of Viscous Fluids*, Eds Y.Giga, and A. Novotný, Springer-Verlag (2017)

4. Galdi, G.P., Kyed, M., Time-periodic flow of a viscous liquid past a body. Partial differential equations in fluid mechanics, 2049, London Math. Soc. Lecture Note Ser., **452**, Cambridge Univ. Press, Cambridge (2018)

5. Galdi, G.P., and Silvestre A.L., Existence of time-periodic solutions to the Navier-Stokes equations around a moving body. *Pacific J. Math.* **223** 251–267 (2006)

6. Galdi, G.P., and Sohr, H., Existence and uniqueness of time-periodic physically reasonable Navier–Stokes flow past a body. *Arch. Ration. Mech. Anal.* **172** 363–406 (2004)

7. Geissert, M., Hieber, M., and Huy, N.-T., A general approach to time periodic incompressible viscous fluid flow problems. *Arch. Ration. Mech. Anal.* **220** 1095–1118 (2016)

8. Heywood, J.G., The Navier-Stokes equations: on the existence, regularity and decay of solutions, *Indiana Univ. Math. J.*, **29**, 639–681 (1980)

9. Hieber, M., Huy, N.-T., and Seyfert, A., On periodic and almost periodic solutions to incompressible viscous fluid flow problems on the whole line. *Mathematics for nonlinear phenomena–analysis and computation*, 51-81, Springer Proc. Math. Stat., **215**, Springer, Cham (2017)

10. Hieber, M., Mahalov, A. and Takada, R., Time periodic and almost time periodic solutions to rotating stratified fluids subject to large forces. *J. Differential Equations* **266**, 977–1002 (2019)

11. Huy, N.-T., Periodic motions of Stokes and Navier-Stokes flows around a rotating obstacle. *Arch. Ration. Mech. Anal.* **213** 689–703 (2014)

12. Kang, K., Miura, H., and Tsai, T–P., Asymptotics of small exterior Navier–Stokes flows with non-decaying boundary data. *Comm. Partial Differential Equations* **37** 1717–1753 (2012)

13. Kozono, H., and Nakao, M., Periodic solutions of the Navier–Stokes equations in unbounded domains. *Tohoku Math. J.* **48** 33–50 (1996)

14. Maremonti, P. Existence and stability of time-periodic solutions to the Navier-Stokes equations in the whole space. *Nonlinearity* **4** 503–529 (1991)

15. Maremonti, P., Padula, M.: Existence, uniqueness and attainability of periodic solutions of the Navier-Stokes equations in exterior Domains. *Zap. Nauchn. Sem. S.-Peterburg. Otdel. Mat. Inst. Steklov* (POMI) **233**, 141–182 (1996)

16. Nguyen, T.-H., Trinh, V.-D., Vu, T.-N. H., and Vu, T.-M., Boundedness, almost periodicity and stability of certain Navier-Stokes flows in unbounded domains, *J. Differential Equations* **263** 8979–9002 (2017)

17. Solonnikov, V.A., Estimates of the solutions of the nonstationary Navier-Stokes system, *Zap. Naucn. Sem. Leningrad. Otdel. Mat. Inst. Steklov* (LOMI), **38** 153–201 (1973)

18. Yamazaki, M., The Navier–Stokes equations in the weak–L_n space with time-dependent external force. *Math. Ann.* **317** 635–675 (2000)

Chapter 5

Compressible Navier-Stokes System on a Moving Domain in the $L_p - L_q$ Framework

Ondřej Kreml (✉) and Šárka Nečasová
Czech Academy of Sciences, Institute of Mathematics, Prague, Czech Republic
e-mail: kreml@math.cas.cz; matus@math.cas.cz

Tomasz Piasecki
University of Warsaw, Institute of Applied Mathematics and Mechanics, Warszawa, Poland
e-mail: tpiasecki@mimuw.edu.pl

We prove the local well-posedness for the barotropic compressible Navier-Stokes system on a moving domain, a motion of which is determined by a given vector field $\mathbf{V}$, in a maximal $L_p - L_q$ regularity framework. Under additional smallness assumptions on the data we show that our solution exists globally in time and satisfies a decay estimate. In particular, for the global well-posedness we do not require exponential decay or smallness of $\mathbf{V}$ in $L_p(L_q)$. However, we require exponential decay and smallness of its derivatives.

5.1 Introduction

We consider a barotropic flow of a compressible viscous fluid in the absence of external forces described by the isentropic compressible Navier-Stokes system

$$\partial_t \varrho + \mathrm{div}_x(\varrho \mathbf{u}) = 0, \tag{5.1}$$

T. Bodnár et al. (eds.), *Waves in Flows*, Advances in Mathematical Fluid Mechanics, https://doi.org/10.1007/978-3-030-68144-9_5

$$\partial_t(\varrho\mathbf{u}) + \mathrm{div}_x(\varrho\mathbf{u}\otimes\mathbf{u}) + \nabla_x\pi(\varrho) = \mathrm{div}_x\mathbb{S}(\nabla_x\mathbf{u}), \tag{5.2}$$

where ϱ is the density of the fluid and $\mathbf{u}$ denotes the velocity. We assume that the stress tensor $\mathbb{S}$ is determined by the standard Newton rheological law

$$\mathbb{S}(\nabla_x\mathbf{u}) = \mu\left(\nabla_x\mathbf{u} + \nabla_x^t\mathbf{u} - \frac{2}{3}\mathrm{div}_x\mathbf{u}\mathbb{I}\right) + \zeta\mathrm{div}_x\mathbf{u}\mathbb{I} \tag{5.3}$$

with constant viscosity coefficients $\mu > 0$ and $\zeta \geq 0$. The pressure $\pi(\varrho)$ is a given sufficiently smooth, non-decreasing function of the density. We assume the fluid occupies a time-dependent bounded domain Ω_t, the motion of which is described by means of a given velocity field $\mathbf{V}(t,x)$, where $t \geq 0$ and $x \in \mathbb{R}^3$. More precisely, we assume that if $\mathbf{X}$ solves the following system of ordinary differential equations

$$\frac{\mathrm{d}}{\mathrm{d}t}\mathbf{X}(t,x) = \mathbf{V}\Big(t, \mathbf{X}(t,x)\Big), \quad t > 0, \;\; \mathbf{X}(0,x) = x,$$

we set

$$\Omega_\tau = \mathbf{X}\left(\tau, \Omega_0\right),$$

where $\Omega_0 \subset \mathbb{R}^3$ is a given bounded domain at initial time $t = 0$. Moreover we denote $\Gamma_\tau = \partial\Omega_\tau$ and

$$Q_\tau = \bigcup_{t\in(0,\tau)} \{t\} \times \Omega_t =: (0,\tau) \times \Omega_t.$$

We consider the system (5.1)–(5.2) supplied with the Dirichlet boundary conditions

$$(\mathbf{u} - \mathbf{V})|_{\Gamma_\tau} = 0 \text{ for any } \tau \geq 0 \tag{5.4}$$

and the initial conditions

$$\varrho(0,\cdot) = \varrho_0, \quad \mathbf{u}(0,\cdot) = \mathbf{u}_0 \quad \text{in } \Omega_0. \tag{5.5}$$

The existence theory for system (5.1)–(5.2) on fixed domains is nowadays quite well developed. The existence of global weak solutions has been first established by Lions [19]. This result has been later extended by Feireisl and coauthors [4–6, 11] to cover larger class of pressure laws. Strong solutions on fixed domains are known to exist locally in time or globally provided certain smallness assumptions on the data. For no-slip boundary conditions see among others [21, 22, 32, 33] for the results in Hilbert spaces, [23–25] in L_p setting and [3] for a maximal $L_p - L_q$ regularity approach. The problem with slip boundary conditions on a fixed domain has been investigated by Zajaczkowski [35], Hoff [13] and, more recently, by Shibata and Murata [20, 29] in the $L_p - L_q$ maximal regularity setting. In [26, 27] the approach from [3] has been adapted to treat a generalization of the compressible Navier-Stokes system describing a flow of a compressible mixture with cross-diffusion. For results on free boundary problems of the system (5.1)–(5.2), we refer to [36, 37] where the global existence of strong solutions in L_2 setting has been shown under the assumption that the domain is close to a ball and to [28] where a free boundary problem is treated in $L_p - L_q$ approach.

The existence theory for the system (5.1)–(5.2) on a moving domain with a given motion of the boundary started to develop with the results for weak solutions obtained using a penalization method in [9] for no-slip boundary conditions and [10] for slip conditions. These results have been recently generalized to the complete system with heat conductivity in [16] and [17]. The first weak-strong uniqueness result on a moving domain has been shown in [2] in case of no-slip boundary condition. A generalization of this to slip conditions as well as a local existence result for strong solution for both types boundary conditions can be found in [18]. There, the authors use the energy approach in L_2 setting for the existence result.

The aim of this paper is to extend the existence theory for strong solutions on a moving domain to $L_p - L_q$ maximal regularity setting. We present a more detailed outline of the proof after stating our main result; however, first let us resume the notation used in the paper.

5.1.1 Notation

We use standard notations for Lebesgue spaces $L_p(\Omega_0)$ and Sobolev spaces $W_p^k(\Omega_0)$ with $k \in \mathbb{N}$ on a fixed domain Ω_0. By $C_B(\Omega_0)$ we denote a space of bounded continuous functions on Ω_0. Furthermore, for a Banach space X, $L_p(0,T;X)$ is a Bochner space of functions for which the norm

$$\|f\|_{L_p(0,T;X)} = \begin{cases} \left(\int_0^T \|f(t)\|_X \,\mathrm{d}t\right)^{1/p}, & 1 \le p < \infty, \\ \operatorname{ess\,sup}_{0 \le t \le T} \|f(t)\|_X, & p = \infty \end{cases}$$

is finite. Then

$$W_p^1(0,T;X) = \{f \in L_p(0,T;X) : \partial_t f \in L_p(0,T;X)\}.$$

For $p \ge 1$ we denote by p' its dual exponent, i.e., $\frac{1}{p} + \frac{1}{p'} = 1$. Next, we recall that for $0 < s < \infty$ and m a smallest integer larger than s we define Besov spaces on domains as intermediate spaces

$$B^s_{q,p}(\Omega_0) = (L_q(\Omega_0), W_q^m(\Omega_0))_{s/m,p}, \tag{5.6}$$

where $(\cdot,\cdot)_{s/m,p}$ is the real interpolation functor, see [1, Chapter 7]. In particular,

$$B^{2(1-1/p)}_{q,p}(\Omega_0) = (L_q(\Omega_0), W_q^2(\Omega_0))_{1-1/p,p} = (W_q^2(\Omega_0), L_q(\Omega_0))_{1/p,p}. \tag{5.7}$$

We shall not distinguish between notation of spaces for scalar and vector valued functions, i.e., we write $L_q(\Omega_0)$ instead of $L_q(\Omega_0)^3$, etc. However, we write vector valued functions in boldface.

For $p, q \in [1, \infty]$ we introduce the vector space $L_p(0, T, L_q(\Omega_t))$ which consists of functions $f : Q_T \mapsto \mathbb{R}$ such that

$$t \mapsto \|f(t, \cdot)\|_{L_q(\Omega_t)}$$

is measurable and L_p integrable on time interval $(0, T)$, i.e., if

$$\|f\|_{L_p(0,T,L_q(\Omega_t))} := \big\|\|f(t, \cdot)\|_{L_q(\Omega_t)}\big\|_{L_p(0,T)} \le C. \tag{5.8}$$

Similarly, for $p, q \in [1, \infty]$ and $k \in \mathbb{N}$ we introduce a vector space $L_p(0, T, W_q^k(\Omega_t))$ which consists of functions $f : Q_T \mapsto \mathbb{R}$ such that

$$\|f\|_{L_p(0,T,W_q^k(\Omega_t))} := \Big\|\|f(t, \cdot)\|_{W_q^k(\Omega_t)}\Big\|_{L_p(0,T)} \le C. \tag{5.9}$$

Let us also introduce a brief notation for the regularity class of the solution. Namely, for a function g and a vector field $\mathbf{f}$ defined on $(0, T) \times \Omega_t$ we define

$$\|g, \mathbf{f}\|_{\mathcal{X}(T)} = \|\mathbf{f}\|_{L_p(0,T;W_q^2(\Omega_t))} + \|\mathbf{f}_t\|_{L_p(0,T;L_q(\Omega_t))} + \|g\|_{L_p(0,T;W_q^1(\Omega_t))} + \|g_t\|_{L_p(0,T;L_q(\Omega_t))}. \tag{5.10}$$

Next, for a pair $\tilde{g}, \tilde{\mathbf{f}}$ defined of $(0, T) \times \Omega_0$ we define a norm

$$\|\tilde{g}, \tilde{\mathbf{f}}\|_{\mathcal{Y}(T)} = \|\tilde{\mathbf{f}}\|_{L_p(0,T;W_q^2(\Omega_0))} + \|\tilde{\mathbf{f}}_t\|_{L_p(0,T;L_q(\Omega_0))} + \|\tilde{g}\|_{W_p^1(0,T;W_q^1(\Omega_0))} \tag{5.11}$$

and a seminorm

$$\|\tilde{g}, \tilde{\mathbf{f}}\|_{\dot{\mathcal{Y}}(T)} = \|\tilde{\mathbf{f}}\|_{L_p(0,T;W_q^2(\Omega_0))} + \|\tilde{\mathbf{f}}_t\|_{L_p(0,T;L_q(\Omega_0))} + \|\nabla\tilde{g}\|_{L_p(0,T;L_q(\Omega_0))} + \|\tilde{g}_t\|_{L_p(0,T;W_q^1(\Omega_0))}. \tag{5.12}$$

Obviously, we denote by $\mathcal{X}(T)$ and $\mathcal{Y}(T)$ spaces for which the norms (5.10) and (5.11), respectively, are finite.

Remark 5.1.1. Notice that the norm (5.11) involves also $\|\tilde{g}_t\|_{L_p(W_q^1(\Omega_0))}$ while in (5.10) we have only $\|g_t\|_{L_p(L_q(\Omega_t))}$. The reason is that in the Lagrangian coordinates we are able to show higher regularity of the density, which does not correspond to equivalent regularity in Eulerian coordinates, see Sect. 5.4.4.

Furthermore we denote

$$\mathcal{H}_q = L_q(\Omega_0) \times W_q^1(\Omega_0), \quad \hat{\mathcal{H}}_q = \{(\mathbf{f}, g) \in \mathcal{H}_q : \int_{\Omega_0} g \, \mathrm{d}x = 0\}, \tag{5.13}$$

$$\mathcal{D}_q = W_q^2(\Omega_0) \times W_q^1(\Omega_0), \quad \hat{\mathcal{D}}_q = \{(\mathbf{f}, g) \in \mathcal{D}_q : \int_{\Omega_0} g \, \mathrm{d}x = 0\}. \tag{5.14}$$

Next, for $0 < \varepsilon < \pi$ and $\lambda_0 > 0$ we introduce

$$\Sigma_\varepsilon = \{\lambda \in \mathbb{C} \setminus \{0\} : |\arg\lambda| \le \pi - \varepsilon\}, \quad \Sigma_{\varepsilon,\lambda_0} = \lambda_0 + \Sigma_\varepsilon, \quad \Lambda_{\varepsilon,\lambda_0} = \{\lambda \in \Sigma_\varepsilon : |\lambda| \ge \lambda_0\}, \tag{5.15}$$

$$\mathbb{C}_+ := \{\lambda \in \mathbb{C} : \mathrm{Re}\lambda \ge 0\}. \tag{5.16}$$

We also recall the definition of $\mathcal{R}$-boundedness:

Definition 5.1.1. Let X and Y be two Banach spaces, and $\|\cdot\|_X$ and $\|\cdot\|_Y$ their norms. A family of operators $\mathcal{T} \subset \mathcal{L}(X,Y)$ is called $\mathcal{R}$-bounded on $\mathcal{L}(X,Y)$ if there exist constants $C > 0$ and $p \in [1,\infty)$ such that for any $n \in \mathbb{T}$, $\{T_j\}_{j=1}^n \subset \mathcal{T}$ and $\{f_j\}_{j=1}^n \subset X$, the inequality

$$\int_0^1 \|\sum_{j=1}^n r_j(u)T_j f_j\|_Y^p \, du \leq C \int_0^1 \|\sum_{j=1}^n r_j(u) f_j\|_X^p \, du,$$

where $r_j : [0,1] \to \{-1,1\}$, $j \in \mathbb{T}$, are the Rademacher functions given by $r_j(t) = \mathrm{sign}(\sin(2^j \pi t))$. The smallest such C is called $\mathcal{R}$-bound of $\mathcal{T}$ on $\mathcal{L}(X,Y)$ which is written by $\mathcal{R}_{\mathcal{L}(X,Y)}\mathcal{T}$.

The Fourier transform and its inverse are defined as

$$\mathcal{F}[f](\tau) = \int_{\mathbb{R}} e^{-it\tau} f(t)dt, \quad \mathcal{F}^{-1}[f](t) = \frac{1}{2\pi}\int_{\mathbb{R}} e^{it\tau} f(\tau)d\tau. \tag{5.17}$$

We shall also need the Laplace transform and its inverse

$$\mathcal{L}[f](\lambda) = \int_{\mathbb{R}} e^{-\lambda t} f(t)dt, \quad \mathcal{L}^{-1}[f](t) = \frac{1}{2\pi}\int_{\mathbb{R}} e^{\lambda t} f(\lambda)d\tau, \quad \text{where } \lambda = \gamma + i\tau\,. \tag{5.18}$$

Finally, by $E(\cdot)$ we shall denote a non-negative non-decreasing continuous function such that $E(0) = 0$. We use the notation $E(T)$ in particular to denote those constants in various inequalities, which can be made arbitrarily small by taking T sufficiently small.

5.1.2 Main Results

The first main result of this paper gives the local well-posedness for system (5.1)–(5.2) with Dirichlet boundary condition.

Theorem 5.1.1. Let $\Omega_0 \subset \mathbb{R}^3$ be a bounded uniform C^2 domain. Assume

$$\varrho_0 \in W_q^1(\Omega_0), \quad \mathbf{u}_0 \in B_{q,p}^{2-2/p}(\Omega_0)$$

and

$$\mathbf{V} \in L_p(0,T; W_q^2(\mathbb{R}^3)) \cap W_p^1(0,T; L_q(\mathbb{R}^3))$$

with $2 < p < \infty$, $3 < q < \infty$ and $\frac{2}{p} + \frac{3}{q} < 1$. Then for any $L > 0$ there exists $T > 0$ such that if

$$\|\varrho_0\|_{W_q^1(\Omega_0)} + \|\mathbf{u}_0\|_{B_{q,p}^{2-2/p}(\Omega_0)} + \|\mathbf{V}\|_{L_p(0,T;W_q^2(\mathbb{R}^3))} + \|\partial_t \mathbf{V}\|_{L_p(0,T;L_q(\mathbb{R}^3))} \leq L, \tag{5.19}$$

then the system (5.1)–(5.5) admits a unique strong solution $(\varrho, \mathbf{u}) \in \mathcal{X}(T)$ and

$$\|\varrho, \mathbf{u}\|_{\mathcal{X}(T)} \leq CL. \tag{5.20}$$

Remark 5.1.2. Let us comment the restrictions on p and q. The condition $q > 3$ is natural as we shall repeatedly use the embedding $W^1_q(\Omega_0) \subset L_\infty(\Omega_0)$. However, a stronger condition $\frac{2}{p} + \frac{3}{q} < 1$ is required since we need the embedding $B^{2(1-1/p)}_{q,p}(\Omega_0) \subset W^1_\infty(\Omega_0)$ to prove Lemma 5.4.3, see Corollary 5.3.1.

The second main result gives the global well-posedness:

Theorem 5.1.2. Let $\Omega_0 \subset \mathbb{R}^3$ be a bounded uniform C^2 domain. Assume that

$$\varrho_0 \in W^1_q(\Omega_0), \quad \mathbf{u}_0 \in B^{2-2/p}_{q,p}(\Omega_0).$$

Furthermore, let $\varrho^*, \gamma > 0$ be given constants. Then there exists $\epsilon > 0$ such that if

$$\|\varrho_0 - \varrho^*\|_{W^1_p(\Omega_0)} + \|\mathbf{u}_0 - \mathbf{V}(0)\|_{B^{2-2/p}_{q,p}(\Omega_0)} + \|e^{\gamma t}(\partial_t \mathbf{V}, \nabla_x \mathbf{V}, \nabla^2_x \mathbf{V})\|_{L_p(0,T;L_q(\mathbb{R}^3))} \leq \epsilon, \tag{5.21}$$

then the unique strong solution to (5.1)–(5.5) is defined globally in time and

$$\begin{aligned} &\|e^{\gamma t}\nabla\varrho\|_{L_p(0,\infty;L_q(\Omega_t))} + \|e^{\gamma t}\partial_t\varrho\|_{L_p(0,\infty;W^1_q(\Omega_t))} + \|\varrho - \varrho^*\|_{L_\infty(0,\infty;W^1_q(\Omega))} \\ &+ \|e^{\gamma t}\partial_t\mathbf{u}\|_{L_p(0,\infty;L_q(\Omega_t))} + \|e^{\gamma t}\nabla_x\mathbf{u}\|_{L_p(0,\infty;W^1_q(\Omega_t))} + \|e^{\gamma t}(\mathbf{u} - \mathbf{V})\|_{L_p(0,\infty;L_q(\Omega_t))} \leq C\epsilon, \end{aligned} \tag{5.22}$$

$$\|\mathbf{u}\|_{L_p(0,\infty;L_q(\Omega_t))} \leq C\epsilon + \|\mathbf{V}\|_{L_p(0,\infty;L_q(\Omega_t))}. \tag{5.23}$$

The paper is structured as follows. In Sect. 5.2 we rewrite the problem on a fixed domain using Lagrangian coordinates. In Sect. 5.3 we prove maximal regularity and exponential decay results for linear problems corresponding to linearization of our system in Lagrangian coordinates. As many similar results are already well known, we skip certain technical details referring to relevant papers. In the same section we also recall some imbedding properties. Section 5.4 is dedicated to the proof of Theorem 5.1.1. We reduce the problem to homogeneous boundary condition and show appropriate estimates of the right hand side of the problem in Lagrangian coordinates and conclude using fixed point argument and linear result recalled in Sect. 5.3. In Sect. 5.5 we prove Theorem 5.1.2. For this purpose we obtain appropriate estimates of the right hand side which allow to show uniform in time estimate for the solution using exponential decay property of the linear problem. This estimate allows to prolong the solution for arbitrarily large time.

5.2 Lagrangian Transformation

Let us start with a following observation

Lemma 5.2.1. Let p and q satisfy the assumptions of Theorem 5.1.1. Then
(i) if $\|f\|_{L_p(0,T;W^2_q(\Omega_0))} \leq M$ for some $M > 0$, then

$$\int_0^T \|\nabla f(t,\cdot)\|_{L_\infty(\Omega_0)} dt \leq ME(T). \tag{5.24}$$

(ii) if $\|e^{\gamma t} f\|_{L_p(0,\infty;W^2_q(\Omega_0))} \leq M$ for some $M, \gamma > 0$, then

$$\int_0^\infty \|\nabla f(t,\cdot)\|_{L_\infty(\Omega_0)} \mathrm{d}t \leq CM. \tag{5.25}$$

Proof: By the imbedding theorem and Hölder inequality we have

$$\int_0^T \|\nabla f(t,\cdot)\|_{L_\infty(\Omega_0)} \mathrm{d}t \leq C \int_0^T \|f(t,\cdot)\|_{W^2_q(\Omega_0)} \mathrm{d}t \leq T^{1/p'} \int_0^T \left(\|f(t,\cdot)\|^p_{W^2_q(\Omega_0)} \right)^{1/p} \mathrm{d}t \leq ME(T),$$

which proves the first assertion, and for the second we have

$$\begin{aligned} \int_0^\infty \|\nabla f(t,\cdot)\|_{L_\infty(\Omega_0)} \mathrm{d}t &\leq C \int_0^\infty e^{-\gamma t} e^{\gamma t} \|f(t,\cdot)\|_{W^2_q(\Omega_0)} \mathrm{d}t \\ &\leq \left(\int_0^\infty e^{-\gamma t p'} \mathrm{d}t \right)^{1/p'} \|e^{\gamma t} f\|_{L_p(0,\infty;W^2_q(\Omega_0))} \leq CM. \end{aligned}$$

□

In order to transform the problem (5.1)–(5.2) to a fixed domain we introduce the change of coordinates

$$\frac{\mathrm{d}}{\mathrm{d}t} \mathbf{X_u}(t,y) = \mathbf{u}\Big(t, \mathbf{X_u}(t,y)\Big) \quad \text{for } t > 0, \quad \mathbf{X_u}(0,y) = y, \tag{5.26}$$

i.e.,

$$\mathbf{X_u}(t,y) = y + \int_0^t \mathbf{u}(s, \mathbf{X_u}(s,y)) \mathrm{d}s. \tag{5.27}$$

Then for any differentiable function f defined on Q_T we have

$$\frac{\mathrm{d}}{\mathrm{d}t} f(t, \mathbf{X_u}(t,y)) = \frac{\partial}{\partial t} f(t, \mathbf{X_u}(t,y)) + \mathbf{u} \cdot \nabla_x f(t, \mathbf{X_u}(t,y)). \tag{5.28}$$

Let us define transformed density and velocities on a fixed domain Ω_0:

$$\tilde{\varrho}(t,y) = \varrho(t, \mathbf{X_u}(t,y)), \quad \tilde{\mathbf{u}}(t,y) = \mathbf{u}(t, \mathbf{X_u}(t,y)), \quad \tilde{\mathbf{V}}(t,y) = \mathbf{V}(t, \mathbf{X_u}(t,y)). \tag{5.29}$$

Lemma 5.2.2. Assume that

$$\int_0^T \|\nabla_y \tilde{\mathbf{u}}\|_\infty \mathrm{d}t \leq \delta \tag{5.30}$$

for sufficiently small $\delta > 0$. Then the inverse to $\mathbf{X_u}$, i.e., $\mathbf{Y}(t,x)$ defined as

$$\mathbf{X_u}(t, \mathbf{Y}(t,x)) = x \quad \forall t \geq 0, \ x \in \Omega_t, \tag{5.31}$$

is well defined and its Jacobian can be expressed in a following way

$$\nabla_x \mathbf{Y}(t, \mathbf{X_u}(t,y)) = [\nabla_y \mathbf{X_u}(t,y)]^{-1} = \mathbf{I} + \mathbf{E}^0(\mathbf{k}_{\tilde{\mathbf{u}}}(t,y)), \tag{5.32}$$

where

$$\mathbf{k}_{\tilde{\mathbf{u}}}(t,y) = \int_0^t \nabla_y \tilde{\mathbf{u}}(s,y)\mathrm{d}s \tag{5.33}$$

and $\mathbf{E}^0(\cdot)$ is a 3×3 matrix of smooth functions with $\mathbf{E}^0(0) = 0$.

Proof: We have

$$\frac{\partial X_i}{\partial y_j}(t,y) = \delta_{ij} + \int_0^t \frac{\partial \tilde{u}_i}{\partial y_j}(s,y)\, ds. \tag{5.34}$$

Therefore, if (5.30) holds for sufficiently small δ, then $\mathbf{Y}(t,x)$ is well defined and we have (5.32)–(5.33) with $\mathbf{E}^0$ as in the statement of the Lemma. Next, by the boundary condition (5.4) we have

$$\mathbf{X}_{\mathbf{u}}(\Gamma_0, t) = \Gamma_t \quad \text{for } t > 0$$

and

$$\mathbf{X}_{\mathbf{u}}(y,t) \subset \Omega_t \quad \text{for } t > 0,\ y \in \Omega_0.$$

Finally, it is well known that $\mathbf{X}_{\mathbf{u}}$ is a diffeomorphism which completes the proof. □

Note that by (5.32) we can write

$$\nabla_x = [\mathbf{I} + \mathbf{E}^0(\mathbf{k}_{\tilde{\mathbf{u}}})]\nabla_y. \tag{5.35}$$

Lemma 5.2.3. Let $(\varrho, \mathbf{u})$ be a solution to (5.1)–(5.5). Then $(\tilde{\varrho}, \tilde{\mathbf{u}})$ solve the following system of equations on the fixed domain Ω_0

$$\tilde{\varrho}\tilde{\mathbf{u}}_t - \mu\Delta_y\tilde{\mathbf{u}} - \left(\frac{\mu}{3} + \zeta\right)\nabla_y \mathrm{div}_y \tilde{\mathbf{u}} + \nabla_y \pi(\tilde{\varrho}) = \mathbf{F}(\tilde{\varrho}, \tilde{\mathbf{u}}), \tag{5.36}$$

$$\tilde{\varrho}_t + \tilde{\varrho}\,\mathrm{div}_y \tilde{\mathbf{u}} = G(\tilde{\varrho}, \tilde{\mathbf{u}}), \tag{5.37}$$

$$\tilde{\mathbf{u}}|_{t=0} = \mathbf{u}_0, \quad \tilde{\mathbf{u}}|_{\partial\Omega_0} = \tilde{\mathbf{V}}. \tag{5.38}$$

The i-th component of $\mathbf{F}(\cdot,\cdot)$ is given by

$$F_i(\tilde{\varrho}, \tilde{\mathbf{u}}) = -E^0_{ij}\partial_{y_j}\pi(\tilde{\varrho}) + R_i(\tilde{\mathbf{u}}), \tag{5.39}$$

and

$$G(\tilde{\varrho}, \tilde{\mathbf{u}}) = -\tilde{\varrho}\mathbf{E}^0_{ij}(\mathbf{k}_{\tilde{\mathbf{u}}})\frac{\partial \tilde{u}_i}{\partial y_j}, \tag{5.40}$$

where the components $R_i(\cdot)$ of $\mathbf{R}(\cdot)$ are expressed as

$$R_i(\tilde{\mathbf{u}}) = \mu[A_{2\Delta}(\mathbf{k}_{\tilde{\mathbf{u}}})\nabla_y^2\tilde{\mathbf{u}} + A_{1\Delta}(\mathbf{k}_{\tilde{\mathbf{u}}})\nabla_y\tilde{\mathbf{u}}]_i + \left(\frac{\mu}{3} + \zeta\right)[A_{2\mathrm{div},i}(\mathbf{k}_{\tilde{\mathbf{u}}})\nabla_y^2\tilde{\mathbf{u}} + A_{1\mathrm{div},i}(\mathbf{k}_{\tilde{\mathbf{u}}})\nabla_y\tilde{\mathbf{u}}], \tag{5.41}$$

with $A_{j\Delta}$ and $A_{j\mathrm{div}}$ $(j = 1, 2)$ given in (5.46), (5.47), (5.49), and (5.50), respectively.

Proof: We have

$$\operatorname{div}_x \mathbf{u} = \operatorname{div}_y \tilde{\mathbf{u}} + \mathbf{E}^0 : \nabla_y \tilde{\mathbf{u}}, \tag{5.42}$$

where $\mathbf{E}^0 : \nabla_y \tilde{\mathbf{u}} = \mathbf{E}^0_{ij}(\mathbf{k}_{\tilde{\mathbf{u}}}) \frac{\partial \tilde{u}_i}{\partial y_j}$, which together with (5.28) gives (5.37).

In order to transform the momentum equation (5.2) it is convenient to rewrite it, using (5.1) and (5.3), as

$$\varrho(\partial_t \mathbf{u} + \mathbf{u} \cdot \nabla_x \mathbf{u}) - \mu \Delta_x \mathbf{u} - (\frac{\mu}{3} + \zeta) \nabla \operatorname{div}_x \mathbf{u} + \nabla_x \pi(\varrho) = 0. \tag{5.43}$$

We have

$$\partial_{x_i} \pi(\varrho) = \partial_{y_i} \pi(\tilde{\varrho}) + E^0_{ij} \partial_{y_j} \pi(\tilde{\varrho}). \tag{5.44}$$

Now we need to transform second order operators. By (5.35), we have

$$\Delta_x \mathbf{u} = \frac{\partial}{\partial x_k} \left(\frac{\partial \mathbf{u}}{\partial x_k} \right) = \left(\delta_{kl} + \mathbf{E}^0_{kl}(\mathbf{k}_{\tilde{\mathbf{u}}})\right) \frac{\partial}{\partial y_l} \left(\left(\delta_{km} + \mathbf{E}^0_{km}(\mathbf{k}_{\tilde{\mathbf{u}}})\right) \frac{\partial \tilde{\mathbf{u}}}{\partial y_m} \right).$$

Therefore

$$\Delta_x \mathbf{u} = \Delta_y \tilde{\mathbf{u}} + A_{2\Delta}(\mathbf{k}_{\tilde{\mathbf{u}}}) \nabla^2_y \tilde{\mathbf{u}} + A_{1\Delta}(\mathbf{k}_{\tilde{\mathbf{u}}}) \nabla_y \tilde{\mathbf{u}} \tag{5.45}$$

with

$$A_{2\Delta}(\mathbf{k}_{\tilde{\mathbf{u}}}) \nabla^2_y \tilde{\mathbf{u}} = 2 \sum_{l,m} \mathbf{E}^0_{kl}(\mathbf{k}_{\tilde{\mathbf{u}}}) \frac{\partial^2 \tilde{\mathbf{u}}}{\partial y_l \partial y_m} + \sum_{k,l,m} \mathbf{E}^0_{kl}(\mathbf{k}_{\tilde{\mathbf{u}}}) \mathbf{E}^0_{km}(\mathbf{k}_{\tilde{\mathbf{u}}}) \frac{\partial^2 \tilde{\mathbf{u}}}{\partial y_l \partial y_m}, \tag{5.46}$$

$$\begin{aligned} A_{1\Delta}(\mathbf{k}_{\tilde{\mathbf{u}}}) \nabla_y \tilde{\mathbf{u}} =& (\nabla_{\mathbf{k}_{\tilde{\mathbf{u}}}} \mathbf{E}^0_{lm})(\mathbf{k}_{\tilde{\mathbf{u}}}) \int_0^t (\partial_l \nabla_y \tilde{\mathbf{u}}) \, \mathrm{d}s \frac{\partial \tilde{\mathbf{u}}}{\partial y_m} \\ &+ \mathbf{E}^0_{kl}(\mathbf{k}_{\tilde{\mathbf{u}}}) (\nabla_{\mathbf{k}_{\tilde{\mathbf{u}}}} \mathbf{E}^0_{km})(\mathbf{k}_{\tilde{\mathbf{u}}}) \int_0^t \partial_l \nabla_y \tilde{\mathbf{u}} \, \mathrm{d}s \frac{\partial \tilde{\mathbf{u}}}{\partial y_m}. \end{aligned} \tag{5.47}$$

Next, by (5.42)

$$\frac{\partial}{\partial x_i} \operatorname{div}_x \mathbf{u} = \sum_{k=1}^3 (\delta_{ik} + \mathbf{E}^0_{ik}(\mathbf{k}_{\tilde{\mathbf{u}}})) \frac{\partial}{\partial y_k} \left(\operatorname{div}_y \tilde{\mathbf{u}} + \sum_{l,m=1}^3 \mathbf{E}^0_{lm}(\mathbf{k}_{\tilde{\mathbf{u}}}) \frac{\partial \tilde{u}_l}{\partial y_m} \right),$$

so we obtain

$$\frac{\partial}{\partial x_i} \operatorname{div}_x \mathbf{u} = \frac{\partial}{\partial y_i} \operatorname{div}_y \tilde{\mathbf{u}} + A_{2\mathrm{div},i}(\mathbf{k}_{\tilde{\mathbf{u}}}) \nabla^2_y \tilde{\mathbf{u}} + A_{1\mathrm{div},i}(\mathbf{k}_{\tilde{\mathbf{u}}}) \nabla_y \tilde{\mathbf{u}}, \tag{5.48}$$

where

$$A_{2\mathrm{div},i}(\mathbf{k}_{\tilde{\mathbf{u}}}) \nabla^2_y \tilde{\mathbf{u}} = \sum_{l,m=1}^3 \mathbf{E}^0_{lm}(\mathbf{k}_{\tilde{\mathbf{u}}}) \frac{\partial^2 \tilde{u}_l}{\partial y_m \partial y_i} + \sum_{k=1}^3 \mathbf{E}^0_{ik}(\mathbf{k}_{\tilde{\mathbf{u}}}) \frac{\partial}{\partial y_k} \operatorname{div}_y \tilde{\mathbf{u}} + \sum_{k,l=1}^3 \mathbf{E}^0_{ik}(\mathbf{k}_{\tilde{\mathbf{u}}}) \mathbf{E}^0_{lm}(\mathbf{k}_{\tilde{\mathbf{u}}}) \frac{\partial^2 \tilde{u}_l}{\partial y_k \partial y_m}, \tag{5.49}$$

$$
\begin{aligned}
A_{1\mathrm{div},i}(\mathbf{k}_{\tilde{\mathbf{u}}})\nabla_y\tilde{\mathbf{u}} &= \sum_{l,m=1}^{3}(\nabla_{\mathbf{k}_{\tilde{\mathbf{u}}}}\mathbf{E}^0_{lm})(\mathbf{k}_{\tilde{\mathbf{u}}})\int_0^t \partial_i\nabla_y\tilde{\mathbf{u}}\,\mathrm{d}s\frac{\partial\tilde{u}_l}{\partial y_m}\\
&+\sum_{k,l,m=1}^{3}\mathbf{E}^0_{ik}(\mathbf{k}_{\tilde{\mathbf{u}}})(\nabla_{\mathbf{k}_{\tilde{\mathbf{u}}}}\mathbf{E}^0_{lm})(\mathbf{k}_{\tilde{\mathbf{u}}})\int_0^t \partial_k\nabla_y\tilde{\mathbf{u}}\,\mathrm{d}s\frac{\partial\tilde{u}_l}{\partial y_m}.
\end{aligned}
\tag{5.50}
$$

Putting together (5.44), (5.45), and (5.48) gives (5.36) with (5.39).

□

5.3 Linear Theory and Auxiliary Results

5.3.1 $L_p - L_q$ Maximal Regularity

This section is dedicated to $L_p - L_q$ maximal regularity results for linear problems on a fixed domain, which will be used in the proof of Theorem 5.1.1. We shall rely on similar already known results, therefore we will be able to avoid most technicalities referring to relevant works and presenting only an outline of the proof. The linearized system of equations on the fixed domain Ω_0 reads as

$$\varrho_0\mathbf{u}_t - \mu\Delta_y\mathbf{u} - (\frac{\mu}{3}+\zeta)\nabla_y\mathrm{div}_y\mathbf{u} + \bar{\gamma}_0\nabla_y\eta = \mathbf{f}, \tag{5.51}$$

$$\eta_t + \varrho_0\mathrm{div}_y\mathbf{u} = g, \tag{5.52}$$

$$\mathbf{u}|_{\partial\Omega_0} = 0, \quad \mathbf{u}|_{t=0} = \mathbf{u}_0, \tag{5.53}$$

where the unknowns are η and $\mathbf{u}$, while $\varrho_0, \bar{\gamma}_0, \mathbf{f}, g$ are given functions such that $\varrho_0 \geq c > 0$ and $\bar{\gamma}_0 \geq 0$. To show the local well-posedness for the Dirichlet boundary condition we will use the following result

Theorem 5.3.1. Let $1 < p, q < \infty$ and $\frac{2}{p}+\frac{1}{q} \neq 1$. Let $\varrho_0, \mathbf{u}_0, \mu$, and ζ satisfy the assumptions of Theorem 5.1.1. Moreover, let $\Omega_0 \subset \mathbb{R}^n$ be a uniform C^2 domain. If $\frac{2}{p}+\frac{1}{q} < 1$, assume additionally that the initial velocity satisfies the compatibility condition $\mathbf{u}_0|_{\partial\Omega_0} = 0$. Finally, assume that for some $T > 0$

$$\mathbf{f} \in L_p(0,T;L_q(\Omega_0)), \quad g \in L_p(0,T;W^1_q(\Omega_0)).$$

Then the problem (5.51)–(5.53) admits a unique solution $(\eta, \mathbf{u}) \in \mathcal{Y}(T)$ such that

$$\|\eta,\mathbf{u}\|_{\mathcal{Y}(T)} \leq C(\mu,\zeta,\|\varrho_0\|_{L_\infty(\Omega_0)})e^{\gamma_0 T}[\|\mathbf{u}_0\|_{B^{2-2/p}_{q,p}(\Omega_0)} + \|\mathbf{f}\|_{L_p(0,T;L_q(\Omega_0)} + \|g\|_{L_p(0,T;W^1_q(\Omega_0)}] \tag{5.54}$$

for some positive constant γ_0.

In order to remove inhomogeneity from the boundary condition we also need to consider separately the linearized momentum equation

$$\varrho_0\mathbf{u}_t - \mu\Delta_y\mathbf{u} - (\frac{\mu}{3}+\zeta)\nabla_y\mathrm{div}_y\mathbf{u} = \mathbf{f}, \tag{5.55}$$

$$\mathbf{u}|_{\partial\Omega_0} = 0, \quad \mathbf{u}|_{t=0} = \mathbf{u}_0. \tag{5.56}$$

For the latter we have

Theorem 5.3.2. Let $\Omega_0, \mathbf{u}_0, p, q, \mu, \zeta$, and $\mathbf{f}$ satisfy the assumptions of Theorem 5.3.1. Then the problem (5.55)–(5.56) admits a unique solution $\mathbf{u} \in L_p(0,T; W_q^2(\Omega_0)) \cap W_p^1(0,T; L_q(\Omega_0))$ such that

$$\|\mathbf{u}\|_{L_p(0,T;W_q^2(\Omega_0))} + \|\mathbf{u}_t\|_{L_p(0,T;L_q(\Omega_0))} \leq C(T, \mu, \zeta, \|\varrho_0\|_{L_\infty(\Omega_0)}) \\ [\|\mathbf{u}_0\|_{B^{2-2/p}_{q,p}(\Omega_0)} + \|\mathbf{f}\|_{L_p(0,T;L_q(\Omega_0))}]. \tag{5.57}$$

The proofs of Theorems 5.3.1 and 5.3.2 are based on the concept of $\mathcal{R}$ - boundedness. For a space X, by $\mathcal{D}(\mathbb{R}, X)$ we shall denote C^∞ functions with compact support and values in X and by $\mathcal{S}(\mathbb{R}, X)$ a Schwartz space of X-valued functions. A fundamental tool enabling application of $\mathcal{R}$-boundedness to maximal regularity of PDE is the following Weis' vector valued Fourier multiplier theorem [34].

Theorem 5.3.3. Let X and Y be UMD spaces and $1 < p < \infty$. Let $M \in C^1(\mathbb{R} \setminus \{0\}, \mathcal{L}(X, Y))$. Let us define the operator $T_M : \mathcal{F}^{-1}\mathcal{D}(\mathbb{R}, X) \to \mathcal{S}'(\mathbb{R}, Y)$:

$$T_M \phi(\tau) = \mathcal{F}^{-1}[M\mathcal{F}[\phi](\tau)]. \tag{5.58}$$

Assume that

$$\mathcal{R}_{\mathcal{L}(X,Y)}(\{M(\tau) : \tau \in \mathbb{R} \setminus \{0\}\}) = \kappa_0 < \infty, \quad \mathcal{R}_{\mathcal{L}(X,Y)}(\{\tau M'(\tau) : \tau \in \mathbb{R} \setminus \{0\}\}) = \kappa_1 < \infty. \tag{5.59}$$

Then, the operator T_M defined in (5.58) is extended to a bounded linear operator $L_p(\mathbb{R}, X) \to L_p(\mathbb{R}, X)$ and

$$\|T_M\|_{\mathcal{L}(L_p(\mathbb{R},X), L_p(\mathbb{R},Y))} \leq C(\kappa_0 + \kappa_1),$$

where $C = C(p, X, Y) > 0$.

Remark 5.3.1. For definitions and properties of UMD spaces we refer the reader, for example, to Chapter 4 in [14]. Here let us only note that L_p spaces and W_p^k spaces are UMD for $1 < p < \infty$.

The following result (Theorem 2.17 in [3]) explains how to obtain L_p - maximal regularity using Theorem 5.3.3 and Laplace transform:

Theorem 5.3.4. Let X and Y be UMD Banach spaces and $1 < p < \infty$. Let $0 < \varepsilon < \frac{\pi}{2}$ and $\gamma_1 \in \mathbb{R}$. Let Φ_λ be a C^1 function of $\tau \in \mathbb{R} \setminus \{0\}$ where $\lambda = \gamma + i\tau \in \Sigma_{\varepsilon,\gamma_1}$ with values in $\mathcal{L}(X, Y)$. Assume that

$$\mathcal{R}_{\mathcal{L}(X,Y)}(\{\Phi_\lambda : \lambda \in \Sigma_{\varepsilon,\gamma_1}\}) \leq M, \quad \mathcal{R}_{\mathcal{L}(X,Y)}\left(\left\{\tau \frac{\partial}{\partial \tau}\Phi_\lambda : \lambda \in \Sigma_{\varepsilon,\gamma_1}\right\}\right) \leq M \tag{5.60}$$

for some $M > 0$. Let us define

$$\Psi f(t) = \mathcal{L}^{-1}[\Phi_\lambda \mathcal{L}[f](\lambda)](t) \quad \text{for} \quad f \in \mathcal{F}^{-1}\mathcal{D}(\mathbb{R}, X), \tag{5.61}$$

where $\mathcal{L}$ and $\mathcal{L}^{-1}$ are the Laplace transform and its inverse defined in (5.18). Then

$$\|e^{-\gamma t}\Psi f\|_{L_p(\mathbb{R},Y)} \leq C(p,X,Y)M\|e^{-\gamma t}f\|_{L_p(\mathbb{R},X)} \quad \forall \gamma \geq \gamma_1. \tag{5.62}$$

Proof: For $\lambda = \gamma + i\tau$ we have the following relation between Laplace and Fourier transforms defined in, respectively, (5.18) and (5.17):

$$\mathcal{L}[f](\lambda) = \int_{\mathbb{R}} e^{-\lambda t} f(t)dt = \mathcal{F}[e^{-\gamma t}f](\tau),$$
$$\mathcal{L}^{-1}[f](t) = \frac{1}{2\pi}\int_{\mathbb{R}} e^{\lambda t} f(\lambda)d\tau = e^{\gamma t}\mathcal{F}^{-1}[f](t).$$

Therefore by (5.61) we have

$$e^{-\gamma t}\Psi f(t) = \mathcal{F}^{-1}[\Psi_{\gamma+i\tau}\mathcal{F}[e^{-\gamma t}f](\tau)](t).$$

Applying Theorem 5.3.3 to the above formula we conclude (5.62)

□

If $(\mathbf{u},\eta)$ solves (5.51)–(5.53), then $\mathbf{u}_\lambda = \mathcal{L}[\mathbf{u}](\lambda)$, $\eta_\lambda = \mathcal{L}[\eta](\lambda)$ satisfy the following resolvent problem

$$\varrho_0\lambda\mathbf{u}_\lambda - \mu\Delta_y\mathbf{u}_\lambda - (\frac{\mu}{3}+\zeta)\nabla_y \mathrm{div}_y\mathbf{u}_\lambda + \gamma_0\nabla_y\eta_\lambda = \hat{\mathbf{f}}, \tag{5.63}$$
$$\lambda\eta_\lambda + \varrho_0 \mathrm{div}_y\mathbf{u}_\lambda = \hat{g}, \tag{5.64}$$
$$\mathbf{u}_\lambda|_{\partial\Omega_0} = 0, \tag{5.65}$$

with $\hat{\mathbf{f}} = \mathcal{L}[\mathbf{f}](\lambda)$ and $\hat{g} = \mathcal{L}[g](\lambda)$. Therefore if we define by Φ_λ a solution operator to (5.63)–(5.65), i.e.,

$$(\mathbf{u}_\lambda, \eta_\lambda) = \Phi_\lambda(\tilde{\mathbf{f}}, \tilde{g}),$$

then $(\mathbf{u},\eta) = \Psi(\mathbf{f},g)$, where Ψ is defined in (5.61) satisfies the original problem (5.51)–(5.53). Hence in order to prove Theorem 5.3.1 it suffices to show $\mathcal{R}$-boundedness of a family of solutions to (5.63)–(5.64). The result is

Theorem 5.3.5. Let $1 < p < \infty$ and $0 < \epsilon < \pi/2$. Assume that Ω_0 is a uniform C^2 domain. Then, there exists a positive constant λ_0 and operator families $\mathcal{A}(\lambda) \in \mathrm{Hol}\,(\Lambda_{\epsilon,\lambda_0}, \mathcal{L}(L_q(\Omega_0)\times W^1_q(\Omega_0), W^1_q(\Omega_0)))$ and $\mathcal{B}(\lambda) \in \mathrm{Hol}\,(\Lambda_{\epsilon,\lambda_0}, \mathcal{L}(L_q(\Omega_0) \times W^1_q(\Omega_0), W^2_q(\Omega_0)^N))$, such that for any $(\hat{\mathbf{f}},\hat{g}) \in L_q(\Omega_0)\times W^1_q(\Omega_0)$ and $\lambda \in \Lambda_{\epsilon,\lambda_0}$, $(\eta_\lambda = \mathcal{A}(\lambda)(\hat{\mathbf{f}},\hat{g}), \mathbf{u}_\lambda = \mathcal{B}(\lambda)(\hat{\mathbf{f}},\hat{g}))$ is a unique solution of (5.63)–(5.65) and

$$\mathcal{R}_{\mathcal{L}(L_q(\Omega_0)\times W^1_q(\Omega_0), W^1_q(\Omega_0))}(\{(\tau\partial_\tau)^\ell\mathcal{A}(\lambda) :\ \lambda\in\Sigma_{\epsilon,\lambda_0}\}) \leq M,$$
$$\mathcal{R}_{\mathcal{L}(L_q(\Omega_0)\times W^1_q(\Omega_0), H^{2-j}_q(\Omega_0)^N)}(\{(\tau\partial_\tau)^\ell(\lambda^{j/2}\mathcal{B}(\lambda)) :\ \lambda\in\Sigma_{\epsilon,\lambda_0}\}) \leq M$$

for $\ell = 0,1$, $j = 0,1,2$ and some constant $M > 0$, where by Hol we denote the space of holomorphic operators.

Remark 5.3.2. We say that the operator valued function $T(\lambda)$ is holomorphic if it is differentiable in norm for all λ in a complex domain. For more details, see [15, Chaper VII,$1.1].

A particular advantage of this approach is a direct treatment of the continuity equation. Namely, in order to prove Theorem 5.3.5 we use (5.64) to get $\eta_\lambda = \lambda^{-1}(g - \varrho_0 \mathrm{div}\,_y \mathbf{u}_\lambda)$. Plugging the latter to (5.63) and skipping the subscripts λ we get

$$\varrho_0\lambda\mathbf{u} - \mu\Delta_y\mathbf{u} - (\frac{\mu}{3} + \zeta + \gamma_0\lambda^{-1}\varrho_0)\nabla_y \mathrm{div}\,_y\mathbf{u} = \mathbf{f} - \gamma_0\lambda^{-1}(\nabla g + \nabla\varrho_0 \mathrm{div}\,\mathbf{u}), \quad \mathbf{u}|_{\partial\Omega_0} = 0 \tag{5.66}$$

and the problem is reduced to showing $\mathcal{R}$-boundedness for the family of solutions to (5.66). With this approach Proposition 5.3.2 becomes a special case of Theorem 5.3.1. Namely, resolvent problem corresponding to (5.55) reads

$$\varrho_0\lambda\mathbf{u} - \mu\Delta_y\mathbf{u} - (\frac{\mu}{3} + \zeta)\nabla_y \mathrm{div}\,_y\mathbf{u} = \mathbf{f}, \quad \mathbf{u}|_{\partial\Omega_0} = 0. \tag{5.67}$$

As the second term on the right hand side of $(5.66)_1$ is of lower order, we can reduce (5.66) to

$$\varrho_0\lambda\mathbf{u} - \mu\Delta_y\mathbf{u} - (\frac{\mu}{3} + \zeta + \gamma_0\lambda^{-1}\varrho_0)\nabla_y \mathrm{div}\,_y\mathbf{u} = \mathbf{F}, \quad \mathbf{u}|_{\partial\Omega_0} = 0. \tag{5.68}$$

Now (5.67) is a particular case of (5.68) with $\gamma_0 = 0$. The following result gives $\mathcal{R}$-boundedness for (5.68):

Proposition 5.3.1. Let $1 < q < \infty$ and $0 < \epsilon < \pi/2$. Assume that Ω_0 is a uniform C^2 domain in $\mathbb{R}^N$. Then, there exists a positive constant λ_0 such that there exists an operator family $\mathcal{C}(\lambda) \in \mathrm{Hol}\,(\Sigma_{\epsilon,\lambda_0}, \mathcal{L}(L_q(\Omega_0)^N, H^2_q(\Omega_0)^N))$ such that for any $\lambda \in \Lambda_{\epsilon,\lambda_0}$ and $\mathbf{F} \in L_q(\Omega_0)^N$, $\mathbf{v} = \mathcal{C}(\lambda)\mathbf{F}$ is a unique solution of (5.68), and

$$\mathcal{R}_{\mathcal{L}(L_q(\Omega_0)^N, H^{2-j}_q(\Omega_0)^N)}(\{(\tau\partial_\tau)^\ell \mathcal{C}(\lambda) \mid \lambda \in \Sigma_{\epsilon,\lambda_0}\}) \leq M$$

for $\ell = 0, 1$, $j = 0, 1, 2$ and some constant $M > 0$.

Proof. An analog of Proposition 5.3.1 has been shown in ([3], Theorem 2.10) for a problem

$$\lambda\mathbf{u} - \mu\Delta_y\mathbf{u} - (\nu + \gamma^2\lambda^{-1}\varrho_0)\nabla_y \mathrm{div}\,_y\mathbf{u} = \mathbf{F}, \quad \mathbf{u}|_{\partial\Omega_0} = 0, \tag{5.69}$$

where μ, ν, and γ are constants satisfying $\mu + \nu > 0$ and $\gamma > 0$. The proof requires only minor modifications in order to prove Proposition 5.3.1, therefore we present only a sketch. First we solve a problem with constant coefficients in the whole space

$$\varrho_0^*\lambda\mathbf{u} - \mu\Delta_y\mathbf{u} - (\frac{\mu}{3} + \zeta + \gamma_0^*\lambda^{-1}\varrho_0^*)\nabla_y \mathrm{div}\,_y\mathbf{u} = \mathbf{F} \quad \in \mathbb{R}^n, \tag{5.70}$$

where $\rho_0^* > 0$ and $\gamma_0^* \geq 0$ are constants. $\mathcal{R}$-boundedness for (5.70) can be shown following the proof of Theorem 3.2 in [3], the only difference is that now we have

$\gamma_0^* \rho_0^*$ instead of γ^2. However, strict positivity of this constant neither its square structure is not necessary, nonnegativity is sufficient.

Next we consider (5.70) in a half-space supplied with the boundary condition $u|_{\partial\mathbb{R}^n_+} = 0$. Here we can follow the proof of Theorem 4.1 in [3] which works without modifications for $\gamma_0^* \rho_0^* \geq 0$.

The third step consists in showing $\mathcal{R}$-boundedness in a bent half-space, the necessary result is Theorem 5.1 in [3], where replacing $\gamma^2 > 0$ with $\gamma_0^* \rho_0^* \geq 0$ is again harmless.

The final step is an introduction of a partition of unity and application of properties of a uniform C^2 domain. Here we have to deal with variable coefficients which is not in the scope of Theorem 2.10 in [3]. However, we can refer to a more recent result, Theorem 4.1 in [26] which gives $\mathcal{R}$-boundedness for a resolvent problem corresponding to more complicated system describing flow of a two-component mixture. It is enough to follow step by step Section 6.3 of [26] omitting all terms which are not relevant here.

□

Proof of Theorem 5.3.5. By definition of $\Lambda_{\varepsilon,\lambda_0}$ we have $|\lambda^{-1}| \leq C$ for $\lambda \in \Lambda_{\varepsilon,\lambda_0}$. Therefore

$$\|\mathbf{f} - \gamma_0 \lambda^{-1}(\nabla g + \nabla \varrho_0 \mathrm{div}\, \mathbf{u})\|_{L_p(\Omega_0)} \leq C[\|f\|_{L_q(\Omega_0)} + \|\nabla \varrho_0\|_{L_q(\Omega_0)} \|\mathrm{div}\, \mathbf{u}\|_{L_\infty(\Omega_0)})]$$
$$\leq C[\|f\|_{L_q(\Omega_0)} + \|\mathbf{u}\|_{W^2_q(\Omega_0)}],\ \|\lambda^{-1}(g - \nabla \varrho_0 \mathrm{div}\, \mathbf{u})\|_{W^1_q} \leq C[\|g\|_{W^1_q(\Omega_0)} + \|\mathbf{u}\|_{W^2_q(\Omega_0)}],$$

and so Theorem 5.3.5 results directly from Proposition 5.3.1.

□

Now it is straightforward to conclude main results of this section.

Proof of Theorems 5.3.1 and 5.3.2.
Theorem 5.3.1 follows from Theorems 5.3.5 and 5.3.4 with $X = L_q(\Omega_0) \times W^1_q(\Omega_0)$ and $Y = W^1_q(\Omega_0) \times W^2_q(\Omega_0)$. Theorem 5.3.2 follows from Proposition 5.3.1 and Theorem 5.3.4, this time with $X = L_q(\Omega_0)$ and $Y = W^2_q(\Omega_0)$.

□

5.3.2 Exponential Decay

Linearization

In order to show the global well-posedness in Theorem 5.1.2 we will linearize the problem around the constant ϱ^*, therefore we consider on the fixed domain Ω_0 a linear problem

$$\varrho^* \mathbf{u}_t - \mu \Delta_y \mathbf{u} - (\frac{\mu}{3} + \zeta) \nabla_y \mathrm{div}\,_y \mathbf{u} + \gamma_0^* \nabla_y \eta = \mathbf{f}, \tag{5.71}$$

$$\eta_t + \varrho^* \mathrm{div}\,_y \mathbf{u} = g, \tag{5.72}$$

$$\mathbf{u}|_{\partial\Omega_0} = 0, \quad \mathbf{u}|_{t=0} = \mathbf{u}_0. \tag{5.73}$$

Here again η and $\mathbf{u}$ are unknowns and f, g are given functions, but this time $\varrho^* > 0$ and $\gamma_0^* \geq 0$ are constants. The main result of this section is the following exponential decay estimate

Proposition 5.3.2. Assume $\varrho^* > 0$ is a constant and $\Omega_0, p, q, \mu, \zeta, \mathbf{u}_0$ satisfy the assumptions of Theorem 5.3.1. Assume moreover that there exists $\gamma > 0$ such that

$$e^{\gamma t}\mathbf{f} \in L_p(0,\infty; L_q(\Omega_0)), \quad e^{\gamma t} g \in L_p(0,\infty; W_q^1(\Omega_0)).$$

Then (5.71)–(5.73) admits a unique solution $\varrho, \mathbf{u}$ such that

$$\begin{aligned}
&\|e^{\gamma t}\mathbf{u}_t\|_{L_p(0,\infty;L_q(\Omega_0))}+\|e^{\gamma t}\mathbf{u}\|_{L_p(0,\infty;W_q^2(\Omega_0))}+\|e^{\gamma t}\nabla\eta\|_{L_p(0,\infty;L_q(\Omega_0))} \\
&\quad +\|e^{\gamma t}\partial_t\eta\|_{L_p(0,\infty;W_q^1(\Omega_0))} \\
&\leq C_{p,q}\left(\|\mathbf{u}_0\|_{B^{2-2/p}_{q,p}(\Omega_0)} + \|e^{\gamma t}\mathbf{f}\|_{L_p(0,\infty;L_q(\Omega_0))} + \|e^{\gamma t} g\|_{L_p(0,\infty;W_q^1(\Omega_0))}\right). \qquad (5.74)
\end{aligned}$$

As before, we also need a decay estimate for the linear momentum equation

$$\varrho^*\mathbf{u}_t - \mu\Delta_y\mathbf{u} - (\frac{\mu}{3}+\zeta)\nabla_y \mathrm{div}_y\mathbf{u} = \mathbf{f}, \qquad (5.75)$$

$$\mathbf{u}|_{\partial\Omega_0} = 0, \quad \mathbf{u}|_{t=0} = \mathbf{u}_0. \qquad (5.76)$$

Proposition 5.3.3. Assume $\varrho^* > 0$ is a constant and $\Omega_0, p, q, \mu, \zeta, \mathbf{u}_0$ satisfy the assumptions of Theorem 5.3.1 and $\mathbf{f}$ satisfies the assumptions of Proposition 5.3.2. Then (5.75) admits a unique solution $\mathbf{u}$ such that

$$\|e^{\gamma t}\mathbf{u}_t\|_{L_p(0,\infty;L_q(\Omega_0))}+\|e^{\gamma t}\mathbf{u}\|_{L_p(0,\infty;W_q^2(\Omega_0))} \leq C_{p,q}\left(\|\mathbf{u}_0\|_{B^{2-2/p}_{q,p}(\Omega_0)}+\|e^{\gamma t}\mathbf{f}\|_{L_p(0,\infty;L_q(\Omega_0))}\right). \qquad (5.77)$$

Proof of Propositions 5.3.2 and 5.3.3

Let us focus on the proof of Proposition 5.3.2, as the proof Proposition 5.3.3 is just a simplification of the latter. Roughly speaking, the point is that the resolvent estimate from the previous section holds for λ bounded away from zero and now we have to extend it to cover some range of negative values of $\mathrm{Re}\,\lambda$ under stronger assumptions (fixed coefficients). The resolvent problem corresponding to (5.71)–(5.73) reads

$$\varrho^*\lambda\mathbf{u} - \mu\Delta_y\mathbf{u} - (\frac{\mu}{3}+\zeta)\nabla_y \mathrm{div}_y\mathbf{u} + \gamma_0^*\nabla_y\eta = \mathbf{f}, \qquad (5.78)$$

$$\lambda\eta + \varrho^*\mathrm{div}_y\mathbf{u} = g, \qquad (5.79)$$

$$\mathbf{u}|_{\partial\Omega_0} = 0. \qquad (5.80)$$

Proposition 5.3.4. Let $q, \varrho_*, \gamma_{0*}, \mu, \zeta$ satisfy the assumptions of Theorem 5.3.1. Then there exists $\delta, \varepsilon_1 > 0$ such that for any

$$\lambda \in \Sigma_{\varepsilon_1,-\delta}$$

and $(\mathbf{f}, g) \in \hat{\mathcal{H}}_q$ the problem (5.78)–(5.80) admits a unique solution $(\mathbf{u}, \eta)$ with the estimate

$$(|\lambda|+1)\|(\mathbf{u},\eta)\|_{\mathcal{H}_q} + \|\mathbf{u}\|_{W^2_q(\Omega_0)} \le C\|(\mathbf{f},g)\|_{\mathcal{H}_q}. \tag{5.81}$$

Proof. Step 1: $\lambda \in \mathbb{C}_+ \cup \Lambda_{\varepsilon,\lambda_0}$**.** The resolvent problem (5.78)–(5.80) is a particular case of (5.63)–(5.65) with fixed coefficients. Therefore from Theorem 5.3.5 it follows that there exists $\lambda_0 > 0$ such that the solution with the estimate (5.81) exists for $\lambda \in \Lambda_{\epsilon,\lambda_0}$. It remains to prove the unique existence and (5.81) for $\{\lambda \in \mathbb{C}_+ : \ |\lambda| \le \lambda_0\}$. Consider first $\lambda \neq 0$. Computing η from (5.79) and plugging to (5.78) we obtain

$$\lambda\mathbf{u} - \varrho_*^{-1}[\mu\Delta\mathbf{u} + (\frac{\mu}{3} + \zeta + \lambda^{-1}\varrho_*)\nabla\mathrm{div}\,\mathbf{u}] = \varrho_*^{-1}(f - \lambda^{-1}\nabla g) =: \tilde{\mathbf{f}}, \quad \mathbf{u}|_{\partial\Omega_0} = 0. \tag{5.82}$$

Let us denote

$$A_\lambda\mathbf{u} = \varrho_*^{-1}[\mu\Delta\mathbf{u} + (\frac{\mu}{3} + \zeta + \lambda^{-1}\varrho_*)\nabla\mathrm{div}\,\mathbf{u}] \tag{5.83}$$

and consider an auxiliary problem with a resolvent parameter $\tau > 0$:

$$\tau\mathbf{v}_\tau - A_\lambda\mathbf{v}_\tau = \tilde{\mathbf{f}}. \tag{5.84}$$

By ([30], Theorem 7.1) there exists $\tau_0 > 0$ such that for $\tau \ge \tau_0$ there exists $(\tau\mathbf{I} - A_\lambda)^{-1} \in \mathcal{L}(L_q(\Omega_0), \mathcal{D}_q)$ and

$$\tau\|\mathbf{v}_\tau\|_{L_q(\Omega_0)} + \tau^{1/2}\|\mathbf{v}_\tau\|_{W^1_q(\Omega_0)} + \|\mathbf{v}_\tau\|_{W^2_q(\Omega_0)} \le C\|\tilde{\mathbf{f}}\|_{L_q(\Omega_0)}.$$

Now we rewrite (5.82) as

$$(\lambda - \tau)\mathbf{u} + (\tau\mathbf{I} - A_\lambda)\mathbf{u} = \tilde{\mathbf{f}}.$$

Applying $(\tau\mathbf{I} - A_\lambda)^{-1}$ to the above equation we get

$$\mathbf{u} + (\lambda - \tau)(\tau\mathbf{I} - A_\lambda)^{-1}\mathbf{u} = (\tau\mathbf{I} - A_\lambda)^{-1}\tilde{\mathbf{f}}. \tag{5.85}$$

Since $(\lambda-\tau)(\tau\mathbf{I}-A_\lambda)^{-1}$ is compact on $L_q(\Omega_0)$, in order to prove unique solvability of (5.85) it is enough to show that

$$\mathrm{Ker}(\mathbf{I} + (\lambda - \tau)(\tau\mathbf{I} - A_\lambda)^{-1}) = \{0\}. \tag{5.86}$$

Assume that

$$\left[\mathbf{I} + (\lambda - \tau)(\tau\mathbf{I} - A_\lambda)^{-1}\right]\mathbf{g} = 0.$$

Setting $\mathbf{v}^0 = (\tau\mathbf{I} - A_\lambda)^{-1}\mathbf{g}$ we get

$$(\lambda\mathbf{I} - A_\lambda)\mathbf{v}^0 = 0.$$

Under the assumptions on the coefficients, by definition of A_λ easily verify that $\mathbf{v}^0 = 0$, therefore $\mathbf{g} = \mathbf{0}$ and so (5.86) holds. This completes the proof of the unique solvability of (5.78)–(5.80) for $\lambda \in \mathbb{C}_+ \setminus \{0\}$.

For $\lambda = 0$ problem (5.78)–(5.80) reduces to inhomogeneous Stokes problem

$$-\mu\Delta_y \mathbf{u} + \gamma_0^* \nabla_y \eta = \mathbf{f} + (\frac{\mu}{3} + \zeta)\varrho_*^{-1}\nabla_y g, \tag{5.87}$$

$$\operatorname{div}_y \mathbf{u} = g, \quad \mathbf{u}|_{\partial\Omega_0} = 0. \tag{5.88}$$

In [7] the unique existence was shown for a problem

$$\kappa\mathbf{u} - \mu\Delta_y \mathbf{u} + \gamma_0^* \nabla_y \eta = \mathbf{f}, \quad \operatorname{div}_y \mathbf{u} = g, \quad \mathbf{u}|_{\partial\Omega_0} = 0$$

with $(\mathbf{f}, g) \in \hat{\mathcal{H}}_q$ for sufficiently large κ. By Fredholm alternative, this result together with uniqueness for (5.87)–(5.88), which is obvious, implies the unique solvability of (5.87)–(5.88). Notice that here we require that g has zero mean. This completes the proof for $\lambda \in \mathbb{C}_+ \cup \Lambda_{\varepsilon,\lambda_0}$

Step 2. Proof for $\lambda \in \Sigma_{\varepsilon_1,-\delta}$. The crucial observation is that the resolvent set is open. Therefore, since the set $\{\lambda \in \mathbb{C} : \mathrm{Re}\lambda = 0,\ \mathrm{Im}\lambda \le \lambda_0\}$ is compact and contained in the resolvent, there exists some $\delta_0 > 0$ such that $\{\lambda \in \mathbb{C} : \mathrm{Re}\lambda \ge -\delta_0,\ \mathrm{Im}\lambda \le \lambda_0\}$ is also in the resolvent. Since we already know that $\Lambda_{\varepsilon,\lambda_0}$ is in the resolvent, it follows that for some $\frac{\pi}{2} > \varepsilon_1 > \varepsilon$ and $0 < \delta < \delta_0$ the resolvent set contains $\Sigma_{\varepsilon_1,-\delta}$.

□

Now we denote

$$AU = \begin{pmatrix} -\varrho^* \operatorname{div} \mathbf{u} \\ \varrho_*^{-1}[\mu\Delta\mathbf{u} + (\frac{\mu}{3} + \zeta)\nabla\operatorname{div}\mathbf{u} + \gamma_{0*}\nabla\eta] \end{pmatrix} \quad \text{for } U = (\eta, \mathbf{u}) \in \mathcal{D}_q, \tag{5.89}$$

$$\hat{A} = A|_{\hat{\mathcal{D}}_q}. \tag{5.90}$$

Let us consider the Cauchy problem

$$\partial_t U - \hat{A}U = 0 \quad \text{for } t > 0, \quad U|_{t=0} = U_0 = (\eta_0, \mathbf{v}_0) \in \hat{\mathcal{H}}_q(\Omega). \tag{5.91}$$

Proposition 5.3.5. $\hat{A}$ generates a C_0 semigroup $\{\dot{T}(t)\}_{t\ge 0}$ on $\hat{\mathcal{H}}_q$ and

$$\|\dot{T}(t)U_0\|_{\mathcal{H}_q} \le Ce^{-\gamma_1 t}\|U_0\|_{\mathcal{H}_q} \tag{5.92}$$

for $U_0 \in \hat{\mathcal{H}}_q$ and some $\gamma_1 > 0$.

Proof. The resolvent problem corresponding to (5.91) is (5.78)–(5.80). Therefore the generation of analytic semigroup results from the general semigroup theory. More precisely, the semigroup can be expressed by the following formula (see (1.50) in Chapter 9 of [15]):

$$\hat{T}(t) = \frac{1}{2\pi i}\int_{-\infty}^{\infty} e^{\lambda t}(\lambda\mathbf{I} - \hat{A})^{-1} d\tau, \tag{5.93}$$

where $\tau = \mathrm{Im}\,\lambda$ and the line along which we integrate is contained in the resolvent of $\hat{A}$. Due to Proposition 5.3.4 we can take $\lambda = -\gamma_1 + i\tau$ for any $\gamma_1 < \delta$ and then the estimate (5.92) follows from (5.81) and (5.93).

□

Now we can start the proof of Proposition 5.3.2. Let $\lambda_1 > 0$ be sufficiently large and $\gamma > 0$ sufficiently small to be precised later. For notational convenience let us denote $U = (\eta, \mathbf{v})$ let us denote

$$P_{\mathbf{v}} U = \mathbf{v} \quad \text{for}\, U = (\eta, \mathbf{v}). \tag{5.94}$$

Now we consider the problem

$$\begin{cases} \partial_t U_1 + \lambda_1 U_1 - AU_1 = G \quad \text{in } \Omega_0 \times (0,T), \\ P_{\mathbf{v}} U_1|_{\partial\Omega_0} = 0, \quad U_1|_{t=0} = U_0. \end{cases} \tag{5.95}$$

Multiplying it by $e^{\gamma t}$ we get

$$\begin{cases} \partial_t(e^{\gamma t}U_1) + (\lambda_1 - \gamma)e^{\gamma t}U_1 - Ae^{\gamma t}U_1 = e^{\gamma t}G \quad \text{in } \Omega_0 \times (0,T), \\ P_{\mathbf{v}} e^{\gamma t}U_1|_{\partial\Omega_0} = 0, \quad e^{\gamma t}U_1|_{t=0} = U_0. \end{cases}$$

Since we need a maximal regularity estimate for $e^{\gamma t}U_1$ independent from the initial condition, we denote by G_0 a zero extension of G on $\mathbb{R}$ and consider a problem

$$\partial_t U_2 + (\lambda_1 - \gamma)U_2 - AU_2 = e^{\gamma t}G_0 \quad \text{in } \Omega_0 \times \mathbb{R}, \qquad P_{\mathbf{v}} U_2|_{\partial\Omega_0} = 0.$$

Applying the Fourier transform in time we obtain

$$(\lambda_1 - \gamma + i\tau)\mathcal{F}[U_2](\cdot,\tau) - A\mathcal{F}[U_2](\cdot,\tau) = \mathcal{F}[e^{\gamma t}G_0](\cdot,\tau) \quad \text{in } \Omega_0, \quad \mathcal{F}[P_{\mathbf{v}}(U_2)]|_{\partial\Omega_0} = 0. \tag{5.96}$$

Choosing λ_1 such that $\lambda_1 - \gamma > \lambda_0$, where λ_0 is from Theorem 5.3.5 we can deduce $\mathcal{R}$ - bounds analogous to those from Theorem 5.3.5, but for (5.96). Therefore

$$M(\tau) := \mathcal{F}[U_2](\cdot,\tau)$$

is well defined for $\tau \in \mathbb{R}$, $M \in C^1(\mathbb{R}, \mathcal{L}(\mathcal{H}_q, \mathcal{D}_q))$ and moreover (5.59) holds. Therefore we can apply directly Theorem 5.3.3 to $M(\cdot)$ to obtain

$$\|\partial_t U_2\|_{L_p(\mathbb{R},\mathcal{H}_q)} + \|U_2\|_{L_p(\mathbb{R},D_q)} \le C\|e^{\gamma t}G_0\|_{L_p(\mathbb{R},\mathcal{H}_q)} \le C\|e^{\gamma t}G\|_{L_p((0,T),\mathcal{H}_q)}. \tag{5.97}$$

Finally consider the Cauchy problem

$$\begin{cases} \partial_t U_3 + (\lambda_1 - \gamma)U_3 - AU_3 = 0 \quad \text{in } \Omega \times (0,\infty) \\ P_{\mathbf{v}} U_3|_{\partial\Omega_0} = 0, \quad U_3|_{t=0} = U_0 - U_2. \end{cases}$$

Let us introduce another constant $\gamma_5 > 0$. Then $e^{\gamma_5 t}U_3$ satisfies

$$\begin{cases} \partial_t e^{\gamma_5 t}U_3 + (\lambda_1 - \gamma - \gamma_5)e^{\gamma_5 t}U_3 - A(e^{\gamma_5 t}U_3) = 0 \quad \text{in } \Omega \times (0,\infty) \\ P_{\mathbf{v}}(e^{\gamma_5 t}U_3)|_{\partial\Omega_0} = 0, \quad e^{\gamma_5 t}U_3|_{t=0} = U_0 - U_2. \end{cases}$$

The resolvent problem corresponding to the latter reads

$$\begin{cases} (\lambda + \lambda_1 - \gamma - \gamma_5)e^{\gamma t}U_3 - A(e^{\gamma t}U_3) = 0 \quad \text{in } \Omega \\ P_{\mathbf{v}}(e^{\gamma t}U_3)|_{\partial\Omega_0} = 0. \end{cases}$$

For λ_1 sufficiently large we can apply Theorem 5.3.1 to obtain the maximal regularity estimate

$$\|e^{\gamma_5 t}\partial_t U_3\|_{L_p((0,T),\mathcal{H}_q)} + \|e^{\gamma_5 t}U_3\|_{L_p((0,T),D_q)} \le Ce^{\gamma_0 T}\|U_0 - U_2\|_{W^1_q(\Omega_0)\times B^{2-2/p}_{q,p}(\Omega_0)}, \tag{5.98}$$

where γ_0 is from Theorem 5.3.1. For sufficiently large λ_1 we can choose γ_5 such that $\gamma_5 - \gamma_0 \ge \gamma$ and then from (5.98) we obtain

$$\|e^{\gamma t}\partial_t U_3\|_{L_p((0,\infty),\mathcal{H}_q)} + \|e^{\gamma t}U_3\|_{L_p((0,\infty),D_q)} \le C\|U_0 - U_2\|_{W^1_q(\Omega_0)\times B^{2-2/p}_{q,p}(\Omega_0)}. \tag{5.99}$$

Since $e^{\gamma t}U_1 = U_2 + U_3$, from (5.97) and (5.99) we obtain

$$\begin{aligned}\|e^{\gamma t}\partial_t U_1\|_{L_p((0,\infty),\mathcal{H}_q)} + \|e^{\gamma t}U_1\|_{L_p((0,\infty),D_q)} \le C\|\eta_0\|_{W^1_q(\Omega_0)} + \|\mathbf{v}_0\|_{B^{2-2/p}_{q,p}(\Omega_0)} \\ +\|e^{\gamma t}G\|_{L_p((0,T),\mathcal{H}_q)}.\end{aligned} \tag{5.100}$$

Now for $U_1 = (\eta_1, \mathbf{v}_1)$ let us define

$$\tilde{U}_1(x,t) = \left(\eta_1(x,t) - \frac{1}{|\Omega_0|}\int_{\Omega_0}\eta_1(y,t)dy,\ \mathbf{v}_1(x,t)\right). \tag{5.101}$$

Consider the problem

$$\begin{cases} \partial_t\tilde{V} - A\tilde{V} = \lambda_1\tilde{U}_1 & \text{in } \Omega_0\times(0,T) \\ P_{\mathbf{v}}\tilde{V}|_{\partial\Omega_0} = 0, & \tilde{V}|_{t=0} = 0. \end{cases} \tag{5.102}$$

Let $\{\hat{T}(t)\}$ be the semigroup on $\hat{\mathcal{H}}_q$ generated by $\hat{A}$. Since $\tilde{U}_1 \in \hat{\mathcal{H}}_q$, we have

$$\tilde{V} = \int_0^t \hat{T}(t-s)\tilde{U}_1(\cdot,s)\mathrm{d}s.$$

Therefore in order to show exponential decay of $\tilde{V}$ we can use (5.92) to obtain

$$\begin{aligned} e^{\gamma t}\|\tilde{V}(\cdot,t)\|_{L_q(\Omega_0)} &\le C\int_0^t e^{\gamma t}e^{-\gamma_1(t-s)}\|\tilde{U}_1(\cdot,s)\|_{\mathcal{H}_q}\mathrm{d}s = C\int_0^t e^{-(\gamma_1-\gamma)(t-s)}e^{\gamma s}\|\tilde{U}_1(\cdot,s)\|_{\mathcal{H}_q}\mathrm{d}s \\ &\le \left(\int_0^t e^{-(\gamma_1-\gamma)(t-s)}\mathrm{d}s\right)^{1/p'}\left(\int_0^t e^{-(\gamma_1-\gamma)(t-s)}[e^{\gamma s}\|\tilde{U}_1(\cdot,s)\|_{\mathcal{H}_q}]^p\right)^{1/p}. \end{aligned}$$

Therefore

$$\int_0^T e^{\gamma t}\|\tilde{V}(\cdot,t)\|^p_{L_q(\Omega_0)} \le C(\gamma_1-\gamma)^{-p/p'}\int_0^T [e^{\gamma s}\|\tilde{U}_1(\cdot,s)\|_{\mathcal{H}_q}]^p\mathrm{d}s\int_s^T e^{-(\gamma_1-\gamma)(t-s)}\mathrm{d}t,$$

which implies

$$\|e^{\gamma t}\tilde{V}\|_{L_p((0,T),\mathcal{H}_q)} \le C(\gamma_1,\gamma,p)\|e^{\gamma t}\tilde{U}_1\|_{L_p((0,T),\mathcal{H}_q)}. \tag{5.103}$$

Adding $\lambda_0 \tilde{V}$ to both sides of (5.102) we see that $\tilde{V}$ satisfies

$$\begin{cases} \partial_t \tilde{V} + \lambda_0 \tilde{V} - A\tilde{V} = \lambda_1 \tilde{U}_1 + \lambda_0 \tilde{V} \quad \text{in } \Omega_0 \times (0,T) \\ P_{\mathbf{v}} \tilde{V}|_{\partial\Omega_0} = 0, \quad \tilde{V}|_{t=0} = 0. \end{cases} \tag{5.104}$$

For sufficiently large λ_0 we have $\mathcal{R}$—boundedness for corresponding resolvent problem and from Theorem 5.3.4 we get

$$\|e^{\gamma t}\partial_t \tilde{V}\|_{L_p(0,T;\mathcal{H}_q)} + \|e^{\gamma t}\tilde{V}\|_{L_p(0,T;D_q)} \le C(\|e^{\gamma t}\tilde{U}_1\|_{L_p(0,T;D_q)} + \|e^{\gamma t}\tilde{V}\|_{L_p(0,T;\mathcal{H}_q)}),$$

which combined with (5.100) and (5.103) gives

$$\begin{aligned} &\|e^{\gamma t}\partial_t \tilde{V}\|_{L_p(0,T;\mathcal{H}_q)} + \|e^{\gamma t}\tilde{V}\|_{L_p(0,T;D_q)} \\ &\quad \le C\left(\|(\eta_0, \mathbf{v}_0)\|_{W^1_q(\Omega_0)\times B^s_{q,p}(\Omega_0)} + \|e^{\gamma t}G\|_{L_p((0,T),\mathcal{H}_q)}\right). \end{aligned} \tag{5.105}$$

Next, if we set

$$V = \tilde{V} - \left(\frac{1}{|\Omega_0|}\int_0^t \int_{\Omega_0} \eta(x,t)\mathrm{d}x, 0\right),$$

then, by (5.101), V satisfies

$$\begin{cases} \partial_t V - AV = \lambda_1 U_1 \quad \text{in } \Omega_0 \times (0,T) \\ P_{\mathbf{v}} \tilde{V}|_{\partial\Omega_0} = 0, \quad \tilde{V}|_{t=0} = 0. \end{cases} \tag{5.106}$$

Now if U_1 solves (5.95) with $G = (\mathbf{f}, g)$, then $(\eta, v) = U_1 + V$ is the unique solution to (5.71) and from (5.100) and (5.105) we conclude the estimate (5.74).

□

5.3.3 Embedding Results

Next we recall some embedding results for Besov spaces. The first one is [1, Theorem 7.34 (c)]:

Lemma 5.3.1. Assume $\Omega \in \mathbb{R}^n$ satisfies the cone condition and let $1 \le p, q \le \infty$ and $sq > n$. Then

$$B^s_{q,p}(\Omega_0) \subset C_B(\Omega_0),$$

where C_B we denote the space of continuous bounded functions.

In particular $u \in B^{2-2/p}_{q,p}(\Omega_0)$ implies $\nabla u \in B^{1-2/p}_{q,p}(\Omega_0)$. Therefore the above Lemma with $s = 1 - 2/p$ yields

Corollary 5.3.1. Assume $\frac{2}{p} + \frac{3}{q} < 1$ and let Ω_0 satisfy the assumptions of Theorem 5.1.1. Then $B^{2-2/p}_{q,p}(\Omega_0) \subset W^1_\infty(\Omega_0)$ and

$$\|f\|_{W^1_\infty(\Omega_0)} \le C\|f\|_{B^{2-2/p}_{q,p}(\Omega_0)}. \tag{5.107}$$

The next result is due to Tanabe (cf. [31, p.10]):

Lemma 5.3.2. Let X and Y be two Banach spaces such that X is a dense subset of Y and $X \subset Y$ is continuous. Then for each $p \in (1,\infty)$

$$W^1_p((0,\infty),Y) \cap L_p((0,\infty),X) \subset C([0,\infty),(X,Y)_{1/p,p})$$

and for every $u \in W^1_p((0,\infty),Y) \cap L_p((0,\infty),X)$ we have

$$\sup_{t\in(0,\infty)} \|u(t)\|_{(X,Y)_{1/p,p}} \leq \left(\|u\|^p_{L_p((0,\infty),X)} + \|u\|^p_{W^1_p((0,\infty),Y)}\right)^{1/p}.$$

□

5.4 Local Well-Posedness

5.4.1 Linearization for the Local Well-Posedness

Let us start with removing inhomogeneity from the boundary condition (5.38). For this purpose we show

Lemma 5.4.1. Let $\mathbf{V}$ satisfy the assumptions of Theorem 5.1.1. Then the problem

$$\begin{aligned}&\varrho_0\partial_t\mathbf{u}_{b1} - \mu\Delta_y\mathbf{u}_{b1} - (\frac{\mu}{3}+\zeta)\nabla_y\mathrm{div}_y\mathbf{u}_{b1} = 0 \qquad \text{in } \Omega_0\times(0,T),\\ &\mathbf{u}_{b1}|_{\Gamma_0} = \tilde{\mathbf{V}}, \quad \mathbf{u}_{b1}|_{t=0} = \mathbf{V}(0)\end{aligned} \tag{5.108}$$

admits a unique solution such that

$$\|\partial_t\mathbf{u}_{b1}\|_{L_p(0,T;L_q(\Omega_0))} + \|\mathbf{u}_{b1}\|_{L_p(0,T;W^2_q(\Omega_0))} \leq C\|\partial_t\tilde{\mathbf{V}}\|_{L_p(0,T;L_q(\Omega_0))} + \|\tilde{\mathbf{V}}\|_{L_p(0,T;W^2_q(\Omega_0))}. \tag{5.109}$$

Proof. Denoting $\tilde{\mathbf{u}}_{b1} = \mathbf{u}_{b1} - \tilde{\mathbf{V}}$ we have

$$\begin{aligned}&\varrho_0\partial_t\tilde{\mathbf{u}}_{b1} - \mu\Delta_y\tilde{\mathbf{u}}_{b1} - (\frac{\mu}{3}+\zeta)\nabla_y\mathrm{div}_y\tilde{\mathbf{u}}_{b1} = \varrho_0\partial_t\tilde{\mathbf{V}} - \mu\Delta_y\tilde{\mathbf{V}} - (\frac{\mu}{3}+\zeta)\nabla_y\mathrm{div}_y\tilde{\mathbf{V}} \quad \text{in } \Omega_0\times(0,T),\\ &\tilde{\mathbf{u}}_{b1}|_{\Gamma_0} = 0, \quad \tilde{\mathbf{u}}_{b1}|_{t=0} = 0.\end{aligned} \tag{5.110}$$

Therefore, if $\mathbf{V}$ satisfies the assumptions of Theorem 5.1.1, then Proposition 5.3.2 gives

$$\|\partial_t\tilde{\mathbf{u}}_{b1}\|_{L_p(0,T;L_q(\Omega_0))} + \|\tilde{\mathbf{u}}_{b1}\|_{L_p(0,T;W^2_q(\Omega_0))} \leq C\|\partial_t\tilde{\mathbf{V}}, \nabla^2_y\tilde{\mathbf{V}}\|_{L_p(0,T;L_q(\Omega_0))}, \tag{5.111}$$

which implies (5.109).

□

As linear system in Theorem 5.3.1 has constant in time coefficients, we linearize (5.36)–(5.37) around the initial condition. Denoting

$$\eta = \tilde{\varrho} - \varrho_0, \quad \mathbf{v} = \tilde{\mathbf{u}} - \mathbf{u}_{b1}$$

we obtain

$$\varrho_0 \mathbf{v}_t - \mu\Delta_y \mathbf{v} - \left(\frac{\mu}{3} + \zeta\right)\nabla_y \mathrm{div}\,_y \mathbf{v} + \gamma_1 \nabla_y \eta = \mathbf{F}_1(\eta, \mathbf{v}) \tag{5.112}$$

$$\eta_t + \varrho_0 \mathrm{div}\,_y \mathbf{v} = G_1(\eta, \mathbf{v}), \tag{5.113}$$

$$\mathbf{v}|_{t=0} = \mathbf{u}_0 - \mathbf{V}(0), \quad \mathbf{v}|_{\partial\Omega_0} = 0, \tag{5.114}$$

where $\gamma_1 = \pi'(\varrho_0)$ and

$$\mathbf{F}_1(\eta, \mathbf{v}) = \mathbf{R}(\eta + \varrho_0, \mathbf{v} + \mathbf{u}_{b1}) - \eta\partial_t(\mathbf{v} + \mathbf{u}_{b1}) - \pi'(\varrho_0)\nabla_y \varrho_0 - [\pi'(\eta + \varrho_0) - \pi'(\varrho_0)]\nabla_y \eta \tag{5.115}$$

$$G_1(\eta, \mathbf{v}) = G(\eta + \varrho_0, \mathbf{v} + \mathbf{u}_{b1}) - (\eta + \varrho_0)\mathrm{div}\,_y \mathbf{u}_{b1} - \eta \mathrm{div}\,_y \mathbf{v}, \tag{5.116}$$

and $\mathbf{R}(\tilde{\varrho}, \tilde{\mathbf{u}})$ is defined in (5.41).

5.4.2 Nonlinear Estimates for the Local Well-Posedness

We start with a following imbedding result:

Lemma 5.4.2. Let $f_t \in L_p(0,T; L_q(\Omega_0))$, $f \in L_p(0,T; W^2_q(\Omega_0))$, $f(0,\cdot) \in B^{2-2/p}_{q,p}(\Omega_0)$. Then

$$\sup_{t\in(0,T)} \|f\|_{B^{2(1-1/p)}_{q,p}(\Omega_0)} \leq C[\|f_t\|_{L_p(0,T;L_q(\Omega_0))} + \|f\|_{L_p(0,T;W^2_q(\Omega_0))} + \|f(0)\|_{B^{2-2/p}_{q,p}(\Omega_0)}] \tag{5.117}$$

$$\sup_{t\in(0,T)} \|f\|_{W^1_\infty(\Omega_0))} \leq C[\|f_t\|_{L_p(0,T;L_q(\Omega_0))} + \|f\|_{L_p(0,T;W^2_q(\Omega_0))} + \|f(0)\|_{B^{2-2/p}_{q,p}(\Omega_0)}]. \tag{5.118}$$

Proof: In order to prove (5.117) we introduce an extension operator

$$e_T[f](\cdot,t) = \begin{cases} f(\cdot,t) & t \in (0,T), \\ f(\cdot, 2T-t) & t \in (T, 2T), \\ 0 & t \in (2T, +\infty). \end{cases} \tag{5.119}$$

If $f|_{t=0} = 0$, then we have

$$\partial_t e_T[f](\cdot,t) = \begin{cases} 0 & t \in (2T, +\infty), \\ (\partial_t f)(\cdot,t) & t \in (0,T), \\ -(\partial_t f)(\cdot, 2T-t) & t \in (T, 2T) \end{cases} \tag{5.120}$$

in a weak sense. Applying Lemma 5.3.2 with $X = H^2_q(\Omega_0)$, $Y = L_q(\Omega_0)$ and using (5.119) and (5.120) we have

$$\begin{aligned} &\sup_{t\in(0,T)} \|f(\cdot,t) - f(0)\|_{B^{2(1-1/p)}_{q,p}(\Omega_0)} \leq \sup_{t\in(0,\infty)} \|e_T[f - f(0)]\|_{B^{2(1-1/p)}_{q,p}(\Omega_0)} \\ &= (\|e_T[f - f(0)]\|^p_{L_p((0,\infty),W^2_q(\Omega_0))} + \|e_T[f - f(0)]\|^p_{W^1_p((0,\infty),L_q(\Omega_0))})^{1/p} \\ &\leq C(\|f - f(0)\|_{L_p(0,\infty;W^2_q(\Omega_0))} + \|\partial_t f\|_{L_p(0,T;L_q(\Omega_0))}). \end{aligned}$$

This gives (5.117), which together with Corollary 5.3.1 imply (5.118).

□

Using the results recalled in the previous section and Lemma 5.4.2 we show the following estimate for functions from the space $\mathcal{Y}(T)$:

Lemma 5.4.3. Let $(z, \mathbf{w}) \in B(0, M) \subset \mathcal{Y}(T)$ and $\|\mathbf{w}(0)\|_{B^{2-2/p}_{q,p}(\Omega_0)} \leq L$. Then

$$\|\mathbf{E}^0(\mathbf{k_w}), \nabla_{\mathbf{k_w}}\mathbf{E}^0(\mathbf{k_w}), \|_{L_\infty((0,T)\times\Omega_0)} \leq C(M, L)E(T), \tag{5.121}$$

$$\sup_{t\in(0,T)}\|z(\cdot, t)\|_{W^1_q(\Omega_0)} \leq C(M, L)E(T), \tag{5.122}$$

$$\sup_{t\in(0,T)}\|\mathbf{w}(\cdot, t) - \mathbf{w}(0)\|_{B^{2(1-1/p)}_{q,p}(\Omega_0)} \leq C(M, L), \tag{5.123}$$

$$\|\mathbf{w}\|_{L_\infty(0,T;W^1_\infty(\Omega_0))} \leq C(M, L), \tag{5.124}$$

where $\mathbf{k_w}$ is defined in (5.33).

Proof. (5.121) follows immediately from Lemma 5.2.1. Next, we have

$$\|z(\cdot, t)\|_{W^1_q(\Omega_0)} \leq \int_0^t \|\partial_t z(\cdot, s)\|_{W^1_q(\Omega_0)}\, ds \leq T^{1/p'}\|\partial_t z\|_{L_p((0,T),W^1_q(\Omega_0))} \leq C(M)E(T),$$

which implies (5.122). Finally, (5.123) and (5.124) follow from (5.117) and (5.118), respectively.

□

Now we can estimate the right hand side of (5.36) in the regularity required by Theorem 5.3.1:

Lemma 5.4.4. Let $\mathbf{F}_1(\eta, \mathbf{v}), G_1(\eta, \mathbf{v})$ be defined in (5.115) and (5.116). Assume that $\varrho_0, \mathbf{u}_0$ and $\mathbf{V}$ satisfy (5.19). Then

$$\|\mathbf{F}_1(\eta, \mathbf{v})\|_{L_p(0,T;L_q(\Omega_0))} + \|G_1(\eta, \mathbf{v})\|_{L_p(0,T;W^1_q(\Omega_0))} \leq E(T)(\|\eta, \mathbf{v}\|_{\mathcal{Y}(T)} + L). \tag{5.125}$$

Proof: The proof relies on the estimates collected in Lemma 5.4.3. By (5.122) and (5.109) we have

$$\begin{aligned} \|\eta\partial_t(\mathbf{v}+\mathbf{u}_{b1})\|_{L_p(0,T;L_q(\Omega_0))} &\leq \|\eta\|_{L_\infty(\Omega_0\times(0,T))}\left(\|\partial_t\mathbf{v}\|_{L_p(0,T;L_q(\Omega_0))}+\|\partial_t\mathbf{u}_{b1}\|_{L_p(0,T;L_q(\Omega_0))}\right) \\ &\leq E(T)[\|\eta, \mathbf{v}\|_{\mathcal{Y}(T)} + L]. \end{aligned} \tag{5.126}$$

In order to estimate the remaining terms notice that all the quantities (5.46)–(5.50) contain either $\mathbf{E}(\mathbf{k}_{\tilde{\mathbf{u}}})$ or $\nabla\mathbf{E}(\mathbf{k}_{\tilde{\mathbf{u}}})$ multiplied by the derivatives of $\tilde{\mathbf{u}}$ with respect to y of at most second order. Therefore (5.121) and (5.109) imply

$$\begin{aligned} \|A_{2\Delta}(\mathbf{k}_{\tilde{\mathbf{u}}})\nabla^2_y\tilde{\mathbf{u}},\ A_{1\Delta}(\mathbf{k}_{\tilde{\mathbf{u}}})\nabla_y\tilde{\mathbf{u}},\ A_{2\mathrm{div}\,,i}(\mathbf{k}_{\tilde{\mathbf{u}}})\nabla^2_y\tilde{\mathbf{u}},\ A_{1\mathrm{div}\,,i}(\mathbf{k}_{\tilde{\mathbf{u}}})\nabla_y\tilde{\mathbf{u}}\|_{L_p(0,T;L_q(\Omega_0))} \\ \leq E(T)[\|\eta, \mathbf{v}\|_{\mathcal{Y}(T)} + L]. \end{aligned} \tag{5.127}$$

Putting together all above estimates we get the estimate for $\mathbf{F}_1$. Next, (5.121) gives immediately

$$\|G(\tilde{\varrho}, \tilde{\mathbf{u}})\|_{L_p(0,T;W^1_q(\Omega_0))} \leq E(T)[\|\eta, \mathbf{v}\|_{\mathcal{Y}(T)} + L],$$

and thus (5.125) follows.

□

5.4.3 Fixed Point Argument

Let us define a solution operator

$$(\eta, \mathbf{v}) = S(\bar{\eta}, \bar{\mathbf{v}}) \iff (\eta, \mathbf{v}) \text{ solves (5.112)–(5.114) with right hand side } \mathbf{F}_1(\bar{\eta}, \bar{\mathbf{v}}), G_1(\bar{\eta}, \bar{\mathbf{v}}).$$

By Theorem 5.3.1 and Lemma 5.4.4, S is well defined on $\mathcal{Y}(T)$ and maps a ball $B(0, M) \subset \mathcal{Y}(T)$ into itself provided T is sufficiently small w.r.t. M and L. Denote

$$(\eta_i, \mathbf{v}_i) = S(\bar{\eta}_i, \bar{\mathbf{v}}_i), \quad i = 1, 2.$$

Then the difference $(\eta_1 - \eta_2, \mathbf{v}_1 - \mathbf{v}_2)$ satisfies

$$\varrho_0 \partial_t(\mathbf{v}_1 - \mathbf{v}_2) - \mu \Delta_y(\mathbf{v}_1 - \mathbf{v}_2) - \left(\frac{\mu}{3} + \zeta\right) \nabla_y \mathrm{div}_y(\mathbf{v}_1 - \mathbf{v}_2) + \gamma_1 \nabla_y(\eta_1 - \eta_2) = \mathbf{F}_1(\bar{\eta}_1, \bar{\mathbf{v}}_1) - \mathbf{F}_1(\bar{\eta}_2, \bar{\mathbf{v}}_2) \tag{5.128}$$

$$\partial_t(\eta_1 - \eta_2) + \varrho_0 \mathrm{div}_y(\mathbf{v}_1 - \mathbf{v}_2) = G_1(\bar{\eta}_1, \bar{\mathbf{v}}_1) - G_1(\bar{\eta}_2, \bar{\mathbf{v}}_2), \tag{5.129}$$

$$(\mathbf{v}_1 - \mathbf{v}_2)|_{t=0} = 0, \quad (\mathbf{v}_1 - \mathbf{v}_2)|_{\partial\Omega_0} = 0, \tag{5.130}$$

and we have

$$\begin{aligned}\mathbf{F}_1(\bar{\eta}_1, \bar{\mathbf{v}}_1) - \mathbf{F}_1(\bar{\eta}_2, \bar{\mathbf{v}}_2) &= \mathbf{R}(\bar{\eta}_1 + \varrho_0, \bar{\mathbf{v}}_1 + \mathbf{u}_{b1}) - \mathbf{R}(\bar{\eta}_2 + \varrho_0, \bar{\mathbf{v}}_2 + \mathbf{u}_{b1}) - (\bar{\eta}_1 - \bar{\eta}_2)\partial_t \mathbf{u}_{b1} - \bar{\eta}_1 \partial_t(\bar{\mathbf{v}}_1 - \bar{\mathbf{v}}_2) \\ &- \partial_t \bar{\mathbf{v}}_2(\bar{\eta}_1 - \bar{\eta}_2) + \pi'(\varrho_0)\nabla_y(\bar{\eta}_1 - \bar{\eta}_2) - \pi'(\bar{\eta}_1 + \varrho_0)\nabla_y(\bar{\eta}_1 - \bar{\eta}_2) - \nabla_y \bar{\eta}_2[\pi'(\bar{\eta}_1 + \varrho_0) - \pi'(\bar{\eta}_2 + \varrho_0)]\end{aligned} \tag{5.131}$$

and

$$\begin{aligned}G_1(\bar{\eta}_1, \bar{\mathbf{v}}_1) - G_1(\bar{\eta}_2, \bar{\mathbf{v}}_2) = &- (\bar{\eta}_1 - \bar{\eta}_2)\mathrm{div}_y \mathbf{u}_{b1} \\ &- \bar{\eta}_2 \mathrm{div}_y(\bar{\mathbf{v}}_1 - \bar{\mathbf{v}}_2) - (\bar{\eta}_1 - \bar{\eta}_2)\mathrm{div}_y \bar{\mathbf{v}}_1 - (\bar{\eta}_1 + \varrho_0) \\ \mathbf{E}^1 : \nabla_y(\bar{\mathbf{v}}_1 - \bar{\mathbf{v}}_2) &- \nabla_y(\bar{\mathbf{v}}_2 + \mathbf{u}_{b1}) : [(\bar{\eta}_1 + \varrho_0)(\mathbf{E}^1 - \mathbf{E}^2) + (\bar{\eta}_1 - \bar{\eta}_2)\mathbf{E}^2],\end{aligned} \tag{5.132}$$

where we have denoted

$$\mathbf{E}^1 = \mathbf{E}^0(\mathbf{k}_{\bar{\mathbf{v}}_1 + \mathbf{u}_{b1}}), \quad \mathbf{E}^2 = \mathbf{E}^0(\mathbf{k}_{\bar{\mathbf{v}}_2 + \mathbf{u}_{b1}}).$$

Since $\mathbf{E}^0(\cdot)$ is smooth, we have

$$|\mathbf{E}^1 - \mathbf{E}^2| \leq C|\mathbf{k}_{\bar{\mathbf{v}}_1 + \mathbf{u}_{b1}} - \mathbf{k}_{\bar{\mathbf{v}}_2 + \mathbf{u}_{b1}}| \leq C\mathbf{k}_{\bar{\mathbf{v}}_1 - \bar{\mathbf{v}}_2}.$$

Therefore, recalling the definition of $\mathbf{R}$ we obtain

$$\|\mathbf{R}(\bar{\eta}_1+\varrho_0,\bar{\mathbf{v}}_1+\mathbf{u}_{b1})-\mathbf{R}(\bar{\eta}_2+\varrho_0,\bar{\mathbf{v}}_2+\mathbf{u}_{b1})\|_{L_p(0,T;L_q(\Omega_0))} \le E(T)\|(\bar{\eta}_1-\bar{\eta}_2,\bar{\mathbf{v}}_1-\bar{\mathbf{v}}_2)\|_{\mathcal{Y}(T)}.$$

Estimating the remaining terms on the right hand side of (5.131) similarly as in the proof of Lemma 5.4.4 we obtain

$$\|\mathbf{F}_1(\bar{\eta}_1,\bar{\mathbf{v}}_1)-\mathbf{F}_1(\bar{\eta}_2,\bar{\mathbf{v}}_2)\|_{L_p(0,T;L_q(\Omega_0))} \le E(T)\|(\bar{\eta}_1-\bar{\eta}_2,\bar{\mathbf{v}}_1-\bar{\mathbf{v}}_2)\|_{\mathcal{Y}(T)}. \quad (5.133)$$

In a similar way we get

$$\|G_1(\bar{\eta}_1,\bar{\mathbf{v}}_1)-G_2(\bar{\eta}_2,\bar{\mathbf{v}}_2)\|_{W^1_p(0,T;L_q(\Omega_0))} \le E(T)\|(\bar{\eta}_1-\bar{\eta}_2,\bar{\mathbf{v}}_1-\bar{\mathbf{v}}_2)\|_{\mathcal{Y}(T)}. \quad (5.134)$$

Applying (5.133), (5.134) and Theorem 5.3.1 to system (5.128)–(5.130) we see that S is a contraction on $B(0,M)\subset\mathcal{Y}(T)$ for sufficiently small times. Therefore it has a unique fixed point $(\eta^*,\mathbf{v}^*)$. Now

$$\tilde{\varrho}=\eta^*+\varrho_0,\quad \tilde{\mathbf{u}}=\mathbf{v}^*+\mathbf{u}_{b1}$$

is a solution to (5.36)–(5.38) and

$$\|\tilde{\varrho},\tilde{\mathbf{u}}\|_{\mathcal{Y}(T)}\le CL.$$

It is quite standard to verify that after coming back to Eulerian coordinates we obtain a solution with the estimate (5.20); however, for the sake of completeness we justify it briefly in the next subsection.

5.4.4 Equivalence of Norms in Lagrangian and Eulerian Coordinates

By (5.124), the Jacobian of the transformation $\mathbf{X}_\mathbf{u}$ is bounded in space-time. Therefore, Lemma 5.2.2 implies the equivalence of $L_p(0,T;L_q)$ norms of a function and its first-order space derivatives. Furthermore, we have

$$\nabla^2_y\tilde{f}(t,y)=\nabla_x f(t,\mathbf{X}_\mathbf{u}(t,y))\nabla^2_y\mathbf{X}_\mathbf{u}+(\nabla_y\mathbf{X}_\mathbf{u})^2\nabla^2_x f(t,\mathbf{X}_\mathbf{u}(t,y)).$$

Again by (5.124), $\nabla_y\mathbf{X}_\mathbf{u}$ is bounded in space-time, which together with embedding $W^1_q(\Omega_t)\subset L_\infty(\Omega_t)$ for $t\in[0,T)$ gives equivalence of $L_p(L_q)$ norms of second space derivatives. However, we have a different situation for the time derivative. The solution constructed in Lagrangian coordinates satisfies

$$\tilde{\varrho}_t\in L_p(0,T;W^1_q(\Omega_0)).$$

However, due to (5.28) this does not imply the same regularity for the density in Eulerian coordinates. Nevertheless, the regularity of $\mathbf{u}$ implies

$$\varrho_t\in L_p(0,T;L_q(\Omega_t)),$$

which is the regularity in the assertion of Theorem 5.1.1.

5.5 Global Well-Posedness

5.5.1 Linearization

Again we first reduce the problem to homogeneous boundary condition.

Lemma 5.5.1. If $\mathbf{V}$ satisfies the assumptions of Theorem 5.1.2, then the problem

$$\varrho^*\partial_t\mathbf{u}_{b2} - \mu\Delta_y\mathbf{u}_{b2} - (\frac{\mu}{3}+\zeta)\nabla_y\mathrm{div}_y\mathbf{u}_{b2} = 0 \qquad \text{in } \Omega_0\times(0,T), \tag{5.135}$$

$$\mathbf{u}_{b2}|_{\Gamma_0} = \tilde{\mathbf{V}}, \quad \mathbf{u}_{b2}|_{t=0} = \mathbf{V}(0)$$

admits a unique global in time solutions $\mathbf{u}_{b2}$ with the decay estimate

$$\|e^{\gamma t}\partial_t\mathbf{u}_{b2}\|_{L_p(0,T;L_q(\Omega_0))} + \|e^{\gamma t}\nabla_y\mathbf{u}_{b2}\|_{L_p(0,T;W^1_q(\Omega_0))} + \|e^{\gamma t}(\mathbf{u}_{b2}-\tilde{\mathbf{V}}))\|_{L_p(0,T;L_q(\Omega_0))} \leq C\|e^{\gamma t}(\partial_t\tilde{\mathbf{V}},\nabla^2_y\tilde{\mathbf{V}})\|_{L_p(0,T;L_q(\Omega_0))}. \tag{5.136}$$

Proof. Let us define $\tilde{\mathbf{u}}_{b_2} = \mathbf{u}_{b2} - \tilde{\mathbf{V}}$. If $\mathbf{V}$ satisfies the assumptions of Theorem 5.1.2, then Theorem 5.3.2 implies

$$\|e^{\gamma t}\partial_t\tilde{\mathbf{u}}_{b2}\|_{L_p(0,T;L_q(\Omega_0))} + \|e^{\gamma t}\tilde{\mathbf{u}}_{b2}\|_{L_p(0,T;W^2_q(\Omega_0))} \leq C\|e^{\gamma t}(\partial_t\tilde{\mathbf{V}},\nabla^2_y\tilde{\mathbf{V}})\|_{L_p(0,T;L_q(\Omega_0))},$$

which gives (5.136).

□

This time we have to linearize the density around the constant ϱ^*. Denoting

$$\sigma = \tilde{\varrho} - \varrho^*, \qquad \mathbf{v} = \tilde{\mathbf{u}} - \mathbf{u}_{b2}$$

we obtain from (5.36)–(5.38)

$$\varrho^*\mathbf{v}_t - \mu\Delta_y\mathbf{v} - \left(\frac{\mu}{3}+\zeta\right)\nabla_y\mathrm{div}_y\mathbf{v} + \gamma_2\nabla_y\sigma = \mathbf{F}_2(\sigma,\mathbf{v}), \tag{5.137}$$

$$\sigma_t + \varrho^*\mathrm{div}_y\mathbf{v} = G_2(\sigma,\mathbf{v}), \tag{5.138}$$

$$\mathbf{v}|_{t=0} = \mathbf{u}_0 - \mathbf{V}(0), \quad \mathbf{v}|_{\partial\Omega_0} = 0, \tag{5.139}$$

where $\gamma_2 = \pi'(\varrho^*)$ and

$$\mathbf{F}_2(\sigma,\mathbf{v}) = \mathbf{F}(\sigma+\varrho^*,\mathbf{v}+\mathbf{u}_{b2}) - \sigma\partial_t(\mathbf{v}+\mathbf{u}_{b2}) - [\pi'(\sigma+\varrho^*) - \pi'(\varrho^*)]\nabla_y\sigma, \tag{5.140}$$

$$G_2(\sigma,\mathbf{v}) = G(\sigma+\varrho^*,\mathbf{v}+\mathbf{u}_{b2}) - (\sigma+\varrho^*)\mathrm{div}_y\mathbf{u}_{b2} - \sigma\mathrm{div}_y\mathbf{v}. \tag{5.141}$$

5.5.2 Nonlinear Estimates for the Global Well-Posedness

We start with an analog of Lemma 5.4.3 which will be used to estimate the nonlinearities for large times. We also recall the definitions (5.11) and (5.12) of norm and seminorm on $\mathcal{Y}(\infty)$.

Lemma 5.5.2. Let $e^{\gamma t}(z, \mathbf{w}) \in \mathcal{Y}(\infty)$ for some $\gamma > 0$ and

$$\|z(0) - \varrho^*\|_{W^1_p(\Omega_0)} + \|\mathbf{w}(0)\|_{B^{2-2/p}_{q,p}(\Omega_0)} \leq \varepsilon.$$

Then

$$\|\mathbf{E}^0(\mathbf{k_w}), \nabla_{\mathbf{k_w}}\mathbf{E}^0(\mathbf{k_w}), \|_{L_\infty((0,\infty)\times\Omega_0)} \leq C\|e^{\gamma t}(z, \mathbf{w})\|_{\dot{\mathcal{Y}}(\infty)}, \tag{5.142}$$

$$\sup_{t\in(0,\infty)}\|z(\cdot,t)\|_{W^1_q(\Omega_0)} \leq C[\epsilon + \|e^{\gamma t}(z, \mathbf{w})\|_{\dot{\mathcal{Y}}(\infty)}], \tag{5.143}$$

$$\sup_{t\in(0,\infty)}\|\mathbf{w}(\cdot,t) - \mathbf{w}(0)\|_{B^{2(1-1/p)}_{q,p}} \leq C\|e^{\gamma t}(z, \mathbf{w})\|_{\dot{\mathcal{Y}}(\infty)}, \tag{5.144}$$

$$\|\mathbf{w}\|_{L_\infty(0,\infty,W^1_\infty(\Omega_0))} \leq C\|e^{\gamma t}(z, \mathbf{w})\|_{\dot{\mathcal{Y}}(\infty)}, \tag{5.145}$$

where $\mathbf{k_v}$ is defined in (5.33).

Proof: We have

$$\int_0^\infty \|\nabla_y \mathbf{w}\|_\infty \mathrm{d}t \leq \left(\int_0^\infty e^{-\gamma t p'}\mathrm{d}t\right)^{1/p'} \left(\int_0^\infty e^{\gamma t p}\|\mathbf{w}\|_{W^2_q(\Omega_0)}\mathrm{d}t\right)^{1/p},$$

which implies (5.142). Next,

$$\|\sigma(\cdot,t)\|_{L_\infty(\Omega_0)} \leq \|z(0) - \varrho^*\|_{L_\infty(\Omega_0)} + \int_0^t \|z_t(s,\cdot)\|_{L_\infty(\Omega_0)}\mathrm{d}t$$

$$\leq \epsilon + C\left(\int_0^t e^{-\gamma s p'}\mathrm{d}s\right)^{1/p'} \left(\int_0^\infty e^{\gamma s p}\|z_t\|_{W^1_q(\Omega_0)}\mathrm{d}s\right)^{1/p},$$

which yields (5.143). Finally, (5.144) and (5.145) follows from Lemma 5.4.2.

□

The following lemma gives estimates for the right hand sides of (5.137)–(5.138).

Lemma 5.5.3. Let $\mathbf{F}_2(\tilde{\varrho}, \tilde{\mathbf{u}}), G_2(\tilde{\varrho}, \tilde{\mathbf{u}})$ be defined in (5.140) and (5.141). Assume that $\varrho_0, \mathbf{u}_0$ and $\mathbf{V}$ satisfy (5.21). Then

$$\|\mathbf{F}_2(\tilde{\varrho}, \tilde{\mathbf{u}})\|_{L_p(0,\infty;L_q(\Omega_0))} + \|G_2(\tilde{\varrho}, \tilde{\mathbf{u}})\|_{L_p(0,\infty;W^1_q(\Omega_0))} \leq C(\|e^{\gamma t}(\sigma, \mathbf{v})\|^2_{\dot{\mathcal{Y}}(\infty)} + \epsilon). \tag{5.146}$$

Proof. First, analogously to (5.127), this time using (5.142) and (5.136) we obtain

$$\begin{aligned}
&\|e^{\gamma t}A_{2\Delta}(\mathbf{k}_{\tilde{\mathbf{u}}})\nabla^2_y\tilde{\mathbf{u}},\ A_{1\Delta}(\mathbf{k}_{\tilde{\mathbf{u}}})\nabla_y\tilde{\mathbf{u}},\ A_{2\mathrm{div}\,,i}(\mathbf{k}_{\tilde{\mathbf{u}}})\nabla^2_y\tilde{\mathbf{u}},\ A_{1\mathrm{div}\,,i}(\mathbf{k}_{\tilde{\mathbf{u}}})\nabla_y\tilde{\mathbf{u}}\|_{L_p(0,\infty;L_q(\Omega_0))}\\
&\leq C\|e^{\gamma t}(\sigma, \mathbf{v})\|_{\dot{\mathcal{Y}}(\infty)}\|e^{\gamma t}(\mathbf{v} + \mathbf{u}_{b2})\|_{L_p(0,\infty;L_q(\Omega_0))}\\
&\leq C\|e^{\gamma t}(\sigma, \mathbf{v})\|_{\dot{\mathcal{Y}}(\infty)}[\|e^{\gamma t}(\sigma, \mathbf{v})\|_{\dot{\mathcal{Y}}(\infty)} + \|e^{\gamma t}(\partial_t\tilde{\mathbf{V}}, \nabla^2_y\tilde{\mathbf{V}})\|_{L_p(0,\infty;L_q(\Omega_0))}]\\
&\leq C\|e^{\gamma t}(\sigma, \mathbf{v})\|_{\dot{\mathcal{Y}}(\infty)}[\|e^{\gamma t}(\sigma, \mathbf{v})\|_{\dot{\mathcal{Y}}(\infty)} + \epsilon].
\end{aligned} \tag{5.147}$$

Next, by (5.143) and (5.136)

$$\|e^{\gamma t}\partial_t(\mathbf{v}+\mathbf{u}_{b2})\|_{L_p(0,\infty;L_q(\Omega_0))} \le \|\sigma\|_{L_\infty((0,\infty)\times\Omega_0)}\|e^{\gamma t}\partial_t(\mathbf{v}+\mathbf{u}_{b2})\|_{L_p(0,\infty;L_q(\Omega_0))} \\ \le C[\varepsilon + \|e^{\gamma t}(\sigma,\mathbf{v})\|_{\dot{\mathcal{Y}}(\infty)}]^2,$$

and

$$\|e^{\gamma t}[\pi'(\tilde{\varrho})-\pi'(\varrho^*)]\nabla_y\sigma\|_{L_p(0,\infty;L_q(\Omega_0))} \le \|\sigma\|_{L_\infty((0,\infty)\times\Omega_0)}\|e^{\gamma t}\nabla_y\sigma\|_{L_p(0,\infty;L_q(\Omega_0))} \\ \le C[\varepsilon + \|e^{\gamma t}(\sigma,\mathbf{v})\|_{\dot{\mathcal{Y}}(\infty)}]^2.$$

Combining all above estimates we get the required estimate for $\|\mathbf{F}_2\|_{L_p(0,\infty;L_q(\Omega_0))}$. Finally, G_2 and its space derivatives are estimated in a similar way using Lemma 5.5.2 and (5.136).

□

5.5.3 Proof of Theorem 5.1.2

It is now easy to verify the following estimate which allows to prolong the local solution for arbitrarily large times.

Lemma 5.5.4. Assume $\sigma,\mathbf{v}$ is solution to (5.137)–(5.139) with $\varrho_0,\mathbf{u}_0$, and $\mathbf{V}$ satisfying the assumptions of Theorem 5.1.2. Then

$$\|e^{\gamma t}(\sigma,\mathbf{v})\|_{\dot{\mathcal{Y}}(\infty)} \le E(\epsilon). \tag{5.148}$$

Proof. Combining Proposition 5.3.2 and Lemma 5.5.3 we obtain

$$\|e^{\gamma t}(\sigma,\mathbf{v})\|_{\dot{\mathcal{Y}}(T)} \le C[\epsilon + \|e^{\gamma t}(\sigma,\mathbf{v})\|^2_{\dot{\mathcal{Y}}(T)}]. \tag{5.149}$$

Note that we derived this inequality for $T=\infty$; however, it is easy to observe that the same arguments yield (5.149) for any $T>0$. Consider the equation

$$x^2 - \frac{x}{C} + \epsilon = 0.$$

Its roots are

$$x_1(\epsilon) = \frac{1}{2C} - \sqrt{\frac{1}{4C^2}-\epsilon}, \quad x_2(\epsilon) = \frac{1}{2C} + \sqrt{\frac{1}{4C^2}-\epsilon}.$$

Notice that the inequality (5.149) implies either $\|e^{\gamma t}(\sigma,\mathbf{v})\|_{\dot{\mathcal{Y}}(T)} \le x_1(\epsilon)$ or $\|e^{\gamma t}(\sigma,\mathbf{v})\|_{\dot{\mathcal{Y}}(T)} \ge x_2(\epsilon)$. However,

$$\|e^{\gamma t}(\sigma,\mathbf{v})\|_{\dot{\mathcal{Y}}(T)} \to 0$$

as $T\to 0$, therefore

$$\|e^{\gamma t}(\sigma,\mathbf{v})\|_{\dot{\mathcal{Y}}(T)} \le x_1(\epsilon)$$

for T small. Finally, $\|e^{\gamma t}(\sigma, \mathbf{v})\|_{\dot{\mathcal{Y}}(T)}$ is continuous in time and therefore

$$\|e^{\gamma t}(\sigma, \mathbf{v})\|_{\dot{\mathcal{Y}}(\infty)} \leq x_1(\epsilon).$$

□

Now it is a standard matter to prolong the local solution for arbitrarily large times. For this purpose it is enough to observe that if the initial data satisfies the smallness assumption from Theorem 5.1.2, then the time of existence from Theorem 5.1.1 satisfies $T > C(\epsilon) > 0$. Therefore, for arbitrarily large T^* we can obtain a solution on $(0, T^*)$ in a finite number of steps. By the estimate (5.148) this solution satisfies (5.22)–(5.23).

Finally, the equivalence of norms can be justified as in Sect. 5.4.4, using (5.145) instead of (5.124).

Acknowledgements

The works of O. Kreml, Š. Nečasová, T.Piasecki were supported by the Czech Science Foundation grant GA19-04243S in the framework of RVO 67985840. The work of T. Piasecki was partially supported by Polish National Science Centre grant 2018/29/B/ST1/00339.

References

1. R. A. Adams, J. F. Fournier, *Sobolev Spaces*, Second edition. Pure and Applied Mathematics (Amsterdam), 140. Elsevier/Academic Press, Amsterdam, (2003).
2. S. Doboszczak. Relative entropy and a weak-strong uniqueness principle for the compressible Navier-Stokes equations on moving domains *Appl. Math. Lett.*, **57**:60–68, 2016
3. Y. Enomoto, Y.Shibata On the R-sectoriality and the Initial Boundary Value Problem for the Viscous Compressible Fluid Flow Funkcialaj Ekvacioj, **56** (2013), 441–505
4. E. Feireisl. Compressible Navier-Stokes equations with a non - monotone pressure law. *J. Differential Equations*, **184**:97–108, 2002.
5. E. Feireisl. *Dynamics of viscous compressible fluids.* Oxford University Press, Oxford, 2004.
6. E. Feireisl. On the motion of a viscous, compressible, and heat conducting fluid. *Indiana Univ. Math. J.*, **53**:1707–1740, 2004.
7. R. Farwig, H. Sohr. Generalized resolvent estimates for the Stokes system in bounded and unbounded domains. J. Math. Soc. Japan, 46 (4), 607–643, 1994.

8. E. Feireisl, B.J. Jin, and A. Novotný. Relative entropies, suitable weak solutions, and weak-strong uniqueness for the compressible Navier-Stokes system. *J. Math. Fluid Mech.* **14**(4):717–730, 2012.

9. E. Feireisl, J. Neustupa, and J. Stebel. Convergence of a Brinkman-type penalization for compressible fluid flows. *J. Differential Equations*, **250**(1):596–606, 2011.

10. E. Feireisl, O. Kreml, Š. Nečasová, J. Neustupa, J. Stebel. Weak solutions to the barotropic Navier-Stokes system with slip boundary conditions in time dependent domains. *J. Differential Equations* **254**(1):125–140, 2013.

11. E. Feireisl, A. Novotný, and H. Petzeltová. On the existence of globally defined weak solutions to the Navier-Stokes equations of compressible isentropic fluids. *J. Math. Fluid Mech.*, **3**:358–392, 2001.

12. E. Feireisl, A. Novotný and Y. Sun. Suitable weak solutions to the Navier-Stokes equations of compressible viscous fluids. *Indiana Univ. Math. J.*, **60**(2):611–631, 2011.

13. D. Hoff. Local solutions of a compressible flow problem with Navier boundary conditions in general three-dimensional domains. *SIAM J. Math. Anal.*, **44**(2):633–650, 2012.

14. T. Hytönen, J. van Neerven, M. Veraar, L. Weis. Analysis in Banach spaces. Vol. I. Martingales and Littlewood-Paley theory. A Series of Modern Surveys in Mathematics, vol. 63. Springer, Cham, 2016.

15. T. Kato. Perturbation Theory for Linear Operators. Springer-Verlag Berlin Heidelberg, 1995.

16. O. Kreml, V. Mácha, Š. Nečasová, A. Wróblewska-Kamińska. Weak solutions to the full Navier-Stokes-Fourier system with slip boundary conditions in time dependent domain *Journal de Mathématiques Pures et Appliquées*, **9**, 109 : 67–92, 2018.

17. O. Kreml, V. Mácha, Š. Nečasová, A. Wróblewska-Kamińska. Flow of heat conducting fluid in a time dependent domain. *Z. Angew. Math. Phys.*, **69**, 5, Art. 119, 2018.

18. O. Kreml, Š. Nečasová, T. Piasecki. Local existence of strong solutions and weak-strong uniqueness for the compressible Navier-Stokes system on moving domains Proc. Roy. Soc. Edinburgh Sect. A, **150**(5):2255–2300, 2020. Edinburgh A, in press, https://doi.org/10.1017/prm.2018.165.

19. P.-L. Lions. *Mathematical topics in fluid dynamics, Vol.2, Compressible models.* Oxford Science Publication, Oxford, 1998.

20. M. Murata On a maximal $L_p - L_q$ approach to the compressible viscous fluid flow with slip boundary condition Nonlinear Analysis 106 (2014), 86–109

21. A. Matsumura, T. Nishida. The initial value problem for the equations of motion of viscous and heat-conductive gases. *J. Kyoto Univ.*, **20**:67–104, 1980.

22. A. Matsumura, T. Nishida. Initial-boundary value problems for the equations of motion of compressible viscous and heat-conductive fluids. *Comm. Math. Phys.*, **89**:445–464, 1983.

23. P.B. Mucha, W.M. Zajączkowski On local-in-time existence for the Dirichlet problem for equations of compressible viscous fluids. Ann. Polon. Math. 78 (2002), no. 3, 227–239.

24. P.B. Mucha, W.M. Zajączkowski On a Lp-estimate for the linearized compressible Navier-Stokes equations with the Dirichlet boundary conditions. J. Differential Equations 186 (2002), no. 2, 377–393.

25. P.B. Mucha, W.M. Zajączkowski Global existence of solutions of the Dirichlet problem for the compressible Navier-Stokes equations. ZAMM Z. Angew. Math. Mech. 84 (2004), no. 6, 417–424.

26. T. Piasecki, Y. Shibata, E. Zatorska On strong dynamics of compressible two-component mixture flow SIAM J. Math. Anal. 51 (2019), no. 4, 2793–2849

27. T. Piasecki, Y. Shibata, E. Zatorska. On the isothermal compressible multi-component mixture flow: the local existence and maximal $L_p - L_q$ regularity of solutions. Nonlinear Analysis 189 (2019) 111571

28. Y. Shibata. On the global well-posedness of some free boundary problem for a compressible barotropic viscous fluid flow. Recent advances in partial differential equations and applications, 341–356, Contemp. Math., 666, Amer. Math. Soc., Providence, RI, 2016

29. Y. Shibata, M. Murata On the global well-posedness for the compressible Navier-Stokes equations with slip boundary condition. J. Differential Equations 260 (2016),

30. Y. Shibata, K. Tanaka. On a resolvent problem for the linearized system from the dynamical system describing the compressible viscous fluid motion. Math. Mech. Appl. Sci. **27**, 1579–1606, 2004.

31. H. Tanabe. *Functional analytic methods for partial differential equations.* Monographs and textbooks in pure and applied mathematics, Vol 204, Marchel Dekker, Inc. New York, Basel, 1997.

32. A. Valli. An existence theorem for compressible viscous fluids. *Ann. Math. Pura Appl. (IV)*, **130**:197–213, 1982. *Ann. Math. Pura Appl. (IV)*, **132**:399–400, 1982.

33. A. Valli. Periodic and stationary solutions for compressible Navier-Stokes equations via a stability method. *Ann. Sc. Norm. Super. Pisa (IV)* **10**:607–647, 1983.

34. L. Weis. Operator-valued Fourier multiplier theorems and maximal L_p-regularity. *Math. Ann. 319, 735–758, 2001.*

35. W.M. Zajączkowski. On nonstationary motion of a compressible barotropic viscous fluid with boundary slip condition. *J. Appl. Anal.* **4**(2):167–204, 1998.

36. W.M. Zajączkowski. On nonstationary motion of a compressible barotropic viscous capillary fluid bounded by a free surface. SIAM J. Math. Anal. 25 no. 1, 1–84, 1994.

37. W.M. Zajączkowski. On nonstationary motion of a compressible barotropic viscous fluid bounded by a free surface. Dissertationes Math. (Rozprawy Mat.) 324 (1993)

Chapter 6

Some New Properties of a Suitable Weak Solution to the Navier–Stokes Equations

Francesca Crispo and Paolo Maremonti (✉)
Università degli Studi della Campania "L. Vanvitelli", Dipartimento di Matematica e Fisica, Caserta, Italy
e-mail: francesca.crispo@unicampania.it; paolo.maremonti@unicampania.it

Carlo Romano Grisanti
Università di Pisa, Dipartimento di Matematica, Pisa, Italy
e-mail: carlo.romano.grisanti@unipi.it

The paper is concerned with the IBVP of the Navier–Stokes equations. The goal is the construction of a weak solution enjoying some new properties. Of course, we look for properties that are global in time. The results hold assuming an initial data $v_0 \in J^2(\Omega)$.

In Memoria di Christian Simader

6.1 Introduction

This note concerns the 3D-Navier–Stokes initial boundary value problem:

$$\begin{aligned} &v_t + v\cdot\nabla v + \nabla\pi_v = \Delta v,\ \nabla\cdot v = 0,\ \text{in } (0,T)\times\Omega,\\ &v = 0 \text{ on } (0,T)\times\partial\Omega,\ v(0,x) = v_0(x) \text{ on } \{0\}\times\Omega. \end{aligned} \tag{6.1}$$

T. Bodnár et al. (eds.), *Waves in Flows*, Advances in Mathematical Fluid Mechanics, https://doi.org/10.1007/978-3-030-68144-9_6

In system (6.1) $\Omega \subseteq \mathbb{R}^3$ is assumed bounded or exterior, and its boundary is smooth. The symbol v denotes the kinetic field, π_v is the pressure field, $v_t := \frac{\partial}{\partial t}v$, and $v \cdot \nabla v := v_k \frac{\partial}{\partial x_k} v$. In several papers, related to the Navier–Stokes initial boundary value problem, the authors give results concerning the partial regularity of a suitable weak solution (see Definition 6.1.2 below). This is made in order to highlight the properties of a weak solution, corresponding to data $v_0 \in L^2(\Omega)$, divergence free, that can be suitable to state the well-posedness of the equations, see, e.g., [4, 5, 7–10, 18, 19, 21, 33].[1] We believe that, in connection with the non-well-posedness of the Navier–Stokes Cauchy or IBVP problem, this kind of investigation achieves a further interest. Actually, in the recent paper [3], it is considered the possibility of non-uniqueness of a weak solution to the Navier–Stokes equations. This is proved for *very weak solutions*, that is, solutions satisfying a variational formulation of the Navier–Stokes equations and simply belonging to $C([0,T); L^2(\Omega))$. As a consequence of the weakness of the solutions, the result of non-uniqueness fails to hold for regular solutions, but a priori it also does not work for a suitable weak solution, that is, a solution verifying an energy inequality. So that, in order to better delimit the validity of a possible counterexample to the uniqueness in the set of weak solutions corresponding to an initial data in $L^2(\Omega)$, it seems of a certain interest to support the energy inequality, or its variants, by means of a wide set of global properties of the weak solutions not necessarily only consequences of the energy inequality, but of the coupling of other a priori estimates.

The aim of this note is to prove some new properties of a weak solution. We investigate two questions. One is related to a sort of energy equality for a suitable weak solution. It is easy to understand that the possible validity of the energy equality achieves a *mechanical* interest that goes beyond the above question concerning the well-posedness. Actually, we construct a weak solution (v, π_v) to the Navier–Stokes initial boundary value problem such that the "energy equalities" of the kind

$$\|v(t)\|_2^2 + 2\int_s^t \|\nabla v(\tau)\|_2^2 d\tau - \|v(s)\|_2^2 = -H(t,s), \text{a.e. in } t \geq s > 0 \text{ and for } s = 0 \tag{6.2}$$

and

$$2\int_s^t \|\nabla v(\tau)\|_2^2 d\tau = F(t,s)\big(\|v(s)\|_2^2 - \|v(t)\|_2^2\big), \text{a.e. in } t \geq s > 0 \text{ and for } s = 0 \tag{6.3}$$

are fulfilled. The functions $H(t,s)$ and $F(t,s)$ have suitable expressions, see formula (6.6) and formula (6.7). If $H(t,s) \leq 0$, then the energy equality holds (that

[1] There is a wide literature concerning extension to the IBVP of results proved for the Cauchy problem. One of the most interesting of these kinds of extensions is sure the energy inequality in strong form, see, e.g., [14, 27].

is, a fortiori $H(t,s)=0$). If $F(t,s)\geq 1$, then the energy equality holds (that is, a fortiori $F(t,s)=1$). These results are a consequence of the fact that we are able to prove that an approximating sequence $\{(v^m,\pi_{v^m})\}$ is strongly converging in $L^r(0,T;W^{1,2}(\Omega))$ for all $r\in[1,2)$. The strong convergence, in turn, is a consequence of the property: $P\Delta v^m\in L^{\frac{2}{3}}(0,T;L^2(\Omega))$ for all $m\in\mathbb{N}$ and $T>0$. Unfortunately we are not able to put $r=2$, that should give the energy equality. For 2D-Navier–Stokes equations, one proves that $H(t,s)=0$. It is important to stress that the term $H(t,s)$ is equal to zero in 2D-case, thanks to our approximating approach, without appealing to the regularity of the limit. Another result proves that $v\in L^{\mu(p)}(0,T;L^p(\Omega))$, with $\mu(p):=\frac{p}{p-2}$ and $p\in(6,\infty]$. This result is not new in the literature. A first contribution in this sense is proved in [12] for a particular geometry and it is reconsidered in [10]. The proof given in [10] for exterior domains is not completely clear to the present authors. However our proof is alternative with respect to the ones of the quoted papers.

In order to better state our result we recall the following definitions. We denote by $J^2(\Omega)$ and $J^{1,2}(\Omega)$ the completion of $\mathcal{C}_0(\Omega)$ in $L^2(\Omega)$ and in $W^{1,2}(\Omega)$, respectively, where $\mathcal{C}_0(\Omega)$ is the set of smooth divergence-free functions. Moreover $(\cdot,\cdot)$ represents the scalar product in $L^2(\Omega)$.

Definition 6.1.1. *Let $v_0\in J^2(\Omega)$. A pair (v,π_v), such that $v:(0,\infty)\times\Omega\to\mathbb{R}^3$ and $\pi_v:(0,\infty)\times\Omega\to\mathbb{R}$, is said to be a weak solution to problem* (6.1) *if*

(i) for all $T>0$, $v\in L^2(0,T;J^{1,2}(\Omega))$, for suitable q,r, $\pi_v\in L^r_{\ell oc}([0,T);L^q_{\ell oc}(\overline{\Omega}))$ and, for all $\psi\in J^2(\Omega)$, $(v(t),\psi)\in C((0,T))$,

(ii) $\lim\limits_{t\to0}(v(t),\psi)=(v_0,\psi)$, for all $\psi\in J^2(\Omega)$,

(iii) for all $t,s\in(0,T)$, the pair (v,π_v) satisfies the equation

$$\int_s^t\Big[(v,\varphi_\tau)-(\nabla v,\nabla\varphi)+(v\cdot\nabla\varphi,v)+(\pi_v,\nabla\cdot\varphi)\Big]d\tau+(v(s),\varphi(s))=(v(t),\varphi(t)),$$

$$\text{for all } \varphi\in C^1_0([0,T)\times\Omega).$$

In order to investigate on the regularity, in [4] and with slight differences in [30] an energy inequality having a local character is introduced. [10pt]

Definition 6.1.2. *A pair (v,π_v) is said to be a suitable weak solution if it is a weak solution in the sense of the Definition 6.1.1 and, moreover,*

$$\begin{aligned}\int_\Omega|v(t)|^2\phi(t)dx+2\int_s^t\int_\Omega|\nabla v|^2\phi\,dxd\tau&\leq\int_\Omega|v(s)|^2\phi(s)dx\\&+\int_s^t\int_\Omega|v|^2(\phi_\tau+\Delta\phi)dxd\tau+\int_s^t\int_\Omega(|v|^2+2\pi_v)v\cdot\nabla\phi dxd\tau,\end{aligned}\tag{6.4}$$

for all $t > s$, for $s = 0$ and a.e. in $s \geq 0$, and for all nonnegative $\phi \in C_0^\infty(\mathbb{R} \times \overline{\Omega})$. We denote by $\Sigma \subseteq [0, \infty)$ the set of the instants s for which inequality (6.4) holds.

Thanks to the properties of the pressure field (in this connection see also Remark 6.1.1) furnished by the existence theorem, from inequality (6.4) one deduces the classical one

$$\|v(t)\|_2^2 + 2 \int_s^t \|\nabla v(\tau)\|_2^2 d\tau \leq \|v(s)\|_2^2, \text{ for all } t > s \text{ and } s \in \Sigma. \tag{6.5}$$

We are going to prove the following result.

Theorem 6.1.1. *For all $v_0 \in J^2(\Omega)$, there exists a suitable weak solution (v, π_v) to problem (6.1) that is the weak limit in $L^2(0,T; J^{1,2}(\Omega)) \times L^q_{\ell oc}([0,T); L^2_{\ell oc}(\overline{\Omega}))$, $q \in (1, \frac{12}{11})$, of a sequence $\{(v^m, \pi_{v^m})\}$ of solutions to (6.16). The sequence $\{v^m\}$ converges strongly to v in $L^p(0,T; W^{1,2}(\Omega))$ for all $p \in [1,2)$. Further, for any $q \in (6, \infty]$, $v \in L^{\mu(q)}(0,T; L^q(\Omega))$ with $\mu(q) := \frac{q}{q-2}$, and v satisfies relation (6.2) with*

$$H(t,s) := \begin{cases} \lim\limits_{\alpha \to 0} \lim\limits_{m \to \infty} \alpha \displaystyle\int_s^t \frac{\|v^m(\tau)\|_2^2}{(K + \|\nabla v^m(\tau)\|_2^2)^{\alpha+1}} \frac{d}{d\tau} \|\nabla v^m(\tau)\|_2^2 \, d\tau, & \text{for } s > 0, \\ \lim\limits_{s \to 0} \lim\limits_{\alpha \to 0} \lim\limits_{m \to \infty} \alpha \displaystyle\int_s^t \frac{\|v^m(\tau)\|_2^2}{(K + \|\nabla v^m(\tau)\|_2^2)^{\alpha+1}} \frac{d}{d\tau} \|\nabla v^m(\tau)\|_2^2 \, d\tau, & \text{for } s = 0, \end{cases} \tag{6.6}$$

for any arbitrary constant $K > 0$, as well as v satisfies relation (6.3) with

$$F(t,s) := \lim_{\alpha \to 0} \lim_{m \to \infty} \frac{1}{\left(K_1 + \|\nabla v^m(t_{\alpha,m})\|_2\right)^\alpha} \tag{6.7}$$

for any arbitrary constant $K_1 \geq 0$. Finally, the following inclusion holds: $\mathcal{G}_1 := \{t, s$ such that (6.2) is true$\} \subseteq \mathcal{G}_2 := \{t, s$ such that (6.3) is true$\}$.

Remark 6.1.1. Since the proprieties of the pressure field π_v are not our main interest in this paper, we limit ourselves to point out the one that allows us to state that (v, π_v) is a suitable weak solution. Actually, in our construction the pressure π_v enjoys the properties that one can deduce by means of Lemma 6.4.3. For more exhaustive properties relative to the pressure field of a suitable weak solution (that is, with an initial data only in $J^2(\Omega)$), a possible reference is [26].

We note that the quantity $H(t,s)$ is independent of the constant K. This fact is intriguing and somehow leads to conjecture that $H(t,s) = 0$.

If $v_0 \in J^2(\Omega) \setminus J^{1,2}(\Omega)$, almost everywhere in $t > 0$, following the proof idea we also get

$$\|v(t)\|_2^2 + 2 \int_0^t \|\nabla v(\tau)\|_2^2 d\tau = - \lim_{\alpha \to 0} \lim_{m \to \infty} \int_0^t \frac{\|v^m(\tau)\|_2^2}{(K + \|\nabla v^m(\tau)\|_2^2)^{\alpha+1}} \frac{d}{d\tau} \|\nabla v^m(\tau)\|_2^2 \, d\tau.$$

One proves that there exists an instant $\theta > 0$ such that $v^m(t,x) \in C([\theta,\infty); J^{1,2})$; in particular, there exists a M such that $\|\nabla v^m(t)\|_2 \leq M$ for all $t \geq \theta$ and $m \in \mathbb{N}$ (one proves this result by repeating the arguments employed for the structure theorem by Leray). Hence, via estimate (6.21) and taking into account the energy inequality, we can deduce

$$\text{for all } t > \theta,\ \lim_{\alpha\to 0}\lim_{m\to\infty} \alpha \int\limits_{\theta}^{t} \frac{\|v^m(\tau)\|_2^2}{(K + \|\nabla v^m(\tau)\|_2^2)^{\alpha+1}} \frac{d}{d\tau}\|\nabla v^m(\tau)\|_2^2\, d\tau = 0\,.$$

Hence we get that function $H(t,s) = H(\theta,s)$ for all $t > \theta$, and H becomes a constant function for $t > \theta$.

Concerning the function F, we remark that its values are independent of $K_1 \geq 0$. The fact that K_1 can be chosen equal to zero makes a difference with K in function H, as well as $\mathcal{G}_1 \subseteq \mathcal{G}_2$ is another difference.

In the introduction we remarked that if $F(t,s) \geq 1$, then the energy equality holds; that is, $F(t,s) = 1$. Since we are not in a position to prove $F(t,s) \geq 1$, a priori we have to consider that $F(t,s) \leq 1$. However, we can claim that almost everywhere in $t > 0$, $F(t,0) > 0$ holds. Actually, more in general, assume that $s \in \Sigma$ and $\|v(s)\|_2 \neq 0$ and that exists a sequence $\{t_p\}$ converging to s such that $F(t_p,s) = 0$. Then, from formula (6.3), we deduce that $\|v(\tau)\|_2 = 0$ a.e. in $\tau \in (s,t_p)$ holds for all $p \in \mathbb{N}$. Hence we can select a new sequence $\{t'_p\} \subset (s,t_p)$ such that $\|v(t'_p)\|_2 = 0$. By virtue of the right-L^2-continuity in s, we get $\lim\limits_{t'_p\to s}\|v(t'_p) - v(s)\|_2 = 0$, which is a contradiction with $\|v(s)\|_2 \neq 0$.

From formulas (6.2)–(6.3), a.e. in $t \in \mathcal{G}_1$, and for $s = 0$, we easily deduce that

$$H(t,0) = (1 - F(t,0))\big(\|v(0)\|_2^2 - \|v(t)\|_2^2\big)\,.$$

Therefore, via (6.2) we deduce

$$H(t,0) = \big(\tfrac{1}{F(t,0)} - 1\big)\int\limits_0^t \|\nabla v(\tau)\|_2^2 d\tau\,. \tag{6.8}$$

Recalling that, for $t \geq \theta$, $H(t,0) = H(\theta,0)$, then from (6.8) we deduce that $F(t,0)$ is a continuous function for $t \geq \theta$.

Remark 6.1.2. In paper [28] a new energy inequality is proposed

$$\|v(t)\|_2^2 + N(t) + 2\int\limits_0^t \|\nabla v(\tau)\|_2^2 d\tau \leq \|v_0\|_2^2, \text{ for all } t > 0\,.$$

Function $N(t) := \limsup\limits_{\delta\to 0}\int\limits_{\delta}^{t}\left\|\dfrac{u(\tau)-u(\tau-\delta)}{\delta^{\frac12}}\right\|_2^2 d\tau \geq 0$ can be interpreted as $\frac12$-time derivative. It is not known if $N(t) > 0$ holds. In the two-dimensional case one proves that $N(t) = 0$. Of course, we are not able to compare the solution furnished in [28] and the one of Theorem 6.1.1.

In paper [21] the compatibility between an energy equality and an initial data $v_0 \in J^2(\Omega)$ is proved. This supports the idea that $H(t,s)$ can be equal to zero.

Remark 6.1.3. We point out that by a proof completely similar to the one of Theorem 6.1.1, one can prove the validity of the following generalized energy equality:

$$\int_\Omega |v(t)|^2\phi(t)dx + 2\int_s^t\int_\Omega |\nabla v|^2\phi\, dxd\tau = \int_\Omega |v(s)|^2\phi(s)dx + \int_s^t\int_\Omega |v|^2(\phi_\tau + \Delta\phi)dxd\tau + \int_s^t\int_\Omega (|v|^2 + 2\pi_v)v\cdot\nabla\phi dxd\tau - \widetilde{H}(t,s), \tag{6.9}$$

a.e. in $t \geq s > 0$ and for $s = 0$, where

$$\widetilde{H}(t,s) := \begin{cases} \lim\limits_{\alpha\to 0}\lim\limits_{m\to\infty} \alpha\displaystyle\int_s^t \frac{\|\phi^{\frac12}(\tau)v^m(\tau)\|_2^2}{(K + \|\nabla v^m(\tau)\|_2^2)^{\alpha+1}}\frac{d}{d\tau}\|\nabla v^m(\tau)\|_2^2\, d\tau, \text{ for } s > 0,\\ \lim\limits_{s\to 0}\lim\limits_{\alpha\to 0}\lim\limits_{m\to\infty} \alpha\displaystyle\int_s^t \frac{\|\phi^{\frac12}(\tau)v^m(\tau)\|_2^2}{(K + \|\nabla v^m(\tau)\|_2^2)^{\alpha+1}}\frac{d}{d\tau}\|\nabla v^m(\tau)\|_2^2\, d\tau, \text{ for } s = 0, \end{cases}$$

for any arbitrary constant $K > 0$ and for all nonnegative $\phi \in C_0^\infty(\mathbb{R}\times\overline{\Omega})$.

The plan of the paper is the following. In Sect. 6.2 we give some preliminaries and auxiliary lemmas. In Sect. 6.3 we give the proof of the theorem. In Appendix we recall some known properties of the pressure field that are employed in Sect. 6.2.

6.2 Some Preliminary Results

For $p \in (1,\infty)$, we set $J^p(\Omega)$:=completion of $\mathcal{C}_0(\Omega)$ in $L^p(\Omega)$. By P_p, we denote the projector from $L^p(\Omega)$ onto $J^p(\Omega)$. In the case of $p = 2$ we write $P_2 \equiv P$. For any $R > 0$, we set $B_R = \{x \in \mathbb{R}^3 : |x| < R\}$.

We start with the following a priori estimate:

Lemma 6.2.1. *Let $\Omega \subseteq \mathbb{R}^n$, and let $u \in W^{2,2}(\Omega)\cap J^{1,2}(\Omega)$. Then there exists a constant c independent of u such that*

$$\|u\|_r \leq c\|P\Delta u\|_2^a\|u\|_q^{1-a},\quad a\big(\tfrac12 - \tfrac2n\big) + (1-a)\tfrac1q = \tfrac1r, \tag{6.10}$$

provided that $a \in [0,1)$.

Proof. The result of the lemma is a special case of a general one proved in [20, 22]. □

Lemma 6.2.2 (Friedrichs's Lemma). *Let Ω be bounded. For all $\varepsilon > 0$, there exists $N \in \mathbb{N}$ such that*

$$\|u\|_2^2 \le (1+\varepsilon)\sum_{j=1}^{N}(u,a^j)^2 + \varepsilon\|\nabla u\|_2^2, \text{ for any } u \in W^{1,2}(\Omega)\,, \tag{6.11}$$

where $\{a^j\}$ is an orthonormal basis of $L^2(\Omega)$.

Corollary 6.2.3. *Assume that $\{u^k(t,x)\}$ is a sequence with*

$$\int_0^T \|u^k(t)\|^2_{W^{1,2}(\Omega)}dt + \operatorname*{ess\,sup}_{(0,T)}\|u^k(t)\|_2 \le M < \infty, \text{ for all } k \in \mathbb{N}\,, \tag{6.12}$$

and

$$\begin{aligned}\|u^k(t)\|^2_{L^2(|x|>R)} \le \|u_0^k\|^2_{L^2(|x|>\frac{R}{2})} + c(t)\psi(R), \text{ for all } k \in \mathbb{N},\\ \text{with } c(t) \in L^\infty((0,T)), \text{ and } \lim_{R\to\infty}\psi(R) = 0.\end{aligned} \tag{6.13}$$

Also, assume that

$$u_0^k \to u_0 \text{ strongly in } L^2(\Omega) \text{ and, a.e. in } t\in(0,T),\ u^k(t) \to u(t) \text{ weakly in } L^2(\Omega)\,. \tag{6.14}$$

Then there exists a subsequence of $\{u^k\}$ strongly converging to u in $L^2(0,T;L^2(\Omega))$.

Proof. The result for Ω bounded is well-known, and a proof is given in [18]. In the case of Ω exterior, a proof is due to Leray in [19]. For the sake of the completeness, we furnish the following proof.

Let u be the weak limit of $\{u^k\}$ in $L^2(0,T;L^2(\Omega))$. By virtue of (6.13), for any $R > 0$ we have

$$\begin{aligned}\int_0^T\int_{|x|>R}|u^k-u|^2\,dx\,dt &\le \int_0^T\int_{|x|>R}|u^k|^2+|u|^2\,dx\,dt\\ &\le \|u_0^k\|^2_{L^2(|x|>\frac{R}{2})} + \psi(R)\int_0^T c(t)\,dt + \int_0^T\int_{|x|>R}|u|^2\,dx\,dt\\ \le 2\big[\|u_0^k-u_0\|_{L^2(|x|>\frac{R}{2})} + \|u_0\|_{L^2(|x|>\frac{R}{2})}\big] &+ \psi(R)\int_0^T c(t)\,dt + \int_0^T\int_{|x|>R}|u|^2\,dx\,dt.\end{aligned}$$

By (6.13) for $\psi(R)$ and (6.14) for u_0^k, and by the absolute continuity of the integral, we get that, for any $\varepsilon > 0$, there exist $\overline{R}$ and $\overline{k}$, such that

$$\int_0^T\int_{|x|>R}|u^k-u|^2\,dx\,dt < \varepsilon \text{ for all } R > \overline{R} \text{ and } k > \overline{k}\,.$$

In the bounded set $\Omega \cap B_R$ we apply Lemma 6.2.2 and we use estimate (6.12), obtaining, for any $k \in \mathbb{N}$,

$$\int_0^T \int_{\Omega\cap B_R} |u^k - u|^2 \, dx \, dt \le (1+\varepsilon) \sum_{j=1}^N \int_0^T \left[\int_{\Omega\cap B_R} (u^k - u) a^j \, dx \right]^2 dt + 2M\varepsilon. \quad (6.15)$$

By the uniform bound (6.12), we have that

$$\left[\int_{\Omega\cap B_R} (u^k - u) a^j \, dx \right]^2 \le 2M^2 \|a^j\|_2^2,$$

and we use the dominated convergence theorem to pass to the limit as $k \to \infty$ in (6.15). The property (6.14) allows us to complete the proof. □

We recall also two basic results.

Lemma 6.2.4. Let Ω be a measurable subset of $\mathbb{R}^n$, and let $v \in L^q(\Omega)$ for any $q \ge \overline{q} \ge 1$. If $\liminf_{q\to\infty} \|v\|_q = l$, then $v \in L^\infty(\Omega)$ and $\|v\|_\infty = l$.

Proof. There exists an increasing sequence $\{q_h\}$ such that $q_h \to \infty$ and $\lim_{h\to\infty} \|v\|_{q_h} = l$. Hence, for any $\varepsilon > 0$ we can find $\bar{h}$ such that $\|v\|_{q_h} \le l + \varepsilon$ for any $h \ge \bar{h}$. Moreover, if $q > q_{\bar{h}}$, we can find $h \ge \bar{h}$ such that $q_h \le q < q_{h+1}$. By interpolation, there exists $\theta_h \in [0,1]$ such that

$$\|v\|_q \le \|v\|_{q_h}^{\theta_h} \|v\|_{q_{h+1}}^{1-\theta_h} \le l + \varepsilon.$$

It follows that $\|v\|_q \le l + \varepsilon$ for any $q > q_{\bar{h}}$. Hence $v \in L^\infty(\Omega)$ and

$$l = \liminf_{q\to\infty} \|v\|_q \le \limsup_{q\to\infty} \|v\|_q \le l + \varepsilon$$

for any $\varepsilon > 0$. It follows that

$$\|v\|_\infty = \lim_{q\to\infty} \|v\|_q = l.$$

□

Lemma 6.2.5. Let $\{g^k\}$ and g be summable functions such that $g^k \to g$ almost everywhere and

$$\lim_{k\to\infty} \int g^k \, dx = \int g \, dx.$$

If $\{f^k\}$ and f are measurable functions such that $|f^k| \le g^k$ almost everywhere and $f^k \to f$ almost everywhere, then

$$\lim_{k\to\infty} \int |f^k - f| \, dx = 0.$$

Proof. The result of the lemma is contained in Theorem 1.20 of [11]. □

It is well-known that in [4] and in [30] it is furnished an existence theorem of suitable weak solutions to the Navier–Stokes Cauchy problem. Here, in order to achieve the same result in the case of problem (6.1), that is, in the case of the initial boundary value problem in bounded or exterior domains Ω, we give the chief steps of the proof in Lemma 6.2.6 and in Appendix. For this goal, we consider a mollified Navier–Stokes system. Hence problem (6.1) becomes

$$\begin{aligned} & v^m_t + J_m[v^m]\cdot\nabla v^m + \nabla\pi_{v^m} = \Delta v^m,\ \nabla\cdot v^m = 0, \text{ in } (0,T)\times\Omega,\\ & v^m = 0 \text{ on } (0,T)\times\partial\Omega,\ v^m(0,x) = v^m_0(x) \text{ on } \{0\}\times\Omega, \end{aligned} \tag{6.16}$$

where $J_m[\cdot]$ is a mollifier and $\{v^m_0\}\subset J^{1,2}(\Omega)$ converges to v_0 in $J^2(\Omega)$. The result of existence is established proving that the sequence of solutions $\{(v^m,\pi_{v^m})\}$ to problem (6.16) converges with respect to the metric stated in Definition 6.1.1, as well as proving that the limit satisfies the energy inequality (6.4). All this is a consequence of the following.

Lemma 6.2.6. *There exists a sequence of solutions $\{(v^m,\pi_{v^m})\}$ such that, for all $m\in\mathbb{N}$ and $T>0$, $v^m\in C([0,T);J^{1,2}(\Omega))\cap L^2(0,T;W^{2,2}(\Omega))$. Moreover, for Ω exterior domain and for $\overline{R}$ sufficiently large, we get*

$$\|v^m(t)\|^2_{L^2(|x|>R)} \le \|v^m_0\|^2_{L^2(|x|>\frac{R}{2})} + c(t)\psi(R) \text{ for any } t>0,\ R>2\overline{R} \text{ and } m\in\mathbb{N}, \tag{6.17}$$

with $c(t)\in L^\infty(0,T)$ and $\psi(R)=o(1)$.

Proof. The above result is well-known. The existence and uniqueness of the solutions and related properties of regularity can be proved as in Theorem 3 of [16] (see also [6]). Concerning estimate (6.17), in the case of the Cauchy problem it was due to Leray in [19]. Subsequently the result is extended to the initial boundary value problem in exterior domains by several authors, in different contexts. Actually, the technique employed by the authors is essentially the same. In this connection, without the aim of being exhaustive, we refer to [14, 27]. In Appendix we give the details of the proof of (6.17). □

Lemma 6.2.7. *For all $T>0$, the sequence of solutions to problem* (6.16) *furnished by Lemma 6.2.6, uniformly in $m\in\mathbb{N}$, satisfies the estimate*

$$\left(\int_0^T \left(\|P\Delta v^m(t)\|_2^2 + \|v^m_t(t)\|_2^2\right)^{\frac{1}{3}} dt\right)^3 \le c\left(\frac{1}{1+\|\nabla v^m(T)\|_2^2} + \|v_0\|_2^6\right). \tag{6.18}$$

Proof. By virtue of the regularity of (v^m,π_{v^m}) stated in Lemma 6.2.6, we multiply equation $(6.16)_1$ by $P\Delta v^m - v^m_t$. Integrating by parts on Ω and applying the Hölder inequality, we get

$$\|P\Delta v^m - v^m_t\|_2 \le \|v^m\cdot\nabla v^m\|_2\,,\ \text{a.e. in } t>0\,. \tag{6.19}$$

Applying inequality (6.10) with $r = \infty$ and $q = 6$, by virtue of the Sobolev inequality, we obtain

$$\|v^m \cdot \nabla v^m\|_2 \le \|v^m\|_\infty \|\nabla v^m\|_2 \le c\|P\Delta v^m\|_2^{\frac{1}{2}} \|\nabla v^m\|_2^{\frac{3}{2}}. \tag{6.20}$$

By inequalities (6.19) and (6.20), we get

$$\begin{aligned}\tfrac{d}{dt}\|\nabla v^m\|_2^2 + \|P\Delta v^m\|_2^2 + \|v_t^m\|_2^2 = \|P\Delta v^m - v_t^m\|_2^2 \le c\|P\Delta v^m\|_2 \|\nabla v^m\|_2^3 \\ \le \frac{1}{2}\|P\Delta v^m\|_2^2 + c\|\nabla v^m\|_2^6,\end{aligned} \tag{6.21}$$

for all $m \in \mathbb{N}$ and a.e. in $t > 0$. We can divide by $(1 + \|\nabla v^m(t)\|_2^2)^2$, and the following holds:

$$\frac{\frac{d}{dt}\|\nabla v^m\|_2^2}{(1 + \|\nabla v^m\|_2^2)^2} + \frac{\frac{1}{2}\|P\Delta v^m\|_2^2 + \|v_t^m\|_2^2}{(1 + \|\nabla v^m\|_2^2)^2} \le c\|\nabla v^m\|_2^2.$$

Integrating on $(0, T)$, we have

$$\frac{1}{1 + \|\nabla v_0^m\|_2^2} - \frac{1}{1 + \|\nabla v^m(T)\|_2^2} + \int_0^T \frac{\frac{1}{2}\|P\Delta v^m\|_2^2 + \|v_t^m\|_2^2}{(1 + \|\nabla v^m\|_2^2)^2}\, dt \le c\int_0^T \|\nabla v^m\|_2^2\, dt.$$

Employing the reverse Hölder inequality (see [1, Theorem 2.12]) with exponents $\frac{1}{3}$ and $-\frac{1}{2}$, we get

$$\int_0^T \frac{\frac{1}{2}\|P\Delta v^m\|_2^2 + \|v_t^m\|_2^2}{(1 + \|\nabla v^m\|_2^2)^2}\, dt \ge \Big[\int_0^T \big[\tfrac{1}{2}\|P\Delta v^m\|_2^2 + \|v_t^m\|_2^2\big]^{\frac{1}{3}} dt\Big]^3 \Big[\int_0^T (1 + \|\nabla v^m\|_2^2) dt\Big]^{-2}.$$

Coupling the above inequalities with the energy inequality (6.5), estimate (6.18) follows. □

6.3 Proof of Theorem 6.1.1

The idea of the proof is the following. We consider the sequence of solutions to problem (6.16) furnished by Lemma 6.2.6. It is well-known that there exists a subsequence $\{(v^m, \pi_{v^m})\}$ whose weak limit (v, π_v) in $L^2(0, T; J^{1,2}(\Omega))$ is a weak solution in the sense of Definition 6.1.2. All this is contained in [19] or, for example, also in [4]. Now, our aim is to prove further estimates on the extract $\{(v^m, \pi_{v^m})\}$ that ensure the thesis of Theorem 6.1.1.

6.3.1 The Strong Convergence in $L^p(0, T; L^2(\Omega))$ for All $p \in [1, 2)$

We start by proving that the sequence $\{v^m\}$ strongly converges in $L^p(0, T; W^{1,2}(\Omega))$, for $p \in [1, 2)$ and for all $T > 0$. We recall that

$$\|\nabla u\|_2 \le \|P\Delta u\|_2^{\frac{1}{2}} \|u\|_2^{\frac{1}{2}}, \text{ for all } u \in W^{2,2}(\Omega) \cap J^{1,2}(\Omega).$$

Hence, integrating on $(0,T)$ and applying the Hölder inequality, we get

$$\int_0^T \|\nabla v^k(t)-\nabla v^m(t)\|_2\,dt \le \Big[\int_0^T \|P\Delta v^k(t)-P\Delta v^m(t)\|_2^{\frac{2}{3}}dt\Big]^{\frac{3}{4}}\Big[\int_0^T \|v^k(t)-v^m(t)\|_2^2\,dt\Big]^{\frac{1}{4}}.$$

By virtue of Lemma 6.2.7, we get the existence of a $M(T)$ such that

$$\int_0^T \|\nabla v^k(t)-\nabla v^m(t)\|_2\,dt \le (2M(T))^{\frac{3}{4}}\Big[\int_0^T \|v^k(t)-v^m(t)\|_2^2\,dt\Big]^{\frac{1}{4}},\ \text{for all } k,m\in\mathbb{N},$$

and, via (6.17), we can apply Corollary 6.2.3 to deduce the strong convergence of the sequence $\{v^m\}$ in $L^1(0,T;W^{1,2}(\Omega))$. Since the energy inequality holds uniformly with respect to $m\in\mathbb{N}$, by interpolation we arrive at the strong convergence of $\{v^m\}$ in $L^p(0,T;W^{1,2}(\Omega))$, for any $p\in[1,2)$. In order to identify the limit point, we remark that $\{v^m\}$ weakly converges to v in $L^2\left(0,T;W^{1,2}(\Omega)\right)$; hence, v has to coincide with the strong limit in each space $L^p\left(0,T;W^{1,2}(\Omega)\right)$. Thus for all $T>0$, we deduce that

$$\begin{aligned}\int_s^t \|\nabla v(\tau)\|_2^2 d\tau &= \lim_{p\to 2^-}\int_s^t \|\nabla v(\tau)\|_2^p d\tau\\ &= \lim_{p\to 2^-}\lim_{m\to\infty}\int_s^t \|\nabla v^m(\tau)\|_2^p d\tau\,,\ \text{for all } t,s\in(0,T)\,.\end{aligned} \tag{6.22}$$

6.3.2 Proof of Formula (6.2)

By the strong convergence in $L^1\left(0,T;W^{1,2}(\Omega)\right)$, there exists a negligible set (for the Lebesgue measure) $\mathcal{I}\subset(0,T)$, such that for any $t\in\mathcal{G}_1:=(0,T)-\mathcal{I}$ the following limits are finite:

$$\lim_{m\to\infty}\|v^m(t)\|_2=\|v(t)\|_2 \ \text{ and } \ \lim_{m\to\infty}\|\nabla v^m(t)\|_2=\|\nabla v(t)\|_2\,. \tag{6.23}$$

From the energy equality for the approximating solutions $\{v^m\}$, we obtain, for any $t\in\mathcal{G}_1$ and any $\alpha,K>0$,

$$\frac{1}{(K+\|\nabla v^m(t)\|_2^2)^\alpha}\frac{d}{dt}\|v^m(t)\|_2^2+\frac{2\|\nabla v^m(t)\|_2^2}{(K+\|\nabla v^m(t)\|_2^2)^\alpha}=0\,. \tag{6.24}$$

Integrating by parts, we get

$$\begin{aligned}\alpha\int_s^t \frac{\|v^m(\tau)\|_2^2}{(K+\|\nabla v^m(\tau)\|_2^2)^{\alpha+1}}\ \frac{d}{d\tau}\|\nabla v^m(\tau)\|_2^2 d\tau+2\int_s^t\frac{\|\nabla v^m(\tau)\|_2^2}{(K+\|\nabla v^m(\tau)\|_2^2)^\alpha}\,d\tau\\ =\frac{\|v^m(s)\|_2^2}{(K+\|\nabla v^m(s)\|_2^2)^\alpha}-\frac{\|v^m(t)\|_2^2}{(K+\|\nabla v^m(t)\|_2^2)^\alpha}\,.\end{aligned}$$

We remark that, for almost every $\tau \in (0,T)$, by (6.23),

$$\frac{\|\nabla v(\tau)\|_2^2}{(K+\|\nabla v(\tau)\|_2^2)^\alpha} \leftarrow \frac{\|\nabla v^m(\tau)\|_2^2}{(K+\|\nabla v^m(\tau)\|_2^2)^\alpha} \le \|\nabla v^m(\tau)\|_2^{2-2\alpha} \to \|\nabla v(\tau)\|_2^{2-2\alpha}$$

and that, for $\alpha \in \left(0,\frac{1}{2}\right]$, by virtue of the strong convergence in $L^{2-2\alpha}\left(0,T;W^{1,2}(\Omega)\right)$,

$$\lim_{m\to\infty}\int_s^t \|\nabla v^m(\tau)\|_2^{2-2\alpha}\,d\tau = \int_s^t \|\nabla v(\tau)\|_2^{2-2\alpha}\,d\tau.$$

Hence we can apply Lemma 6.2.5 to obtain that, for any $t,s\in\mathcal{G}_1$,

$$\lim_{m\to\infty}\alpha\int_s^t \frac{\|v^m(\tau)\|_2^2}{(K+\|\nabla v^m(\tau)\|_2^2)^{\alpha+1}}\frac{d}{d\tau}\|\nabla v^m(\tau)\|_2^2\,d\tau + 2\int_s^t \frac{\|\nabla v(\tau)\|_2^2}{(K+\|\nabla v(\tau)\|_2^2)^\alpha}\,d\tau$$
$$= \frac{\|v(s)\|_2^2}{(K+\|\nabla v(s)\|_2^2)^\alpha} - \frac{\|v(t)\|_2^2}{(K+\|\nabla v(t)\|_2^2)^\alpha}\,. \tag{6.25}$$

Applying once again Lemma 6.2.5, we get

$$\lim_{\alpha\to 0}\int_s^t \frac{\|\nabla v(\tau)\|_2^2}{(K+\|\nabla v(\tau)\|_2^2)^\alpha}\,d\tau = \int_s^t \|\nabla v(\tau)\|_2^2\,d\tau.$$

Then, letting $\alpha\to 0$ in (6.25), we deduce (6.2) with

$$H(t,s) := \lim_{\alpha\to 0}\lim_{m\to\infty}\alpha\int_s^t \frac{\|v^m(\tau)\|_2^2}{(K+\|\nabla v^m(\tau)\|_2^2)^{\alpha+1}}\frac{d}{d\tau}\|\nabla v^m(\tau)\|_2^2\,d\tau\,. \tag{6.26}$$

6.3.3 Proof of Formula (6.3)

We denote by $\mathcal{G}_2$ the set of $t\ge 0$ such that the estract $\{v^m\}$ is strongly convergent in $L^2(\Omega)$. Recalling the definition of $\mathcal{G}_1$, we have $\mathcal{G}_1\subseteq\mathcal{G}_2$. By virtue of Lemma 6.4.5, we claim that $\|\nabla v^m(t)\|_2\ne 0$ for all $t>0$ and $m\in\mathbb{N}$. Hence we consider formula (6.24) that we rewrite with K_1 as

$$\frac{1}{(K_1+\|\nabla v^m(t)\|_2^2)^\alpha}\frac{d}{dt}\|v^m(t)\|_2^2 + \frac{2\|\nabla v^m(t)\|_2^2}{(K_1+\|\nabla v^m(t)\|_2^2)^\alpha} = 0\,,$$

where, by the above claim, we can consider $K_1\ge 0$. Integrating on (s,t), for $s,t\in\mathcal{G}_2$, and applying the mean value theorem for the integrals, we get

$$\frac{1}{(K_1+\|\nabla v(t_{\alpha,m})\|_2^2)^\alpha} = \left(\|v^m(s)\|_2^2-\|v^m(t)\|_2^2\right)^{-1} 2\int_s^t \frac{\|\nabla v^m(\tau)\|_2^2}{(K_1+\|\nabla v^m(t)\|_2^2)^\alpha}\,d\tau\,.$$

Since the right-hand side admits limit as $m \to \infty$ and as $\alpha \to 0$, the limit $F(t,s) := \lim\limits_{\alpha\to 0}\lim\limits_{m\to\infty} \dfrac{1}{(K_1 + \|\nabla v^m(t_{\alpha,m})\|_2^2)^\alpha}$ is well-posed and (6.3) is proved.

6.3.4 The $L^{\mu(q)}(0,T;L^q(\Omega))$ Property

We use the symbol $C(v_0)$ to denote a positive constant depending only on $\|v_0\|_2$, whose numerical value is unessential for our aims, and we can use the same symbol to denote different constants even in the same equation. By virtue of estimate (6.10), we get

$$\|v^m(t)\|_\infty \le c\|P\Delta v^m\|_2^{\frac{1}{2}}\|\nabla v^m\|_2^{\frac{1}{2}}\,.$$

Employing the energy relation (6.5) and estimate (6.18) and applying Hölder inequality, for all $T > 0$, we deduce that, for any $m \in \mathbb{N}$,

$$\int\limits_0^T \|v^m(\tau)\|_\infty d\tau \le c\Big[\int\limits_0^T \|P\Delta v^m(\tau)\|_2^{\frac{2}{3}} d\tau\Big]^{\frac{3}{4}}\Big[\int\limits_0^T \|\nabla v^m(\tau)\|_2^2 d\tau\Big]^{\frac{1}{4}} \le C(v_0),$$

Therefore, by L^p-interpolation and recalling that $v^m \in L^\infty\big(0,T;J^2(\Omega)\big)$, uniformly in $m \in \mathbb{N}$ and $T > 0$, we arrive at

$$\int\limits_0^T \|v^m(\tau)\|_q^{\frac{q}{q-2}} d\tau \le \sup_{(0,T)} \|v^m(\tau)\|_2^{\frac{2}{q-2}} \int\limits_0^t \|v^m(\tau)\|_\infty d\tau \le C(v_0).$$

This allows us to claim that, for all $T > 0$, the weak solution v to problem (6.1), limit of the sequence $\{v^m\}$, belongs to $L^{\frac{q}{q-2}}(0,T;L^q(\Omega))$, for all $q \in (6,\infty)$, with

$$\int\limits_0^T \|v(\tau)\|_q^{\frac{q}{q-2}} d\tau \le C(v_0).$$

This limit property and Fatou's lemma ensure that, for all $T > 0$ and for any sequence $q_h \to \infty$, the following estimate holds true:

$$\int\limits_0^T \liminf_{h\to\infty}\|v(\tau)\|_{q_h} \le \lim_{h\to\infty} T^{\frac{2}{q_h}} C(v_0)^{\frac{q_h-2}{q_h}} = C(v_0)\,. \tag{6.27}$$

The thesis of the theorem in the case $q = \infty$ follows straightforward by Lemma 6.2.4.

□

Remark 6.3.1. We verify that $H(t,s)=0$ in the case of $2D$-Navier–Stokes equations. It is important to realize the result in the framework of the construction given in the above proof, that is, not relying on the regularity of the limit solution v. We start remarking that estimate (6.20), for $\Omega\subset\mathbb{R}^2$, via (6.10), becomes

$$\|v^m\cdot\nabla v^m\|_2\le c\|v^m\|_2^{\frac12}\|P\Delta v^m\|_2^{\frac12}\|\nabla v^m\|_2\,.$$

Hence, in place of (6.21), we deduce the differential inequality

$$\frac{d}{dt}\|\nabla v^m\|_2^2+\|P\Delta v^m\|_2^2+\|v_t^m\|_2^2\le c\|v^m\|_2^2\|\nabla v^m\|_2^4\le c\|v_0^m\|_2^2\|\nabla v^m\|_2^4\,. \tag{6.28}$$

Hence we achieve the result of Lemma 6.2.7 also for $\Omega\subset\mathbb{R}^2$, with the only difference that on the right-hand side of estimate (6.18) we have $\frac{1}{(1+\|\nabla v^m(T)\|_2^2)}+c\|v_0\|_2^2$. By the same arguments of the three-dimensional case, we obtain that $\{v^m\}$ strongly converges in $L^p(0,T;W^{1,2}(\Omega))$, for all $p\in[1,2)$, that is the key ingredient to arrive at the identity (6.2).

Now we prove that $H(t,s)\le 0$, which implies, by virtue of the energy inequality (6.5), that $H(t,s)=0$. By (6.28), (6.5), and the Hölder inequality, we have

$$\begin{aligned}
&\alpha\int_s^t\frac{\|v^m(\tau)\|_2^2}{(K+\|\nabla v^m(\tau)\|_2^2)^{\alpha+1}}\,\frac{d}{d\tau}\|\nabla v^m(\tau)\|_2^2d\tau\\
&\le\alpha c\|v_0^m\|_2^2\int_s^t\|v^m(\tau)\|_2^2\frac{\|\nabla v^m\|_2^4}{(K+\|\nabla v^m\|_2^2)^{\alpha+1}}\,d\tau\le\alpha c\|v_0^m\|_2^4\int_s^t\|\nabla v^m(\tau)\|_2^{2-2\alpha}\,d\tau\\
&\qquad\le\alpha c\|v_0^m\|_2^4(t-s)^{\frac1\alpha}\Big(\int_s^t\|\nabla v^m(\tau)\|_2^2\,d\tau\Big)^{1-\alpha}\le\alpha c\|v_0^m\|_2^{6-2\alpha}(t-s)^{\frac1\alpha}.
\end{aligned}$$

Passing to the limit for $m\to\infty$ and then for $\alpha\to 0$, we get that $H(t,s)\le 0$.

6.4 Appendix

6.4.1 Some Results Related to the Construction of the Weak Solution

In this section we recall some results that are fundamental in order to construct a suitable weak solution. These results essentially concern estimates of the pressure field π_{v^m} that appears in (6.16). Of course we look for estimates that are uniform with respect to $m\in\mathbb{N}$. Our aim is to justify estimate (6.17).

We start by recalling that the energy relation holds uniformly in $m\in\mathbb{N}$:

$$\|v^m(t)\|_2^2+2\int_0^t\|\nabla v^m(\tau)\|_2^2d\tau=\|v_0^m\|_2^2\le\|v_0\|_2^2\ \text{ for all } t>0\,.$$

We introduce the following functionals:

$$\lambda \in (0,1),\ q\in(1,\infty),\ < a >_q^\lambda \quad := \Big[\int\limits_{\partial\Omega}\int\limits_{\partial\Omega} \frac{|a(x)-a(y)|^q}{|x-y|^{2+\lambda q}} d\sigma_y d\sigma_x\Big]^{\frac{1}{q}},$$

$$\|a\|_{W^{1-\frac{1}{q},q}(\partial\Omega)} := \|a\|_{L^q(\partial\Omega)} + < a >_q^{1-\frac{1}{q}} .$$

We consider the following Neumann problem:

$$\Delta\pi = 0,\ \pi \to 0 \text{ for } |x|\to\infty,\ \frac{d\pi}{d\nu} = \nu\cdot\nabla\times a \text{ on } \partial\Omega. \tag{6.29}$$

Lemma 6.4.1. *In* (6.29) *assume* $a \in W^{1-\frac{1}{q},q}(\partial\Omega)$. *Then for all* $\lambda\in(0,1-\frac{1}{q}]$ *and* R_0 *sufficiently large, there exists a constant* c *independent of* a *such that*

$$\|\pi\|_{L^q(\Omega\cap B_{R_0})} \le c < a >_q^\lambda . \tag{6.30}$$

The lemma is due to Solonnikov in [31, 32]. A recent proof of the same result, by similar techniques, can be found, for example, in [23].

Applying the Hölder inequality and the Gagliardo trace theorem, one gets

$$\begin{aligned}\|\pi\|_{L^q(\Omega\cap B_{R_0})} &\le c < a >_q^\lambda \le c\|a\|^\beta_{L^q(\partial\Omega)}\Big[< a >_q^{1-\frac{1}{q}}\Big]^{1-\beta}\\ &\le c\Big[\|a\|_{L^q(\Omega\cap B_{R_0})} + \|a\|^{\frac{1}{q'}}_{L^q(\Omega)}\|\nabla a\|_q^{\frac{1}{q}}\Big]^\beta \|\nabla a\|_q^{1-\beta}\end{aligned} \tag{6.31}$$

with $\beta := \frac{q(1-\lambda)-1}{1+q}$. Now, we consider (U,π) as a solution to the Stokes problem

$$\begin{aligned}&U_t + \nabla\pi = \Delta U\,,\ \nabla\cdot U = 0\,,\ \text{ in } (0,T)\times\Omega\,,\\ &U = 0 \text{ on } (0,T)\times\partial\Omega\,,\ U = v_0 \text{ on } \{0\}\times\Omega\,.\end{aligned} \tag{6.32}$$

We estimate π by means of (6.31). That is, we set $a := \operatorname{curl} U$, we assume $v_0 \in J^2(\Omega)$, and, via the semigroup properties of U (see, e.g., [24]), for $q=2$, for all $T>0$, we get

$$\|\pi(t)\|_{L^2(\Omega\cap B_{R_0})} \le c(T)\|v_0\|_2\Big[t^{-1+\frac{\beta}{2}} + t^{-1+\frac{\beta}{4}}\Big],\ \text{ for all } t\in(0,T)\,, \tag{6.33}$$

with $\beta = \frac{1-2\lambda}{3}$. Keeping this in hand, we can also deduce an estimate in the exterior of B_{R_0}. Actually, by means of a cut of Eq. (6.29) in B_{R_0}, we get

$$\Delta(\pi h_{R_0}) = \pi\Delta h_{R_0} + 2\nabla\pi\cdot\nabla h_{R_0},$$

with h_{R_0} smooth function such that $h_{R_0}(x) = 1$ for $|x| > R_0$, $h_{R_0}(x) = 0$ for $|x| < \frac{R_0}{2}$. Then, by the representation formula of the solution, we obtain

$$\pi(t,x) = -\int\limits_{\mathbb{R}^3}\mathcal{E}(x-y)\pi\Delta h_{R_0}dy - 2\int\limits_{\mathbb{R}^3}\nabla\mathcal{E}(x-y)\nabla h_{R_0}\pi dy,$$

with $\mathcal{E}$ fundamental solution. So that, for $\overline{r} > 3$ and for $|x| > 2R_0$, we easily get

$$\begin{aligned} \|\pi(t)\|_{L^{\overline{r}}(|x|>R_0)} &\le c\|\pi(t)\|_{L^2(\Omega\cap B_{R_0})} \\ &\le c(T)\|v_0\|_2\Big[t^{-1+\frac{\beta}{2}} + t^{-1+\frac{\beta}{4}}\Big], \text{ for all } t \in (0,T). \end{aligned} \tag{6.34}$$

Consider the following initial boundary value problem for the Stokes system:

$$\begin{aligned} &W_t - \Delta W + \nabla \pi_W = F\,,\ \nabla\cdot W = 0\,,\ \text{ on } (0,T)\times\Omega\,,\\ &W = 0 \text{ on } (0,T)\times\Omega\,,\ W = 0 \text{ on } \{0\}\times\Omega\,. \end{aligned} \tag{6.35}$$

Lemma 6.4.2. *In problem* (6.35) *assume* $F \in L^r(0,T;L^s(\Omega))$, $\frac{3}{s}+\frac{2}{r} = 4$, $s \in (1,\frac{3}{2})$. *Then there exists a unique solution to problem* (6.35) *such that*

$$\int_0^T \Big[\|D^2W(\tau)\|_s^r + \|\nabla\pi_W(\tau)\|_s^r + \|W_\tau(\tau)\|_s^r\Big]d\tau \le c\int_0^T \|F(\tau)\|_s^r d\tau\,, \tag{6.36}$$

with c *independent of* F *and* T.

Proof. This result is well-known, and a proof can be found in [24, 25]. □

Lemma 6.4.3. *Let* $\{(v^m,\pi_{v^m})\}$ *be the sequence of solutions to problem* (6.16) *furnished by Lemma 6.2.6. Then there exist functions* $\pi^1_{v^m}$, $\pi^2_{v^m}$ *such that* $\pi_{v^m} = \pi^1_{v^m} + \pi^2_{v^m}$, *and, for all* $\overline{r} > 3$, $R_0 > 0$ *and* $\lambda \in (0,\frac{1}{2})$, *we also obtain*

$$\begin{aligned} &\|\pi^1_{v^m}(t)\|_{L^2(\Omega\cap B_{R_0})} + \|\pi^1_{v^m}(t)\|_{L^{\overline{r}}(|x|>R_0)} \le c(T)\|v_0\|_2 t^{-1+\frac{\beta}{4}}\,,\ \textit{with } \beta := \tfrac{1-2\lambda}{3}\,,\\ &\textit{and}\\ &\int_0^T \|\nabla\pi^2_{v^m}(\tau)\|_s^r d\tau \le c\|v_0\|_2^{2r}\,,\ \tfrac{3}{s}+\tfrac{2}{r} = 4\,. \end{aligned} \tag{6.37}$$

Proof. The result of the lemma is an immediate consequence of the following decomposition:

$$\text{for all } m \in \mathbb{N}, v^m = U^m + W^m \quad \text{and} \quad \pi_{v^m} = \pi_{U^m} + \pi_{W^m}\,,$$

with (U_m,π_{U^m}) solution to problem (6.32) with initial data $U^m = v_0^m$, and (W^m,π_{W^m}) solution to problem (6.35) with $F^m = J_m(v^m)\cdot\nabla v^m$. Since $v^m \in L^2(0,T;J^{1,2}(\Omega))$, one easily deduces that $F^m \in L^r(0,T;L^s(\Omega))$ provided that $\frac{3}{s}+\frac{2}{r} = 4$. □

From estimate $(6.37)_2$, via the Sobolev embedding theorem (cf. Lemma 5.2 of [13]), there exists a function $\pi^0_{v^m}(\tau)$ such that

$$\int_0^T \|\pi^2_{v^m}(\tau) - \pi^0_{v^m}(\tau)\|^r_{\frac{3s}{3-s}} d\tau \le c\int_0^T \|\nabla\pi^2_{v^m}(\tau)\|_s^r d\tau \le c\|v_0\|_2^{2r}\,, \tag{6.38}$$

for $\frac{3}{s}+\frac{2}{r}=4$. In the sequel we assume that $\pi_{v^m} := \pi^1_{v^m}(t,x)+\pi^2_{v^m}(t,x)-\pi^0_{v^m}(t)$.

Now, we are in a position to prove estimate (6.17).

We consider $R>0$ such that $R>2R_0$. We denote by h_R a smooth function such that $h_R(x)=1$ for $|x|>R$, $h_R=0$ for $|x|<\frac{R}{2}$ with $|D^2h_R|+|\nabla h_R|\le\frac{c}{R}$. We multiply equation $(6.16)_1$ by $v^m(t,x)h_R(x)$. Integrating by parts on $(0,T)\times\Omega$, we get

$$\begin{aligned}\|v^m(t)\|^2_{L^2(|x|>R)} &\le \|v^m_0\|^2_{L^2(|x|>\frac{R}{2})} + cR^{-1}\int_0^t\Big[\|v^m\|_2^2+\|v^m\|_3^3+\|\pi_{v^m}|v^m|\|_{L^1(R<|x|<2R)}\Big]d\tau\\ &=: \|v^m_0\|^2_{L^2(|x|>\frac{R}{2})} + cR^{-1}\int_0^t\big[I_1(\tau)+I_2(\tau)+I_3(\tau)\big]d\tau.\end{aligned}$$

Applying the Gagliardo–Nirenberg inequality and the energy relation, we deduce

$$I_1(\tau)+I_2(\tau)\le\|v^m_0\|_2^2+c\|v^m_0\|_2^{\frac{3}{2}}\|\nabla v^m(\tau)\|_2^{\frac{3}{2}}.$$

By virtue of Lemma 6.4.3, assuming $\overline{r}\in(3,6)$ and $\frac{3}{s}+\frac{2}{r}=4$, and applying the Hölder inequality, for I_3 we get

$$\begin{aligned}\|\pi_{v^m}|v^m|\|_{L^1(R<|x|<2R)} &\le \Big[cR^{3\frac{\overline{r}-2}{2\overline{r}}}\|\pi^1_{v^m}(\tau)\|_{L^{\overline{r}}(|x|>\frac{R}{2})}+cR^{\frac{5s-6}{2s}}\|\pi^2_{v^m}-\pi^0_{v^m}\|_{\frac{3s}{3-s}}\Big]\|v^m\|_2\\ &\le c\Big[R^{3\frac{\overline{r}-2}{2\overline{r}}}\|\pi^1_{v^m}(\tau)\|_{L^{\overline{r}}(|x|>\frac{R}{2})}+R^{\frac{5s-6}{2s}}\|\nabla\pi^2_{v^m}\|_s\Big],\end{aligned}$$

for all $R>2R_0$. Increasing the terms I_i, $i=1,2,3$, by means of the above estimates, we arrive at

$$\begin{aligned}\|v^m(t)\|^2_{L^2(|x|>R)} \le \|v^m_0\|^2_{L^2(|x|>\frac{R}{2})} + cR^{-1}\int_0^t\Big[&\|v^m_0\|_2^2+c\|v^m_0\|_2^{\frac{3}{2}}\|\nabla v^m(\tau)\|_2^{\frac{3}{2}}\\ &+R^{3\frac{\overline{r}-2}{2\overline{r}}}\|\pi^1_{v^m}(\tau)\|_{L^{\overline{r}}(|x|>\frac{R}{2})}+R^{\frac{5s-6}{2s}}\|\nabla\pi^2_{v^m}\|_s\Big]d\tau.\end{aligned}$$

Via estimates (6.37) and applying the Hölder inequality and the energy relation, we prove that

$$\|v^m(t)\|^2_{L^2(|x|>R)}\le\|v^m_0\|^2_{L^2(|x|>\frac{R}{2})}+cR^{-1}\big[t+t^{\frac{1}{4}}\|v^m_0\|_2+t^{\frac{\beta}{4}}R^{3\frac{\overline{r}-2}{2\overline{r}}}+t^{1-\frac{1}{r}}\|v^m_0\|_2R^{\frac{5s-6}{2s}}\big]\|v^m_0\|_2^2,$$

which furnishes (6.17).

6.4.2 Uniqueness Backward in Time for the Sequence $\{v^m\}$

For the sake of the completeness, we prove a result concerning the uniqueness backward in time for solutions (v^m,π^m) to the IBVP (6.16), whose existence is furnished by Lemma 6.2.6. A wide literature on the topic can be found in [29] and [2]. Here we employ the logarithmic convexity method developed in [15, 17]. In order to prove the result we premise a result.

Lemma 6.4.4. Let $v_0^m \in \mathcal{C}_0(\Omega)$ and (v^m, π_{v^m}) be the solution to problem (6.16). Then, for all $T > 0$, there exists a constant A_m such that

$$\|v^m(t)\|_\infty \leq A_m \text{ for all } t \in [0, T]. \tag{6.39}$$

Proof. The result of the lemma is classical in the case of a solution to problem (6.1), provided that one considers $[0, T]$ as a subset of the local interval of existence of the solution. In the case of a solution to problem (6.16), by employing the properties of the mollifier, one can prove property (6.39) on $[0, T]$ for all $T > 0$ with a bound depending on m. Actually, for all $T > 0$, one proves Ladyzhenskaya's estimate (see [18] or [16]), that is $\|v^m(t)\|_{2,2} \leq A_m$ for any $t \in [0, T]$. These considerations allow us to omit further details related to estimate (6.39). □

We are going to prove the following.

Lemma 6.4.5. *If $v_0^m \neq 0$, then the solution (v^m, π^m) to problem (6.16) enjoys the property $\|\nabla v^m(t)\|_2 > 0$ for all $t > 0$.*

Proof. We start from (6.19), that furnishes

$$\tfrac{d}{dt}\|\nabla v^m\|_2^2 + \|P\Delta v^m\|_2^2 + \|v_t^m\|_2^2 \leq \|v\|_\infty^2 \|\nabla v^m\|_2^2, \text{ a.e. in } t>0. \tag{6.40}$$

We recall that the following estimates hold:

$$\|\nabla v^m\|_2^2 \leq \|P\Delta v^m\|_2 \|v^m\|_2 \text{ and } \|\nabla v^m\|_2^2 \leq \|v_t^m\|_2 \|v^m\|_2. \tag{6.41}$$

By virtue of the energy equation for (v^m, π^m), we get

$$\dot{E}(t) = -2\|\nabla v^m\|_2^2, \ \ddot{E}(t) = -2\frac{d}{dt}\|\nabla v^m\|_2^2, \tag{6.42}$$

where we set $E(t) := \|v^m\|_2^2$. Therefore, by (6.39) and (6.40)-(6.42) we arrive at

$$-\ddot{E} + \frac{\dot{E}^2}{E} \leq 2A_m^2 \|\nabla v^m\|_2^2. \tag{6.43}$$

We prove the result of the lemma claiming that if $\overline{t} > 0$ is the first instant such that $\|\nabla v^m(\overline{t})\|_2 = 0$, then $\|v^m(t)\|_2 = 0$ for all $t \in [0, \overline{t}]$, that is a contradiction. Since, for all $T > 0$, $v^m \in C([0, T); J^{1,2}(\Omega))$, if there exists $\overline{t} > 0$ such that $\|v^m(\overline{t})\|_2 = 0$, then there exists $\delta > 0$ such that $\|v^m(t)\|_2 \leq 1$ for all $t \in [\overline{t} - \delta, \overline{t}]$. So that, for a suitable $h > 0$, the inequality (6.43) can be written as

$$\ddot{E} - \frac{\dot{E}^2}{E} \geq h\dot{E} \Rightarrow \frac{\ddot{E}E - \dot{E}^2}{E^2} \geq h\frac{\dot{E}}{E} \Leftrightarrow \frac{d}{dt}\Big[e^{-ht}\frac{\dot{E}}{E}\Big] \geq 0, \text{ for all } t \in [\overline{t} - \delta, \overline{t}].$$

Set $\sigma := e^{ht}$ and $\widetilde{E}(\sigma) := E \circ t^{-1}(\sigma)$, we deduce

$$\frac{d}{d\sigma}\Big[\frac{1}{\widetilde{E}}\frac{d}{d\sigma}\widetilde{E}\Big] \geq 0.$$

This last implies that $\log \widetilde{E}$ is a convex function. That is, for all $\lambda \in [0,1]$,

$$\log \widetilde{E}(\lambda\sigma + (1-\lambda)\sigma_0) \leq \lambda \log \widetilde{E}(\sigma) + (1-\lambda)\log \widetilde{E}(\sigma_0), \text{ with } \sigma_0 := e^{h(\bar{t}-\delta)}. \tag{6.44}$$

Since in $\bar{t} > 0$ it is $\|v^m(\bar{t})\|_2 = 0$, then we arrive at $\|v^m(t)\|_2 = 0$ for all $t < \bar{t}$, which is a contradiction with the hypothesis $v_0^m \neq 0$. □

Acknowledgments

The research activity of C.R. Grisanti is performed under the auspices of GNAMPA-INdAM and partially supported by the Research Project of the University of Pisa "Energia e regolarità: tecniche innovative per problemi classici di equazioni alle derivate parziali." The research activity of F. Crispo and P. Maremonti is performed under the auspices of GNFM-INdAM. F. Crispo has been supported by the Program (Vanvitelli per la Ricerca: VALERE) 2019 financed by the University of Campania "L. Vanvitelli".

References

1. R.A. Adams and J.J.F. Fournier, *Sobolev spaces*, second ed., Pure and Applied Mathematics (Amsterdam), vol. 140, Elsevier/Academic Press, Amsterdam, 2003.

2. K.A. Ames and B. Straughan, *Non-Standard and Improperly Posed Problems*, Academic Press, San Diego - Toronto, 1997.

3. T. Buckmaster and V. Vicol, *Nonuniqueness of weak solutions to the Navier-Stokes equation*, Ann. of Math. (2) **189** (2019), no. 1, 101–144.

4. L. Caffarelli, R. Kohn, and L. Nirenberg, *Partial regularity of suitable weak solutions of the Navier-Stokes equations*, Comm. Pure Appl. Math. **35** (1982), no. 6, 771–831.

5. K. Choi and A.F. Vasseur, *Estimates on fractional higher derivatives of weak solutions for the Navier-Stokes equations*, Ann. Inst. H. Poincaré Anal. Non Linéaire **31** (2014), no. 5, 899–945.

6. P. Constantin and C. Foias, *Navier-Stokes equations*, Chicago Lectures in Mathematics, University of Chicago Press, Chicago, IL, 1988.

7. F. Crispo and P. Maremonti, *On the spatial asymptotic decay of a suitable weak solution to the Navier-Stokes Cauchy problem*, Nonlinearity **29** (2016), no. 4, 1355–1383.

8. F. Crispo and P. Maremonti, *A remark on the partial regularity of a suitable weak solution to the Navier-Stokes Cauchy problem*, Discrete Contin. Dyn. Syst. **37** (2017), no. 3, 1283–1294.

9. F. Crispo and P. Maremonti, *Some remarks on the partial regularity of a suitable weak solution to the Navier-Stokes Cauchy problem*, Zap. Nauchn. Sem. S.-Petersburg. Otdel. Mat. Inst. Steklov. (POMI) **477** (2018).

10. G.F.D. Duff, *Derivative estimates for the Navier-Stokes equations in a three-dimensional region*, Acta Math. **164** (1990), no. 3-4, 145–210.

11. L.C. Evans and R.F. Garlepy, *Measure theory and fine properties of functions*, Textbooks in Mathematics, CRC Press, (2015).

12. C. Foiaş, C. Guillopé, and R. Temam, *New a priori estimates for Navier-Stokes equations in dimension* 3, Comm. Partial Differential Equations **6** (1981), no. 3, 329–359.

13. G.P. Galdi, *An Introduction to the Mathematical Theory of the Navier-Stokes Equations*, Springer Monographs in Mathematics, vol. 1 New-York (2011).

14. G.P. Galdi and P. Maremonti, *Monotonic decreasing and asymptotic behavior of the kinetic energy for weak solutions of the Navier-Stokes equations in exterior domains*, Arch. Rational Mech. Anal. **94** (1986), no. 3, 253–266.

15. G.P. Galdi and B. Straughan, *Stability of solutions of the Navier-Stokes equations backward in time*, Arch. Ration. Mech. Anal. **101** (1988), 107–114.

16. J.G. Heywood, *The Navier-Stokes equations: on the existence, regularity and decay of solutions*, Indiana Univ. Math. J. **29** (1980), no. 5, 639–681.

17. R.J. Knops and L.E. Payne, *On the stability of solutions of the Navier-Stokes equations backward in time*, Arch. Rational Mech. Anal. **29** 1968 331–335.

18. O.A. Ladyzhenskaya, *The mathematical theory of viscous incompressible flow*, Revised English edition. Translated from the Russian by Richard A. Silverman, Gordon and Breach Science Publishers, New York-London, 1963.

19. J. Leray, *Sur le mouvement d'un liquide visqueux emplissant l'espace*, Acta Math. **63** (1934), no. 1, 193–248.

20. P. Maremonti, *Some interpolation inequalities involving Stokes operator and first order derivatives*, Ann. Mat. Pura Appl. (4) **175** (1998), 59–91.

21. P. Maremonti, *A note on Prodi-Serrin conditions for the regularity of a weak solution to the Navier-Stokes equations*, J. Math. Fluid Mech. **20** (2018), no. 2, 379–392.

22. P. Maremonti, *On an interpolation inequality involving the Stokes operator*, Mathematical analysis in fluid mechanics—selected recent results, Contemp. Math., vol. 710, Amer. Math. Soc., Providence, RI, 2018, pp. 203–209.

23. P. Maremonti, *On the L^p Helmholtz decomposition: A review of a result due to Solonnikov*, Lithuanian Mathematical J., **58** (2018) 268–283, https://doi.org/10.1007/s10986-018-9403-6

24. P. Maremonti and V.A. Solonnikov, *On nonstationary Stokes problem in exterior domains*, Ann. Scuola Norm. Sup. Pisa Cl. Sci. **24** (1997) 395–449.

25. P. Maremonti and V. A. Solonnikov, *An estimate for the solutions of a Stokes system in exterior domains* (Russian, with English summary), Zap. Nauchn. Sem. Leningrad. Otdel. Mat. Inst. Steklov. (LOMI) **180** (1990), no. Voprosy Kvant. Teor. Polya i Statist. Fiz. 9, 105–120, 181; English transl., J. Math. Sci. **68** (1994), no. 2, 229–239.

26. J.A. Mauro, *Some Analytic Questions in Mathematical Physics Problems*, Ph.D. Thesis, University of Pisa, Italy, 2010. http://etd.adm.unipi.it/t/etd-12232009-161531/

27. T. Miyakawa and H. Sohr, *On energy inequality, smoothness and large time behavior in L^2 for weak solutions of the Navier-Stokes equations in exterior domains*, Math. Z. **199** (1988), no. 4, 455–478.

28. T. Nagasawa, *A new energy inequality and partial regularity for weak solutions of Navier-Stokes equations*, J. Math. Fluid Mech. **3** (2001), no. 1, 40–56.

29. L.E. Payne, *Improperly Posed Problems in Partial Differential Equations*, SIAM, 1975.

30. V. Scheffer, *Hausdorff measure and the Navier-Stokes equations*, Comm. Math. Phys., **55** (1977), 97–112.

31. V.A.Solonnikov, *Estimates of the solutions of the nonstationary Navier-Stokes system*, in Boundary Value Problems of Mathematical Physics and Related Questions in the Theory of Functions. Part 7, Zap. Nauchn. Semin. Leningr. Otd. Mat. Inst. Steklova, **38**, Nauka, Leningrad, (1973) pp. 153–231 (in Russian).

32. V.A. Solonnikov, *Estimates for solutions of nonstationary Navier–Stokes equations*, J. Sov. Math., **8** (1977) 467–529.

33. A. Vasseur, *Higher derivatives estimate for the 3D Navier-Stokes equation*, Ann. Inst. H. Poincaré Anal. Non Linéaire **27** (2010), no. 5, 1189–1204.

Chapter 7

Existence, Uniqueness, and Regularity for the Second–Gradient Navier–Stokes Equations in Exterior Domains

Marco Degiovanni, Alfredo Marzocchi (✉), and Sara Mastaglio
Università Cattolica del Sacro Cuore, Dipartimento di Matematica e Fisica, Brescia, Italy
e-mail: marco.degiovanni@unicatt.it; alfredo.marzocchi@unicatt.it; sara.mastaglio@gmail.com

We study the well-posedness of the problem

$$\begin{cases} \dfrac{\partial u}{\partial t} + (Du)\, u + \nabla p = \nu \Delta u - \tau \Delta \Delta u & \text{in }]0, +\infty[\times \Omega \,, \\ \operatorname{div} u = 0 & \text{in }]0, +\infty[\times \Omega \,, \\ u(t,x) = \dfrac{\partial u}{\partial n}(t,x) = 0 & \text{on }]0, +\infty[\times \partial\Omega \,, \\ u(0,x) = u_0(x) & \text{in } \Omega \,, \end{cases}$$

where $u :]0, +\infty[\times \Omega \to \mathbb{R}^n$ is the velocity field, $p :]0, +\infty[\times \Omega \to \mathbb{R}$ is the pressure, ν is the kinematical viscosity, τ the so-called *hyperviscosity* and Ω is a general domain as for existence and uniqueness of the solution, and an exterior domain as for regularity results.

T. Bodnár et al. (eds.), *Waves in Flows*, Advances in Mathematical Fluid Mechanics, https://doi.org/10.1007/978-3-030-68144-9_7

This problem has been physically well motivated in the recent years as the simplest case of an isotropic second-order fluid, i.e. a fluid whose power expended depends on second derivatives of the velocity field.

7.1 Introduction

In the last years a considerable interest has been put on the study on the so-called *second-gradient materials*, which can be characterized as continua where the inner expended power depends also on second derivatives of the velocity field, rather than, as customary, only on the first ones. Starting from the seminal paper by GERMAIN [13] on microstructured continua, the setting has been studied by FRIED and GURTIN [9] and many others. For these materials, the expression of the expended inner power reads as

$$P(M,u) = -\int_M T \cdot \operatorname{Sym} Du \, d\mathcal{L} + \int_M C \cdot D^2 u \, d\mathcal{L},$$

where $u : M \to \mathbb{R}^3$ is the velocity field, $D^j u$ its successive gradients, T is the Cauchy stress tensor field, and C is a third-order tensor field called *hyperstress*. The momentum balance law is then generalized to

$$\varrho a = \varrho b + \operatorname{div} T - \operatorname{div} \operatorname{div} C,$$

where $a = \partial u/\partial t + (Du)u$ is the acceleration field, ϱ is the density, and b is the external force field density.

T and C must then be constitutively expressed in terms of the descriptors of the problems (the "state variables"). For fluids, the simplest material turns out to be the isotropic incompressible one (see [23]) and gives the expression (see [16])

$$T = -p_0 I + 2\mu \operatorname{Sym} Du$$

$$\begin{aligned} C =& (\eta_1 - \eta_2)\nabla\nabla u + 3\eta_2 \operatorname{Sym} \nabla\nabla u \\ &- (\eta_2 + 5\eta_3)\Delta u \otimes I + 3\eta_3 \operatorname{Sym}(\Delta u \otimes I) - I \otimes P\,, \end{aligned}$$

where I is the identity tensor, p_0 is a scalar field, the reaction to the incompressibility constraint, called *pressure*, and P is a vector field called *hyperpressure*, depending on the density. The parameters μ and η_k must satisfy appropriate conditions for the dissipation to be positive. Since the density in fluids is generally supposed constant in space, we have $\varrho = \varrho_0 > 0$, so that P is constant and the balance law reads then

$$\frac{\partial u}{\partial t} + (Du)u = b - \nabla p + \nu\Delta u - \tau\Delta\Delta u,$$

where $p = p_0 - \operatorname{div} P$, $\nu = \mu/\varrho_0$, and $\tau = \zeta(\eta_k)/\varrho_0 > 0$, where $\zeta(\eta_k)$ is a strictly positive expression containing the coefficients η_k (see [16]). Incompressibility implies $\operatorname{div} u = 0$ and boundary and initial conditions are prescribed. For the sake of simplicity we chose boundary conditions of Dirichlet type, i.e. $u = 0$ and $\partial u/\partial n=0$ on the boundary. We will also suppose b to be conservative, i.e. $b = -\nabla V$ and absorb V into the pressure term.

This problem has been considered e.g. in [14], [17], [18] in unbounded domains, while its linear steady counterpart has been studied in [3], but only for bounded domains.

In this sight, this problem is no more a mere perturbation of the classical Navier–Stokes one: yet, it is considerably simpler in view of the regularizing property of the double Laplacian. Anyway, a complete theory concerning existence, uniqueness, and regularity for external domains is to our knowledge not available and we present it here making use of the theory of maximal monotone operators. We believe that these results could be helpful in the theory of slender bodies as extensively studied by Galdi ([12],[10]).

7.2 Notations and Preliminaries

Let Ω be an open subset of $\mathbb{R}^n$.

7.2.1 Sobolev Spaces

We will denote by $\|\cdot\|_q$ the usual norm in $L^q(\Omega)$, $1 \le q \le \infty$, and, for every $m \ge 1$, by $\|\cdot\|_{m,2}$ the $W^{m,2}(\Omega)$-norm defined by

$$\|u\|_{m,2} := \left(\sum_{|\alpha|\le m} \|D^\alpha u\|_2^2 \right)^{1/2} .$$

We also denote by $W_0^{m,2}(\Omega)$ the closure of $C_c^\infty(\Omega)$ in $W^{m,2}(\Omega)$ and by $W^{-m,2}(\Omega)$ the dual space of $W_0^{m,2}(\Omega)$, whose norm will be denoted by $\|\ \|_{-m,2}$. Since $C_c^\infty(\Omega)$ is dense in $W_0^{m,2}(\Omega)$, we can identify $W^{-m,2}(\Omega)$ with a linear subspace of the distributions $\mathcal{D}'(\Omega)$. We also denote by $W_{loc}^{-m,2}(\Omega)$ the set of distributions $f \in \mathcal{D}'(\Omega)$ such that $\vartheta f \in W^{-m,2}(\Omega)$ for all $\vartheta \in C_c^\infty(\Omega)$. Finally, we set $W^{0,2}(\Omega) = L^2(\Omega)$ and $W_{loc}^{0,2}(\Omega) = L_{loc}^2(\Omega)$.

Since, for all $u \in W_0^{2,2}(\Omega)$,

$$\int_\Omega |\nabla u|^2\, dx = -\int_\Omega u \cdot \Delta u\, dx \le \|u\|_2 \|\Delta u\|_2 ,$$

$$\int_\Omega \sum_{j,k=1}^n D_{jk}^2 u \cdot D_{jk}^2 u\, dx = \int_\Omega \sum_{j,k=1}^n D_{jj}^2 u \cdot D_{kk}^2 u\, dx = \int_\Omega |\Delta u|^2\, dx ,$$

it turns out that

$$\left(\|u\|_2^2 + \|\Delta u\|_2^2\right)^{1/2}$$

is a norm in $W_0^{2,2}(\Omega)$ equivalent to the canonical one.

7.2.2 Hydrodynamic Spaces

We set

$$\mathcal{C}_{div}(\Omega) = \{u \in C_c^\infty(\Omega) : \ \operatorname{div} u = 0\}$$

and then

$$\begin{aligned} J^2(\Omega) &= \text{the closure of } \mathcal{C}_{div}(\Omega) \text{ in } L^2(\Omega)\,, \\ G^2(\Omega) &= \left\{v \in L^2(\Omega) : \ \langle v, u\rangle = 0 \text{ for all } u \in \mathcal{C}_{div}(\Omega)\right\}\,, \\ J^{m,2}(\Omega) &= \text{the closure of } \mathcal{C}_{div}(\Omega) \text{ in } W_0^{m,2}(\Omega)\,, \\ G^{-m,2}(\Omega) &= \left\{v \in W^{-m,2}(\Omega) : \ \langle v, u\rangle = 0 \text{ for all } u \in \mathcal{C}_{div}(\Omega)\right\}\,. \end{aligned}$$

We clearly have

$$L^2(\Omega) = J^2(\Omega) \oplus G^2(\Omega)\,.$$

Moreover, since

$$L^2(\Omega) \cap G^{-m,2}(\Omega) = G^2(\Omega)\,,$$

we also have

$$L^2(\Omega) \subseteq J^2(\Omega) \oplus G^{-m,2}(\Omega) \subseteq W^{-m,2}(\Omega)\,.$$

We will denote by

$$\pi : \left(J^2(\Omega) \oplus G^{-m,2}(\Omega)\right) \to J^2(\Omega)$$

the projection associated with the direct sum.

According to de Rham's Theorem (see e.g. [25]), it turns out that

$$\begin{aligned} G^2(\Omega) &= \left\{v \in L^2(\Omega) : \ v = \nabla p \text{ for some } p \in W_{loc}^{1,2}(\Omega)\right\}\,, \\ G^{-m,2}(\Omega) &= \left\{v \in W^{-m,2}(\Omega) : \ v = \nabla p \text{ for some } p \in W_{loc}^{-m+1,2}(\Omega)\right\}\,. \end{aligned}$$

Finally, for every $u \in W_{loc}^{1,2}(\Omega)$, we define

$$(u \cdot \nabla)u = (Du)u \in L_{loc}^1(\Omega)$$

by

$$[(u \cdot \nabla)u]_k = [(Du)u]_k := \sum_{j=1}^n u_j D_j u_k\,.$$

7.3 The Nonlinear Term

Let Ω be an open subset of $\mathbb{R}^n$ with $n \le 5$. According to [6], there exist $C > 0$ and $\alpha, \beta \in]0,1[$ such that

$$\|u\|_6 \le C\left(\|u\|_2 + \|u\|_2^\alpha \|\Delta u\|_2^{1-\alpha}\right)\,, \qquad \|Du\|_3 \le C\left(\|u\|_2 + \|u\|_2^\beta \|\Delta u\|_2^{1-\beta}\right)\,, \tag{7.1}$$

for all $u \in W_0^{2,2}(\Omega)$.

Proposition 7.3.1. The following facts hold:

(a) for every $u \in W_0^{2,2}(\Omega)$ we have $(Du)u \in L^2(\Omega)$ and the map

$$\begin{array}{ccc} W_0^{2,2}(\Omega) & \longrightarrow & L^2(\Omega) \\ u & \longmapsto & (Du)u \end{array}$$

is of class C^∞;

(b) there exists $\gamma > 0$ and, for every $\varepsilon > 0$, $C_\varepsilon > 0$ such that

$$\int_\Omega [(Du)u - (Dv)v] \cdot (u - v)\, dx \geq -\varepsilon \|\Delta u - \Delta v\|_2^2 - C_\varepsilon (1 + \|u\|_2 + \|v\|_2)^\gamma \, \|u - v\|_2^2$$

for all $u, v \in W_0^{2,2}(\Omega)$ with $\operatorname{div} u = \operatorname{div} v$.

Proof. Combining Hölder's inequality with (7.1), we get

$$\|(Du)v\|_2 \leq C\|Du\|_3 \|v\|_6 \leq C\|u\|_{2,2}\, \|v\|_{2,2} \qquad \text{for all } u, v \in W_0^{2,2}(\Omega)$$

so that the map $(u, v) \mapsto (Du)v$ is bilinear and continuous. Then assertion (a) follows.

If $\operatorname{div} u = \operatorname{div} v$, we also have

$$\begin{aligned} \int_\Omega [(Du)u - (Dv)v] \cdot (u - v)\, dx &= \int_\Omega [(Du - Dv)u] \cdot (u - v)\, dx \\ &\qquad + \int_\Omega [(Dv)(u - v)] \cdot (u - v)\, dx \\ &= \int_\Omega [(Du - Dv)u] \cdot (u - v)\, dx \\ &\qquad - \int_\Omega [(Du - Dv)(u - v)] \cdot v\, dx \\ &\geq -C(\|u\|_2 + \|v\|_2)\|Du - Dv\|_3 \|u - v\|_6 \,. \end{aligned}$$

Combining this fact with (7.1) and Young's inequality, assertion (b) also follows. □

7.4 Steady-State Variational Inequalities

Let Ω be an open subset of $\mathbb{R}^n$ with $n \leq 5$, let $\tau > 0$ and $\nu \geq 0$. For every $R > 0$, we set

$$\begin{aligned} \widehat{K}_R &:= \left\{ u \in W_0^{2,2}(\Omega) : \ \|u\|_2 \leq R \right\}, \\ K_R &:= \left\{ u \in J^{2,2}(\Omega) : \ \|u\|_2 \leq R \right\}. \end{aligned}$$

Proposition 7.4.1. For every $R > 0$, the following facts hold:

(*a*) the map $F : \widehat{K}_R \longrightarrow W^{-2,2}(\Omega)$ defined by

$$F(u) = \tau\Delta\Delta u - \nu\Delta u + (Du)u$$

is continuous;

(*b*) there exists $\omega_R > 0$ such that

$$\langle F(u) - F(v), u - v\rangle \geq \frac{\tau}{2}\|\Delta u - \Delta v\|_2^2 - \omega_R\|u - v\|_2^2$$

for all $u, v \in \widehat{K}_R$ with $\operatorname{div} u = \operatorname{div} v$.

Proof. Assertion (*a*) easily follows from (*a*) of Proposition 7.3.1. Taking into account (*b*) of Proposition 7.3.1, we also have

$$\begin{aligned}\langle F(u) - F(v), u - v\rangle &= \tau\|\Delta u - \Delta v\|_2^2 + \nu\|Du - Dv\|_2^2 \\ &\quad + \int_\Omega [(Du)u - (Dv)v] \cdot (u - v)\, dx \\ &\geq \tau\|\Delta u - \Delta v\|_2^2 - \frac{\tau}{2}\|\Delta u - \Delta v\|_2^2 \\ &\quad - C_\tau(1 + \|u\|_2 + \|v\|_2)^\gamma\, \|u - v\|_2^2\end{aligned}$$

and assertion (*b*) follows. □

Theorem 7.4.1. Let $R > 0$ and let ω_R be as in Proposition 7.4.1. Then, for every $f \in W^{-2,2}(\Omega)$, there exists one and only one $u \in K_R$ satisfying

$$\langle F(u), v - u\rangle + \omega_R \int_\Omega u(v - u)\, dx + \int_\Omega u(v - u)\, dx \geq \langle f, v - u\rangle \quad \text{for all } v \in K_R\,.$$

Proof. If we define $\widetilde{F} : K_R \longrightarrow W^{-2,2}(\Omega)$ by

$$\widetilde{F}(u) = F(u) + \omega_R u + u - f\,,$$

from Proposition 7.4.1 we infer that $\widetilde{F}$ is continuous and satisfies

$$\langle \widetilde{F}(u) - \widetilde{F}(v), u - v\rangle \geq \frac{\tau}{2}\|\Delta u - \Delta v\|_2^2 + \|u - v\|_2^2 \qquad \text{for all } u, v \in K_R\,.$$

Moreover, K_R is a nonempty, closed, and convex subset of $W_0^{2,2}(\Omega)$. Then the assertion follows from [20, Corollary III.1.8]. □

Now we aim to reformulate the above result in a different way. For every $R > 0$ and $u \in K_R$, we set

$$N_{K_R}(u) = \left\{f \in W^{-2,2}(\Omega) :\ \langle f, v - u\rangle \leq 0 \text{ for all } v \in K_R\right\}\,.$$

The next result readily follows.

Corollary 7.4.2. Let $R > 0$ and let ω_R be as in Proposition 7.4.1. Then, for every $f \in W^{-2,2}(\Omega)$, there exists one and only one $u \in K_R$ satisfying

$$F(u) + \omega_R\, u + u + N_{K_R}(u) \ni f\,.$$

Let us provide a more explicit description of $N_{K_R}(u)$.

Proposition 7.4.3. For every $u \in K_R$, we have

$$N_{K_R}(u) = \begin{cases} G^{-2,2}(\Omega) & \text{if } \|u\|_2 < R\,, \\ [0,+\infty[\, u + G^{-2,2}(\Omega) & \text{if } \|u\|_2 = R\,. \end{cases}$$

In particular, for every $u_1, u_2 \in K_R$ and $f_j \in N_{K_R}(u_j)$ for $j = 1, 2$, we have

$$\langle f_1 - f_2, u_1 - u_2 \rangle \geq 0\,.$$

Proof. Let $u \in K_R$ and $f \in N_{K_R}(u)$. If $v \in \mathcal{C}_{div}(\Omega)$ and $\langle u, v\rangle = 0$, then for every $\varepsilon > 0$ there exists $\tau > 0$ such that $\|u + \tau(v - \varepsilon u)\|_2 < R$, whence

$$\langle f, v - \varepsilon u\rangle \leq 0\,.$$

From the arbitrariness of ε and the fact that we can exchange v with $-v$ we infer that

$$\langle f, v\rangle = 0 \qquad \text{for all } v \in \mathcal{C}_{div}(\Omega) \text{ with } \langle u, v\rangle = 0\,.$$

If $\|u\|_2 < R$, we also have $\|u \pm \tau u\|_2 < R$, provided that $\tau > 0$ is small enough, whence

$$\langle f, u\rangle = 0\,.$$

It follows

$$\langle f, v\rangle = 0 \qquad \text{for all } v \in \mathcal{C}_{div}(\Omega)\,,$$

whence $f \in G^{-2,2}(\Omega)$.

If $\|u\|_2 = R$, let

$$t = \frac{\langle f, u\rangle}{\|u\|_2^2}\,,$$

so that $t \geq 0$ and $\langle f - tu, u\rangle = 0$. Then, for every $v \in \mathcal{C}_{div}(\Omega)$ with $\langle u, v\rangle = 0$, we have

$$\langle f - tu, v\rangle = \langle f, v\rangle = 0\,.$$

As before, we infer that $(f - tu) \in G^{-2,2}(\Omega)$.

Therefore, we have

$$\begin{aligned} N_{K_R}(u) &\subseteq G^{-2,2}(\Omega) && \text{if } \|u\|_2 < R\,, \\ N_{K_R}(u) &\subseteq [0,+\infty[\, u + G^{-2,2}(\Omega) && \text{if } \|u\|_2 = R\,. \end{aligned}$$

The opposite inclusions are easy to prove. □

Now we aim to consider the Hilbert space $J^2(\Omega)$ as framework. Let us recall from [5, 24] some basic notions.

Definition 7.4.2. Let H be a Hilbert space and let $\omega \geq 0$. A multivalued operator

$$A : D(A) \longrightarrow \mathfrak{P}(H)\,,$$

where $\mathfrak{P}(H)$ denotes the power set of H, is said to be

(*a*) *monotone* if

$$\langle f_1 - f_2, u_1 - u_2 \rangle \geq 0$$

whenever $u_1, u_2 \in D(A)$, $f_1 \in A(u_1)$, and $f_2 \in A(u_2)$;

(*b*) *maximal monotone* if it is monotone and cannot be properly extended to a monotone operator;

(*c*) $\mathcal{M}(\omega)$ if

$$\langle f_1 - f_2, u_1 - u_2 \rangle \geq -\omega \|u_1 - u_2\|^2$$

whenever $u_1, u_2 \in D(A)$, $f_1 \in A(u_1)$, and $f_2 \in A(u_2)$;

(*d*) *maximal* $\mathcal{M}(\omega)$ if it is $\mathcal{M}(\omega)$ and cannot be properly extended to a $\mathcal{M}(\omega)$ operator.

We set

$$D(A) = \left\{ u \in K_R : \; \Delta\Delta u \in J^2(\Omega) \oplus G^{-2,2}(\Omega) \right\}$$

and define a multivalued operator in $J^2(\Omega)$

$$A : D(A) \longrightarrow \mathfrak{P}(J^2(\Omega))\,,$$

by

$$A(u) = [F(u) + N_{K_R}(u)] \cap J^2(\Omega) = [\tau\Delta\Delta u - \nu\Delta u + (Du)u + N_{K_R}(u)] \cap J^2(\Omega)\,.$$

Theorem 7.4.3. Let $R > 0$ and ω_R as in Proposition 7.4.1. Then the following facts hold:

(*a*) the operator $A + \omega_R I$ is maximal monotone in $J^2(\Omega)$;

(*b*) the operator A is maximal $\mathcal{M}(\omega_R)$ in $J^2(\Omega)$;

(*c*) for every $u \in D(A)$ and $f \in A(u)$, we have

$$\tau \|\Delta u\|_2^2 + \nu \|Du\|_2^2 \leq \|f\|_2 \|u\|_2\,;$$

(*d*) for every $u \in D(A)$ with $\|u\|_2 < R$, the set $A(u)$ is a singleton $\{f(u)\}$ and we have

$$\tau \|\pi(\Delta\Delta u)\|_2 \leq (\|f(u)\|_2 + \nu\|\Delta u\|_2 + \|(Du)u\|_2)\,.$$

Proof.
(a) First of all, it is easily seen that

$$[\tau\Delta\Delta u - \nu\Delta u + (Du)u + N_{K_R}(u)] \cap J^2(\Omega) \neq \emptyset \iff \Delta\Delta u \in J^2(\Omega) \oplus G^{-2,2}(\Omega).$$

Let $u_1, u_2 \in D(A)$ and $f_j \in A(u_j)$ for $j = 1, 2$. Then let $g_j \in N_{K_R}(u_j)$ be such that $f_j = F(u_j) + g_j$.

We have

$$\int_\Omega (f_1 - f_2) \cdot (u_1 - u_2)\, dx + \omega_R \int_\Omega (u_1 - u_2) \cdot (u_1 - u_2)\, dx$$
$$= \langle F(u_1) - F(u_2), u_1 - u_2\rangle + \langle g_1 - g_2, u_1 - u_2\rangle + \omega_R \|u_1 - u_2\|_2^2 \geq 0$$

by Proposition 7.4.1. Therefore, the operator $A + \omega_R I$ is monotone.

Now let $f \in J^2(\Omega) \subseteq W^{-2,2}(\Omega)$. By Corollary 7.4.2 there exists $u \in K_R$ such that

$$F(u) + N_{K_R}(u) + \omega_R u + u \ni f\,.$$

From [5, Proposition 2.2] we infer that $A + \omega_R I$ is also maximal.
(b) According to [24, Lecture 4], the assertion is equivalent to (a).
(c) Since

$$\langle (Du)u, u\rangle = 0 \qquad \text{for all } u \in J^{2,2}(\Omega)\,,$$
$$\langle g, u\rangle \geq 0 \qquad \text{for all } u \in K_R \text{ and } g \in N_{K_R}(u)\,,$$

we have

$$\|f\|_2\|u\|_2 \geq \langle f, u\rangle \geq \tau\|\Delta u\|_2^2 + \nu\|Du\|_2^2\,.$$

(d) The assertion follows from Proposition 7.4.3 and the fact that

$$J^2(\Omega) \cap G^{-2,2}(\Omega) = \{0\}\,.$$

□

In the following, we will also consider the lower semicontinuous and convex functionals $\varphi, \varphi_R : J^2(\Omega) \to [0, +\infty]$ defined as

$$\varphi(u) = \begin{cases} \displaystyle\int_\Omega \left(\frac{\tau}{2}\,|\Delta u|^2 + \frac{\nu}{2}\,|Du|^2\right) dx & \text{if } u \in J^{2,2}(\Omega)\,, \\ +\infty & \text{otherwise}\,, \end{cases}$$

$$\varphi_R(u) = \begin{cases} \varphi(u) & \text{if } u \in K_R\,, \\ +\infty & \text{otherwise}\,. \end{cases}$$

It is easily seen that

$$f - \pi[(Du)u] \in \partial\varphi_R(u) \qquad \text{for all } u \in D(A) \text{ and } f \in A(u)\,,$$

where

$$\partial\varphi_R(u) = \left\{g \in J^2(\Omega) : \ \varphi_R(v) \geq \varphi_R(u) + \langle g, v - u\rangle \text{ for all } v \in K_R\right\}\,.$$

7.5 Uniqueness and Existence

Let Ω be an open subset of $\mathbb{R}^n$ with $n \leq 5$, let $\tau > 0$ and $\nu \geq 0$. The next concept is an adaptation of [5, Définition 3.1] to our setting.

Definition 7.5.1. We say that u is a *strong solution* of

$$\begin{cases} \dfrac{\partial u}{\partial t} + (Du)u + \nabla p = \nu \Delta u - \tau \Delta\Delta u & \text{in }]0,+\infty[\times\Omega \\ \operatorname{div} u = 0 & \text{in }]0,+\infty[\times\Omega \\ u(t,x) = \dfrac{\partial u}{\partial n}(t,x) = 0 & \text{on } \partial\Omega \end{cases} \tag{7.2}$$

if $u : [0,+\infty[\to J^2(\Omega)$ is continuous and the following facts hold:

- $u :]0,+\infty[\to J^2(\Omega)$ is absolutely continuous on compact subsets,
- $u(t) \in J^{2,2}(\Omega)$ for a.a. $t > 0$,
- $u' + \tau\Delta\Delta u - \nu\Delta u + (Du)u \in G^{-2,2}(\Omega)$ for a.a. $t > 0$.

In particular, it follows that for a.a. $t > 0$ there exists $p \in W^{-1,2}_{loc}(\Omega)$ such that

$$u' + (Du)u + \nabla p = \nu\Delta u - \tau\Delta\Delta u \qquad \text{in } W^{-2,2}(\Omega)$$

and we have

$$\Delta\Delta u(t) \in J^2(\Omega) \oplus G^{-2,2}(\Omega) \quad \text{for a.a. } t > 0\,.$$

Theorem 7.5.2. Let u_1, u_2 be two strong solutions of (7.2) and let

$$R \geq \max\left\{\|u_1(0)\|_2, \|u_2(0)\|_2\right\}.$$

Then we have

$$\|u_j(T)\|_2^2 + 2\tau \int_0^T \|\Delta u_j\|_2^2\, dt \leq \|u_j(0)\|_2^2 \qquad \text{for all } j = 1,2 \text{ and } T \geq 0\,,$$

$$\|u_1(T) - u_2(T)\|_2 \leq \|u_1(0) - u_2(0)\|_2 \exp\left(\omega_R T\right) \qquad \text{for all } T \geq 0\,,$$

where ω_R is given by Proposition 7.4.1.

Proof. First of all, for a.a. $t > 0$ we have

$$\frac{d}{dt}\|u_j\|_2^2 = 2\left\langle u_j', u_j\right\rangle = -2\tau\|\Delta u_j\|_2^2 - 2\nu\|Du_j\|_2^2 \leq -2\tau\|\Delta u_j\|_2^2\,,$$

whence

$$\|u_j(T)\|_2^2 + 2\tau \int_{\underline{t}}^T \|\Delta u_j\|_2^2\, dt \leq \|u_j(\underline{t})\|_2^2 \qquad \text{whenever } 0 < \underline{t} \leq T\,.$$

By the continuity of u_j and the monotone convergence theorem, it follows that

$$\|u_j(T)\|_2^2 + 2\tau \int_0^T \|\Delta u_j\|_2^2 \, dt \le \|u_j(0)\|_2^2 \le R^2 \qquad \text{for all } T \ge 0\,.$$

Then, for a.a. $t > 0$ we also have

$$\frac{d}{dt}\|u_1 - u_2\|_2^2 = 2\,\langle u_1' - u_2', u_1 - u_2\rangle = -2\,\langle F(u_1) - F(u_2), u_1 - u_2\rangle \le 2\omega_R \|u_1 - u_2\|_2^2\,,$$

whence

$$\|u_1(t) - u_2(t)\|_2 \le \|u_1(\tau) - u_2(\tau)\|_2 \exp\left(\omega_R (t - \tau)\right) \qquad \text{whenever } 0 < \tau < t < T$$

and the second inequality also follows from the continuity of u_1 and u_2. □

Theorem 7.5.3. For every $u_0 \in J^{2,2}(\Omega)$ with

$$\Delta\Delta u_0 \in J^2(\Omega) \oplus G^{-2,2}(\Omega)\,,$$

there exists one and only one strong solution u of (7.2) such that $u(0) = u_0$.

Moreover, for every $T > 0$ the following facts hold:

- $u : [0, T] \to J^2(\Omega)$ is Lipschitz continuous;
- $u(t) \in J^{2,2}(\Omega)$ for all $t \ge 0$ and

$$\sup\left\{\|\Delta u(t)\|_2 : \; 0 \le t \le T\right\} < +\infty\,;$$

- $\Delta\Delta u(t) \in J^2(\Omega) \oplus G^{-2,2}(\Omega)$ for all $t \ge 0$ and

$$\sup\left\{\|\pi(\Delta\Delta u(t))\|_2 : \; 0 \le t \le T\right\} < +\infty\,;$$

- $\varphi \circ u : [0, T] \to \mathbb{R}$ is Lipschitz continuous and

$$\|u'\|_2^2 + (\varphi \circ u)' = \langle (Du)u, u'\rangle \qquad \text{for a.a. } t > 0\,.$$

Proof. Let $R > \|u_0\|_2$ and ω_R as in Proposition 7.4.1. Since $u_0 \in D(A)$, combining Theorem 7.4.3 with [5, Théorème 3.17] or [24, Lecture 6], we infer that there exists a locally Lipschitz map $u : [0, +\infty[\to J^2(\Omega)$ such that $u(0) = u_0$, $u(t) \in D(A)$ for all $t \ge 0$, namely

$$\begin{aligned} & u(t) \in J^{2,2}(\Omega) && \text{for all } t \ge 0\,,\\ & \Delta\Delta u(t) \in J^2(\Omega) \oplus G^{-2,2}(\Omega) && \text{for all } t \ge 0\,, \end{aligned}$$

and

$$\sup_{0\le t\le T}\ \inf_{f\in A(u(t))}\|f\|_2<+\infty \qquad \text{for all } T>0\,,$$
$$-\,u'(t)\in A(u(t)) \qquad \text{for a.a. } t>0\,.$$

Then we also have

$$-u'(t)-\pi[(Du(t))u(t)]\in\partial\varphi_R(u(t)) \qquad \text{for a.a. } t>0\,.$$

Since $\langle f,u(t)\rangle\ge 0$ for all $t\ge 0$ and $f\in N_{K_R}(u(t))$, we have

$$\frac{d}{dt}\|u\|_2^2=2\,\langle u',u\rangle\le -2\,\langle F(u),u\rangle=-2\tau\|\Delta u\|_2^2-2\nu\|Du\|_2^2\le 0\,,$$

whence $\|u(t)\|_2\le\|u_0\|_2<R$ for all $t\ge 0$. By Proposition 7.4.3 it follows that $N_{K_R}(u(t))=G^{-2,2}(\Omega)$ for all $t\ge 0$, so that u is a strong solution of (7.2) satisfying

$$\sup\{\|\Delta u(t)\|_2:\ 0\le t\le T\}<+\infty\,,$$
$$\sup\{\|\pi(\Delta\Delta u(t))\|_2:\ 0\le t\le T\}<+\infty\,,$$

for all $T>0$, by Theorem 7.4.3 and Proposition 7.3.1.

In particular the map $t\mapsto\pi[(Du(t))u(t)]$ belongs to $L^\infty(0,T;J^2(\Omega))$. By [5, Théorème 3.6] the function $\varphi\circ u$ is Lipschitz continuous on $[0,T]$ with

$$\|u'\|_2^2+(\varphi\circ u)'=\langle(Du)u,u'\rangle \qquad \text{for a.a. } t>0\,.$$

Uniqueness follows from Theorem 7.5.2. □

Theorem 7.5.4. For every $u_0\in J^{2,2}(\Omega)$, there exists one and only one strong solution u of (7.2) such that $u(0)=u_0$.

Moreover, for every $T>0$ the following facts hold:

- $u:[0,T]\to J^2(\Omega)$ is absolutely continuous;
- $u(t)\in J^{2,2}(\Omega)$ for all $t\ge 0$ and
$$\sup\{\|\Delta u(t)\|_2:\ 0\le t\le T\}<+\infty\,;$$
- $\varphi\circ u:[0,T]\to\mathbb{R}$ is absolutely continuous and
$$\|u'\|_2^2+(\varphi\circ u)'=\langle(Du)u,u'\rangle \qquad \text{for a.a. } t>0\,;$$
- $\Delta\Delta u(t)\in J^2(\Omega)\oplus G^{-2,2}(\Omega)$ for all $t>0$;

– for every $t_0 > 0$, the map $v(t) = u(t_0 + t)$ is the strong solution of (7.2) such that $v(0) = u(t_0)$, according to Theorem 7.5.3.

Proof. Let us first observe that, if u is given by Theorem 7.5.3 and $R > \|u(0)\|$, we have

$$\|u'\|_2^2 + (\varphi \circ u)' = \langle (Du)u, u' \rangle \le \frac{1}{2}\,\|(Du)u\|_2^2 + \frac{1}{2}\,\|u'\|_2^2\,,$$

whence

$$\|u'\|_2^2 + 2(\varphi \circ u)' \le \|(Du)u\|_2^2\,.$$

On the other hand, by Proposition 7.3.1 there exists $C > 0$ such that

$$\|(Dv)v\|_2 \le C\left(\tau\|\Delta v\|_2^2 + \|v\|_2^2\right) \le 2C\varphi(v) + CR^2 \qquad \text{for all } v \in K_R\,.$$

In particular, we have

$$\|u'\|_2^2 + 2(\varphi \circ u)' \le \left(2C(\varphi \circ u) + CR^2\right)^2 \qquad \text{for a.a. } t > 0\,. \tag{7.3}$$

Now let $R > \|u_0\|_2$ and let $(u_0^{(k)})$ be a sequence in $\mathcal{C}_{div}(\Omega)$ converging to u_0 in $J^{2,2}(\Omega)$. Then $\Delta\Delta u_0^{(k)} \in J^2(\Omega)$ for all $k \in \mathbb{N}$. Let $u^{(k)}$ be the strong solution of (7.2) such that $u^{(k)}(0) = u_0^{(k)}$ given by Theorem 7.5.3, so that

$$-\left(u^{(k)}\right)'(t) \in A(u^{(k)}(t)) \qquad \text{for a.a. } t > 0\,.$$

According to [24], there exists a continuous map $u : [0, +\infty[\to J^2(\Omega)$ such that $(u^{(k)})$ is uniformly convergent to u on $[0, T]$ for all $T > 0$, as $k \to \infty$. In particular, $u(0) = u_0$.

Moreover, from (7.3) we infer that there exists $\delta > 0$ such that

$$\sup\left\{\varphi(u^{(k)}(t)) : \ k \in \mathbb{N}\,,\ 0 \le t \le \delta\right\} < +\infty\,,$$

whence $u(t) \in J^{2,2}(\Omega)$ for all $t \in [0, \delta]$ with

$$\sup\{\|\Delta u(t)\|_2 : \ 0 \le t \le \delta\} < +\infty$$

and $\pi[(Du)u] \in L^\infty(0, \delta; J^2(\Omega))$.

Again by (7.3) it follows that

$$\sup_k \int_0^\delta \|\left(u^{(k)}\right)'\|_2^2\, dt < +\infty\,.$$

Therefore $u : [0, \delta] \to J^2(\Omega)$ is absolutely continuous and, for every $t_0 > 0$, there exists $\underline{t} \in]0, t_0]$ such that each $u^{(k)}$ is differentiable at $\underline{t}$ with

$$-\left(u^{(k)}\right)'(\underline{t}) \in A(u^{(k)}(\underline{t})) \qquad \text{for all } k \in \mathbb{N}\,,$$

$$\liminf_k \|\left(u^{(k)}\right)'(\underline{t})\|_2 < +\infty\,.$$

From [24, Lecture 4, Lemma 1] we infer that $u(\underline{t}) \in D(A)$. Again by [24] it follows that $u(t) \in D(A)$ for all $t \geq \underline{t}$ and $w(t) = u(\underline{t} + t)$ is the strong solution of (7.2) such that $w(0) = u(\underline{t})$, according to Theorem 7.5.3. In particular, $u(t_0) \in D(A)$ and $v(t) = u(t_0+t)$ is the strong solution of (7.2) such that $v(0) = u(t_0)$, according to Theorem 7.5.3. Since

$$-u'(t) - \pi[(Du(t))u(t)] \in \partial\varphi_R(u(t)) \qquad \text{for a.a. } t > 0\,,$$

by [5, Théorème 3.6] the function $\varphi \circ u$ is absolutely continuous on $[0, \delta]$.

Then the other assertions follow from Theorem 7.5.3. □

Theorem 7.5.5. For every $u_0 \in J^2(\Omega)$, there exists one and only one strong solution u of (7.2) such that $u(0) = u_0$.

Moreover, we have that

$$u(t) \in J^{2,2}(\Omega)\,, \qquad \Delta\Delta u(t) \in J^2(\Omega) \oplus G^{-2,2}(\Omega) \qquad \text{for all } t > 0$$

and, for every $t_0 > 0$, the map $v(t) = u(t_0 + t)$ is the strong solution of (7.2) such that $v(0) = u(t_0)$, according to Theorem 7.5.3.

Proof. Let $R > \|u_0\|_2$ and let $(u_0^{(k)})$ be a sequence in $\mathcal{C}_{div}(\Omega)$ converging to u_0 in $J^2(\Omega)$. As before, let $u^{(k)}$ be the strong solution of (7.2) such that $u^{(k)}(0) = u_0^{(k)}$ given by Theorem 7.5.3 and let $u : [0, +\infty[\to J^2(\Omega)$ be the continuous map such that $(u^{(k)})$ is uniformly convergent to u on $[0, T]$ for all $T > 0$, as $k \to \infty$. In particular, $u(0) = u_0$.

Moreover, by Theorem 7.5.2 we have

$$\sup_k \int_0^T \|\Delta u^{(k)}\|_2^2\, dt < +\infty \qquad \text{for all } T > 0\,.$$

Therefore, for every $t_0 > 0$, there exists $\underline{t} \in]0, t_0]$ such that

$$\liminf_k \|\Delta u^{(k)}(\underline{t})\|_2 < +\infty\,,$$

whence $u(\underline{t}) \in J^{2,2}(\Omega)$. As before, $w(t) = u(\underline{t} + t)$ is the strong solution of (7.2) such that $w(0) = u(\underline{t})$, according to Theorem 7.5.4 and the assertion follows. □

7.6 Regularity in Space for Exterior Domains

Lemma 7.6.1. Let $u \in J^{2,2}(\mathbb{R}^n)$, $p \in W^{-1,2}_{loc}(\mathbb{R}^n)$ with $\nabla p \in W^{-2,2}(\mathbb{R}^n)$ and $f \in W^{-1,2}(\mathbb{R}^n)$ be such that

$$\int_{\mathbb{R}^n} (\Delta u \cdot \Delta v + u \cdot v)\, dx = \langle f + \nabla p, v\rangle \qquad \text{for all } v \in W^{2,2}(\mathbb{R}^n)\,.$$

Then $u \in J^{3,2}(\mathbb{R}^n)$, $p \in L^2_{loc}(\mathbb{R}^n)$ with $\nabla p \in W^{-1,2}(\mathbb{R}^n)$ and

$$\|u\|_{3,2} + \|\nabla p\|_{-1,2} \leq c\|f\|_{-1,2}\,.$$

Proof. We have

$$\int_{\mathbb{R}^n} (\Delta u \cdot \Delta v + u \cdot v)\, dx = \langle f, v \rangle \qquad \text{for all } v \in J^{2,2}(\mathbb{R}^n)\,. \tag{7.4}$$

Then $u \in J^{3,2}(\mathbb{R}^n)$ and

$$\|u\|_{3,2} \le c\|f\|_{-1,2}\,. \tag{7.5}$$

To see this, let $\hat{u}$, $\hat{f}$ be the Fourier transforms of u and f, respectively. Then

$$\|f\|_{-1,2}^2 = \int_{\mathbb{R}^n} \frac{|\hat{f}(\xi)|^2}{|\xi|^2+1}\, d\xi$$

and (7.4) is equivalent to

$$\int_{\mathbb{R}^n} (|\xi|^4+1)\hat{u}(\xi) \cdot v(\xi)\, d\xi = \int_{\mathbb{R}^n} v(\xi) \cdot \hat{f}(\xi)\, d\xi \tag{7.6}$$

for all $v \in L^2(\mathbb{R}^n)$ with

$$\int_{\mathbb{R}^n} (|\xi|^4+1)|v(\xi)|^2\, d\xi < +\infty \qquad \text{and} \qquad \xi \cdot v(\xi) = 0 \quad \text{a.e. in } \mathbb{R}^n.$$

Fix $k \in \mathbb{N}$ and take

$$v(\xi) = (|\xi|^2+1)\chi_{\{|\xi|<k\}}(\xi)\, \hat{u}(\xi)$$

in (7.6). Then

$$\int_{\{|\xi|<k\}} (|\xi|^4+1)(|\xi|^2+1)|\hat{u}(\xi)|^2\, d\xi = \int_{\{|\xi|<k\}} (|\xi|^2+1)^{3/2}\hat{u}(\xi) \cdot \frac{\hat{f}(\xi)}{(|\xi|^2+1)^{1/2}}\, d\xi\,,$$

which implies

$$\int_{\{|\xi|<k\}} (|\xi|^2+1)^3|\hat{u}(\xi)|^2\, d\xi \le c\left(\int_{\mathbb{R}^n} \frac{|\hat{f}(\xi)|^2}{|\xi|^2+1}\right)^{1/2} \left(\int_{\{|\xi|<k\}} (|\xi|^2+1)^3|\hat{u}(\xi)|^2\, d\xi\right)^{1/2},$$

namely

$$\left(\int_{\{|\xi|<k\}} (|\xi|^2+1)^3|\hat{u}(\xi)|^2\, d\xi\right)^{1/2} \le c\left(\int_{\mathbb{R}^n} \frac{|\hat{f}(\xi)|^2}{|\xi|^2+1}\right)^{1/2}.$$

Passing to the limit as $k \to +\infty$, we get formula (7.5).

Since $\nabla p = \tau\Delta^2 u - \nu\Delta u + u - f$, the assertion follows. □

Lemma 7.6.2. Let $u \in J^{2,2}(\mathbb{R}^n)$, $p \in W^{-1,2}_{loc}(\mathbb{R}^n)$ with $\nabla p \in W^{-2,2}(\mathbb{R}^n)$ and $f \in L^2(\mathbb{R}^n)$ be such that

$$\int_{\mathbb{R}^n} (\Delta u \cdot \Delta v + u \cdot v)\, dx = \langle f + \nabla p, v \rangle \qquad \text{for all } v \in W^{2,2}(\mathbb{R}^n)\,.$$

Then $u \in J^{4,2}(\mathbb{R}^n)$, $p \in W^{1,2}_{loc}(\mathbb{R}^n)$ with $\nabla p \in L^2(\mathbb{R}^n)$ and

$$\|u\|_{4,2} + \|\nabla p\|_2 \le c\|f\|_2\,.$$

Proof. The argument is similar to the previous one. This time one takes

$$v(\xi) = (|\xi|^2 + 1)^2 \chi_{\{|\xi|<k\}}(\xi)\, \hat{u}(\xi)\,.$$

□

Lemma 7.6.3. Let $u \in W^{2,2}(\mathbb{R}^n)$, $p \in W^{-1,2}_{loc}(\mathbb{R}^n)$ with $\nabla p \in W^{-2,2}(\mathbb{R}^n)$, $f \in W^{-1,2}(\mathbb{R}^n)$ and $g \in W^{2,2}(\mathbb{R}^n)$ be such that

$$\begin{cases} \displaystyle\int_{\mathbb{R}^n} (\Delta u \cdot \Delta v + u \cdot v)\, dx = \langle f + \nabla p, v \rangle & \text{for all } v \in W^{2,2}(\mathbb{R}^n)\,, \\ \operatorname{div} u = g & \text{a.e. in } \mathbb{R}^n\,. \end{cases}$$

Assume also that there exists $R > 0$ such that $g = 0$ a.e. outside $B_R(0)$ and that

$$\int_{\mathbb{R}^n} g\, dx = 0\,.$$

Then $u \in W^{3,2}(\mathbb{R}^n)$, $p \in L^2_{loc}(\mathbb{R}^n)$ with $\nabla p \in W^{-1,2}(\mathbb{R}^n)$ and

$$\|u\|_{3,2} + \|\nabla p\|_{-1,2} \le c_R\, (\|f\|_{-1,2} + \|g\|_{2,2,})\,.$$

Proof. Since $g \in W^{2,2}_0(B_R(0))$, by [11, Theorem III.3.3] there exists $w \in W^{3,2}_0(B_R(0))$ such that $\operatorname{div} w = g$ a.e. in $B_R(0)$ and

$$\|w\|_{3,2} \le c_R \|g\|_{2,2}\,. \tag{7.7}$$

We can extend w to $\mathbb{R}^n$ with value 0 outside $B_R(0)$ and we get $w \in W^{3,2}(\mathbb{R}^n)$ with $\operatorname{div} w = g$ a.e. in $\mathbb{R}^n$.

Then $(u - w) \in J^{2,2}(\mathbb{R}^n)$ and

$$\int_{\mathbb{R}^n} (\Delta(u-w) \cdot \Delta v + (u-w) \cdot v)\, dx \\ = \langle (f - \Delta\Delta w - w) + \nabla p, v \rangle \qquad \text{for all } v \in W^{2,2}(\mathbb{R}^n)$$

with $\Delta\Delta w + w \in W^{-1,2}(\mathbb{R}^n)$ and

$$\|f - \Delta\Delta w - w\|_{-1,2} \le c\, (\|f\|_{-1,2} + \|w\|_{3,2})\,. \tag{7.8}$$

From Lemma 7.6.1, (7.7), and (7.8) the assertion follows. □

Lemma 7.6.4. Let $u \in W^{2,2}(\mathbb{R}^n)$, $p \in W^{-1,2}_{loc}(\mathbb{R}^n)$ with $\nabla p \in W^{-2,2}(\mathbb{R}^n)$, $f \in L^2(\mathbb{R}^n)$ and $g \in W^{3,2}(\mathbb{R}^n)$ be such that

$$\begin{cases} \displaystyle\int_{\mathbb{R}^n} (\Delta u \cdot \Delta v + u \cdot v)\, dx = \langle f + \nabla p, v \rangle & \text{for all } v \in W^{2,2}(\mathbb{R}^n)\,, \\ \operatorname{div} u = g & \text{a.e. in } \mathbb{R}^n\,. \end{cases}$$

Assume also that there exists $R > 0$ such that $g = 0$ a.e. outside $B_R(0)$ and that

$$\int_{\mathbb{R}^n} g\,dx = 0\,.$$

Then $u \in W^{4,2}(\mathbb{R}^n)$, $p \in W^{1,2}_{loc}(\mathbb{R}^n)$ with $\nabla p \in L^2(\mathbb{R}^n)$ and

$$\|u\|_{4,2} + \|\nabla p\|_2 \le c_R\left(\|f\|_2 + \|g\|_{3,2,}\right)\,.$$

Proof. It is a simple variant of the previous proof, with Lemma 7.6.1 replaced by Lemma 7.6.2. □

Now let Ω be an open subset of $\mathbb{R}^n$ such that $\mathbb{R}^n \setminus \Omega$ is bounded. Let $R > 0$ be such that $\mathbb{R}^n \setminus \Omega \subseteq B_R(0)$ and let $\vartheta_1, \vartheta_2 \in C_c^\infty(\mathbb{R}^n)$ be such that $0 \le \vartheta_j \le 1$ on $\mathbb{R}^n$, $\vartheta_1 = 1$ on $B_{2R}(0)$, $\vartheta_1 = 0$ outside $B_{3R}(0)$, $\vartheta_2 = 1$ on $B_{3R}(0)$, and $\vartheta_2 = 0$ outside $B_{4R}(0)$. We can also agree that, from now on, we will only consider $p \in W^{m,2}_{loc}(\Omega)$ satisfying

$$\langle p, \vartheta_2(1-\vartheta_2)\rangle = 0\,,$$

so that

$$\|p\|_{W^{m,2}(\{2R<|x|<4R\})} \le c_{R,m}\|\nabla p\|_{W^{m-1,2}(\Omega)} \qquad \text{for all } m \in \mathbb{Z}\,.$$

Lemma 7.6.5. Let $u \in J^{2,2}(\Omega)$, $p \in W^{-1,2}_{loc}(\Omega)$ with $\nabla p \in W^{-2,2}(\Omega)$ and $f \in W^{-1,2}(\Omega)$ be such that

$$\int_\Omega (\Delta u \cdot \Delta v + u \cdot v)\,dx = \langle f + \nabla p, v\rangle \qquad \text{for all } v \in W^{2,2}_0(\Omega)\,.$$

Then $(1-\vartheta_1)u \in W^{3,2}(\Omega)$, $(1-\vartheta_1)p \in L^2_{loc}(\Omega)$ with $\nabla[(1-\vartheta_1)p] \in W^{-1,2}(\Omega)$ and

$$\|(1-\vartheta_1)u\|_{3,2} + \|\nabla[(1-\vartheta_1)p]\|_{-1,2} \le c_R\|f\|_{-1,2}\,.$$

Proof. We have $\vartheta_1 u \in W^{2,2}_0(\Omega \cap B_{3R}(0))$ with

$$\int_\Omega \operatorname{div}(\vartheta_1 u)\,dx = \int_{\Omega\cap B_{3R}(0)} \operatorname{div}(\vartheta_1 u)\,dx = 0\,.$$

We can extend u to $\mathbb{R}^n$ with value 0 outside Ω and then $(1-\vartheta_1)u \in W^{2,2}(\mathbb{R}^n)$ with

$$\operatorname{div}[(1-\vartheta_1)u] = -\operatorname{div}(\vartheta_1 u) = -u \cdot \nabla\vartheta_1 = 0 \qquad \text{outside } B_{3R}(0)\,,$$

$\operatorname{div}[(1-\vartheta_1)u] \in W^{2,2}(\mathbb{R}^n)$ and

$$\int_{\mathbb{R}^n} \operatorname{div}[(1-\vartheta_1)u]\,dx = 0\,.$$

We can also define $(1-\vartheta_1)f \in W^{-1,2}(\mathbb{R}^n)$ by

$$\langle (1-\vartheta_1)f, v\rangle = \langle f, (1-\vartheta_1)v\rangle \qquad \text{for all } v \in W^{1,2}(\mathbb{R}^n)$$

and $(1-\vartheta_1)p \in W^{-1,2}_{loc}(\mathbb{R}^n)$ in a similar way, so that $\nabla[(1-\vartheta_1)p] \in W^{-2,2}(\mathbb{R}^n)$. Moreover, for every $v \in W^{2,2}(\mathbb{R}^n)$ we can test the equation on $(1-\vartheta_1)v \in W^{2,2}_0(\Omega)$, obtaining

$$\int_{\mathbb{R}^n} \{\Delta[(1-\vartheta_1)u]\cdot \Delta v + [(1-\vartheta_1)u]\cdot v\}\, dx$$
$$= \langle (1-\vartheta_1)f + \check{f} + \nabla[(1-\vartheta_1)p], v\rangle \qquad \text{for all } v \in W^{2,2}(\mathbb{R}^n),$$

where $\check{f} \in W^{-1,2}(\mathbb{R}^n)$ incorporates several terms from u, p and ϑ_1 and

$$\|\check{f}\|_{-1,2} \le c_R \left(\|u\|_{W^{2,2}(\{2R<|x|<3R\})} + \|p\|_{W^{-1,2}(\{2R<|x|<3R\})}\right).$$

Then the assertion follows from Lemma 7.6.3, as we know from the beginning that

$$\|u\|_{2,2} + \|\nabla p\|_{-2,2} \le c\|f\|_{-2,2}.$$

□

Lemma 7.6.6. Assume that $\partial\Omega$ is of class $C^{3,1}$. Let $u \in J^{2,2}(\Omega)$, $p \in W^{-1,2}_{loc}(\Omega)$ with $\nabla p \in W^{-2,2}(\Omega)$ and $f \in W^{-1,2}(\Omega)$ be such that

$$\int_\Omega (\Delta u\cdot \Delta v + u\cdot v)\, dx = \langle f + \nabla p, v\rangle \qquad \text{for all } v \in W^{2,2}_0(\Omega).$$

Then $\vartheta_1 u \in W^{3,2}(\Omega)$, $\vartheta_1 p \in L^2_{loc}(\Omega)$ with $\nabla(\vartheta_1 p) \in W^{-1,2}(\Omega)$ and

$$\|\vartheta_1 u\|_{3,2} + \|\nabla(\vartheta_1 p)]\|_{-1,2} \le c_\Omega \|f\|_{-1,2}.$$

Proof. Arguing as in the previous proof, now we have $\vartheta_1 u \in W^{2,2}_0(\Omega\cap B_{3R}(0))$, with $\operatorname{div}(\vartheta_1 u) = u\cdot\nabla\vartheta_1 \in W^{2,2}_0(\Omega\cap B_{3R}(0))$, and

$$\int_{\Omega\cap B_{3R}(0)} \{\Delta(\vartheta_1 u)\cdot\Delta v + (\vartheta_1 u)\cdot v\}\, dx$$
$$= \langle \vartheta_1 f + \check{f} + \nabla(\vartheta_1 p), v\rangle \qquad \text{for all } v \in W^{2,2}_0(\Omega\cap B_{3R}(0)),$$

with

$$\|\check{f}\|_{-1,2} \le c_R \left(\|u\|_{W^{2,2}(\{2R<|x|<3R\})} + \|p\|_{W^{-1,2}(\{2R<|x|<3R\})}\right).$$

Now the assertion follows from [3, Théorème B] in the case $m = -1$. □

Theorem 7.6.1. Assume that $\partial\Omega$ is of class $C^{3,1}$. Let $u \in J^{2,2}(\Omega)$, $p \in W^{-1,2}_{loc}(\Omega)$ with $\nabla p \in W^{-2,2}(\Omega)$ and $f \in W^{-1,2}(\Omega)$ be such that

$$\int_\Omega (\Delta u\cdot \Delta v + u\cdot v)\, dx = \langle f + \nabla p, v\rangle \qquad \text{for all } v \in W^{2,2}_0(\Omega).$$

Then $u \in W^{3,2}(\Omega)$, $p \in L^2_{loc}(\Omega)$ with $\nabla p \in W^{-1,2}(\Omega)$ and

$$\|u\|_{3,2} + \|\nabla p\|_{-1,2} \le c_\Omega \|f\|_{-1,2}.$$

Proof. It follows from Lemmas 7.6.5 and 7.6.6. □

Lemma 7.6.7. Let $u \in J^{2,2}(\Omega)$, $p \in W^{-1,2}_{loc}(\Omega)$ with $\nabla p \in W^{-2,2}(\Omega)$ and $f \in L^2(\Omega)$ be such that

$$\int_\Omega (\Delta u \cdot \Delta v + u \cdot v)\, dx = \langle f + \nabla p, v\rangle \qquad \text{for all } v \in W^{2,2}_0(\Omega)\,.$$

Then $(1-\vartheta_2)u \in W^{4,2}(\Omega)$, $(1-\vartheta_2)p \in W^{1,2}_{loc}(\Omega)$ with $\nabla[(1-\vartheta_2)p] \in L^2(\Omega)$ and

$$\|(1-\vartheta_2)u\|_{4,2} + \|\nabla[(1-\vartheta_2)p]\|_2 \le c_R \|f\|_2\,.$$

Proof. From Lemma 7.6.5 we first infer that $(1-\vartheta_1)u \in W^{3,2}(\Omega)$, $(1-\vartheta_1)p \in L^2_{loc}(\Omega)$ with $\nabla[(1-\vartheta_1)p] \in W^{-1,2}(\Omega)$ and

$$\|(1-\vartheta_1)u\|_{3,2} + \|\nabla[(1-\vartheta_1)p]\|_{-1,2} \le c_R \|f\|_{-1,2}\,.$$

Since $(1-\vartheta_2)u = (1-\vartheta_2)(1-\vartheta_1)u$ and $(1-\vartheta_2)p = (1-\vartheta_2)(1-\vartheta_1)p$, the proof is a simple variant of that of Lemma 7.6.5. □

Lemma 7.6.8. Assume that $\partial\Omega$ is of class $C^{3,1}$. Let $u \in J^{2,2}(\Omega)$, $p \in W^{-1,2}_{loc}(\Omega)$ with $\nabla p \in W^{-2,2}(\Omega)$ and $f \in L^2(\Omega)$ be such that

$$\int_\Omega (\Delta u \cdot \Delta v + u \cdot v)\, dx = \langle f + \nabla p, v\rangle \qquad \text{for all } v \in W^{2,2}_0(\Omega)\,.$$

Then $\vartheta_2 u \in W^{4,2}(\Omega)$, $\vartheta_2 p \in W^{1,2}_{loc}(\Omega)$ with $\nabla(\vartheta_2 p) \in L^2(\Omega)$ and

$$\|\vartheta_2 u\|_{4,2} + \|\nabla(\vartheta_2 p)]\|_2 \le c_\Omega \|f\|_2\,.$$

Proof. From Theorem 7.6.1 we first infer that $u \in W^{3,2}(\Omega)$, $p \in L^2_{loc}(\Omega)$ with $\nabla p \in W^{-1,2}(\Omega)$ and

$$\|u\|_{3,2} + \|\nabla p\|_{-1,2} \le c_\Omega \|f\|_{-1,2}\,.$$

Now the proof is a simple variant of that of Lemma 7.6.6, where we appeal now to [3, Théorème B] in the case $m = 0$. □

Theorem 7.6.2. Assume that $\partial\Omega$ is of class $C^{3,1}$. Let $u \in J^{2,2}(\Omega)$, $p \in W^{-1,2}_{loc}(\Omega)$ with $\nabla p \in W^{-2,2}(\Omega)$ and $f \in L^2(\Omega)$ be such that

$$\int_\Omega (\Delta u \cdot \Delta v + u \cdot v)\, dx = \langle f + \nabla p, v\rangle \qquad \text{for all } v \in W^{2,2}_0(\Omega)\,.$$

Then $u \in W^{4,2}(\Omega)$, $p \in W^{1,2}_{loc}(\Omega)$ with $\nabla p \in L^2(\Omega)$ and

$$\|u\|_{4,2} + \|\nabla p\|_2 \le c_\Omega \|f\|_2\,.$$

Proof. It follows from Lemmas 7.6.7 and 7.6.8. □

Now let $n \leq 5$ and let Ω be an open subset of $\mathbb{R}^n$ with $\partial\Omega$ of class $\mathcal{C}^{3,1}$ such that $\mathbb{R}^n \setminus \Omega$ is bounded. Let $\tau > 0$, $\nu \geq 0$, $R > 0$ and let

$$A : D(A) \longrightarrow \mathfrak{P}(J^2(\Omega)),$$

be the operator considered in Sect. 7.4.

Theorem 7.6.3. For every $u \in D(A)$ with $\|u\|_2 < R$, we have $u \in W^{4,2}(\Omega)$ and

$$\|u\|_{4,2} \leq c_\Omega \left(\|f(u)\|_2 + \|u\|_{2,2} + \|u\|_{2,2}^2\right),$$

where $A(u) = \{f(u)\}$ according to Theorem 7.4.3.

Proof. From Theorem 7.4.3 and Proposition 7.3.1 we infer that $\Delta\Delta u \in J^2(\Omega) \oplus G^{-2,2}(\Omega)$, so that

$$\Delta\Delta u = \pi(\Delta\Delta u) + \nabla p$$

with $p \in W^{-1,2}_{loc}(\Omega)$, and that

$$\begin{aligned}\|\pi(\Delta\Delta u)\|_2 &\leq c\,(\|f(u)\|_2 + \|\Delta u\|_2 + \|(Du)u\|_2) \\ &\leq c\left(\|f(u)\|_2 + \|u\|_{2,2} + \|u\|_{2,2}^2\right).\end{aligned}$$

Of course, we also have

$$\Delta\Delta u + u = \pi(\Delta\Delta u) + u + \nabla p$$

and the assertion follows from Theorem 7.6.2. □

Theorem 7.6.4. Let u be the solution given by Theorem 7.5.3. Then $u(t) \in W^{4,2}(\Omega)$ for all $t \geq 0$ and

$$\sup\{\|u(t)\|_{4,2} :\ 0 \leq t \leq T\} < +\infty \qquad \text{for all } T > 0.$$

Proof. From Theorem 7.5.3 we know that $u(t) \in J^{2,2}(\Omega)$ and $\Delta\Delta u(t) \in J^2(\Omega) \oplus G^{-2,2}(\Omega)$ for every $t > 0$. Moreover, for every $T > 0$ we have

$$\begin{aligned}&\sup\{\|u(t)\|_2 :\ 0 \leq t \leq T\} < +\infty, \\ &\sup\{\|\pi(\Delta\Delta u(t))\|_2 :\ 0 \leq t \leq T\} < +\infty.\end{aligned}$$

Then the argument is similar to that of the previous result. □

References

1. R.A. Adams, J.J.F. Fournier. *Sobolev Spaces.* Academic Press (2003).
2. S. Agmon, A. Douglis, L. Nirenberg. *Estimates near the boundary for solutions of elliptic partial differential equations satisfying general boundary conditions.* Communications on pure and applied mathematics, 12 (1959), 623–727.

3. C. Amrouche, V. Girault. *Problèmes généralisés de Stokes.* Potugal. Math. 49(4) (1992), 463–503.

4. S.S. Antman and J.E. Osborn. *The principle of virtual work and integral laws of motion.* Arch. Rational Mech. Anal., 69(3) (1979) 231–262.

5. H. Brezis. *Opérateurs maximaux monotones et semi-groupes de contractions dans les espaces de Hilbert,* North-Holland Publishing Co., Amsterdam (1973).

6. H. Brezis. *Functional analysis, Sobolev spaces and partial differential equations,* Springer, New York (2011).

7. F. Crispo, P. Maremonti. *An Interpolation Inequality in Exterior Domains.* Rend. Sem. Mat. Univ. Padova, 112 (2004).

8. M. Degiovanni, A. Marzocchi, A. Musesti. *Virtual powers on diffused subbodies and normal traces of tensor-valued measures.* M. Šilhavý ed., Mathematical modeling of bodies with complicated bulk and boundary behavior, Quaderni di Matematica, 20, 21–53.

9. E. Fried, M.E. Gurtin. *Tractions, Balances, and Boundary Conditions for Nonsimple Materials with Application to Liquid Flow at Small-Length Scales.* Arch. Rational Mech. Anal., 182 (2006) 513–554.

10. G. P. Galdi, On the motion of a rigid body in a viscous liquid: a mathematical analysis with applications, in *Handbook of mathematical fluid dynamics* Volume 1, North-Holland, Amsterdam (2002), 653–791.

11. G. P. Galdi, An introduction to the mathematical theory of the Navier-Stokes equations, Steady-state problems, Springer, New York, 2011.

12. Galdi, G.P., Neustupa, J. (2018). Steady-state Navier-Stokes flow around a moving body. In Handbook of Mathematical Analysis in Mechanics of Viscous Fluids. (pp. 341–417). doi: 10.1007/978-3-319-13344-7-7.

13. P. Germain. *La méthode des puissances virtuelles en mécanique des milieux continus. I. Théorie du second gradient.* Mécanique, 12 (1973) 235–274.

14. G.G. Giusteri, E. Fried. *Slender-body theory for viscous flow via dimensional reduction and hyperviscous regularization.* Meccanica, 49 (2014) 2153–2167.

15. G.G. Giusteri, A. Marzocchi, A. Musesti. *Three-dimensional nonsimple viscous liquids dragged by one-dimensional immersed bodies.* Mechanics Research Communications, 37 (2010), 642–646.

16. G. G. Giusteri, A. Marzocchi, A. Musesti. *Nonsimple isotropic incompressible linear fluids surrounding one-dimensional structures,* Acta Mechanica, 217 (3) (2011), 191–204.

17. G.G. Giusteri, A. Marzocchi, A. Musesti. *Steady free fall of one-dimensional bodies in a hyperviscous fluid at low Reynolds number,* Evol. Equat. Control Theory, 3(3) (2014), 429–445.

18. G.G. Giusteri, A. Marzocchi, A. Musesti. *Nonlinear free fall of one-dimensional rigid bodies in hyperviscous fluids*, Discret Contin. Dyn. Syst. Ser. B, 19(7) (2014), 2145–2157.

19. J.G. Heywood. *The Navier-Stokes Equations: On the Existence, Regularity and Decay of solutions.* Indiana Univ. Math. J., 29 (1980) 639–681.

20. D. Kinderlehrer, G. Stampacchia, *An introduction to variational inequalities and their applications*, Classics in Applied Mathematics, 31, SIAM, Philadelphia, PA, (2000).

21. A. Kufner, O. John, S. Fučík, *Function Spaces.* Noordhoff Int. Publishing Leyden (1977).

22. P. Maremonti. *Navier-Stokes initial boundary value problem: a short review of the chief results.* Unpublished.

23. A. Musesti. *Isotropic linear constitutive relations for nonsimple fluids.* Acta Mech., 204 (2009) 81–88.

24. A. Pazy. *Semi-groups of nonlinear contractions in Hilbert space.* G. Prodi ed., Problems in non-linear analysis, C.I.M.E. 1970, 343–430, Edizioni Cremonese, Rome, 1971.

25. R. Temam. *Navier-Stokes Equations: Theory and Numerical Analysis.* North-Holland Pub. Co., Amsterdam-New York-Tokyo.

Chapter 8

A Review on Rigorous Derivation of Reduced Models for Fluid–Structure Interaction Systems

Mario Bukal
University of Zagreb, Faculty of Electrical Engineering and Computing, Zagreb, Croatia
e-mail: mario.bukal@fer.hr

Boris Muha (✉)
University of Zagreb, Faculty of Science, Department of Mathematics, Zagreb, Croatia
e-mail: borism@math.hr

In this paper we review and systematize the mathematical theory on justification of sixth-order thin-film equations as reduced models for various fluid–structure interaction systems in which fluids are lubricating underneath elastic structures. Justification is based on careful examination of energy estimates, weak convergence results of solutions of the original fluid–structure interaction systems to the solution of the sixth-order thin-film equation, and quantitative error estimates that provide even strong convergence results.

T. Bodnár et al. (eds.), *Waves in Flows*, Advances in Mathematical Fluid Mechanics, https://doi.org/10.1007/978-3-030-68144-9_8

8.1 Introduction

In 1886 Reynolds derived the fundamental equation of the lubrication approximation [45], which serves until nowadays in many engineering applications. It is an elliptic equation for the pressure distribution in relatively thin viscous fluid in laminar flow between two rigid parallel plates in a relative motion of constant velocity. The equation can be understood as a reduced model for the basic Navier–Stokes equations describing the fluid motion. In Sect. 8.2 we outline the heuristic derivation of the Reynolds equation. However, often in nature and in engineering applications, those "parallel plates" are not rigid but have their own dynamics that also affect the fluid quantities. Such systems are widely known as fluid–structure interaction (FSI) systems. They appear in medicine, in particular in modelling of cardio-vascular systems [5, 9], then in aero-elasticity [21], marine engineering [54], etc., and as interesting systems of partial differential equations, they also gained a huge attention in applied mathematics community.

In this review paper we are particularly interested in FSI systems in which fluids are lubricating underneath elastic structures. Such models describe, for instance, the growth of magma intrusions [34, 36], the fluid-driven opening of fractures in the Earth's crust [8, 26], subglacial floods [18, 51], the passage of air flow in the lungs [27], and the operation of vocal cords [50]. They are also inevitable in engineering, for example, in manufacturing of silicon wafers [30, 31], suppression of viscous fingering [43, 44], and in an emerging area of microfluidics [29, 32, 49] with particular applications to the so-called lab-on-a-chip technologies [19, 47]. Describing such systems with their true physical models: Navier–Stokes equations coupled with elasticity equations both on relatively thin domains is inappropriate from analytical and numerical points of view and, thus, inappropriate for engineering applications. Therefore, reduced and simplified models are sought that will maintain the essence of the fluid–structure interaction. Depending on the original problem at hand, many such reduced models have been derived, especially in the engineering literature. We emphasize at this point that our aim is not to cover all those examples, which would be impossible, but we consider several important model examples and concentrate on rigorous derivation of the reduced models and identification of the necessary scaling assumption in system parameters, which "sees" the interaction between the two subsystems.

Using asymptotic expansion techniques, several reduced models of Biot type describing the flow through a long elastic axially symmetric channel have been derived in [15, 37, 48]. A rigorous justification of those reduced models by means of weak convergence results and the corresponding error estimates was provided in [15]. Periodic flow in thin channel with visco-elastic walls was analyzed in [41], where starting from a linear 2D (fluid)/1D (structure) FSI model and under particular assumption on the ratio of the channel height and rigidity of the wall, a linear sixth-order thin-film equation describing the wall displacement emanated as a reduced model. A similar problem has been also addressed in [16], and the reduced model in terms of a linear sixth-order equation arose again. The reduced models in both papers have been justified by the corresponding weak convergence results.

In this review paper we mainly focus on linear FSI problems, meaning that equations for both subsystems are linear and, moreover, the interaction between the fluid and the structure is realized through a fixed interface. The main reason for such simple FSI models is that they provide global-in-time existence of weak solutions that possess sufficient regularity for passing from the FSI to the reduced model in a rigorous way by means of weak convergence results. We discuss two types of the linear FSI problems. First, a 3D/2D FSI problem, analogous to the FSI problem in [41], is discussed in Sect. 8.3.1. In this model the structure is originally described by a lower-dimensional elasticity model. Hence, the dimension reduction only applies to the fluid part. Additional horizontal dimension in our case does not bring any conceptual novelty. In contrast to the asymptotic expansion techniques employed in [41], our approach relies on careful examination of energy estimates. Based on these, quantitative a priori estimates in terms of the fluid thickness have been derived for weak solutions, and weak convergence results have been established. Moreover, these quantitative a priori estimates provide the necessary scaling assumption on model parameters, which ensures the nontrivial reduced model. The same ideas have been recently employed in [6], where the authors analyzed a linear 3D/3D FSI problem, in which the simultaneous dimension reduction in the fluid and the structure has been performed and again a linear sixth-order thin-film equation has been derived as the reduced model under particular scaling assumptions. We briefly report on this problem and results in Sect. 8.3.2. Finally, we address the physically most relevant nonlinear FSI problem in which equations of fluid motion are nonlinear and the fluid domain is also unknown in the system. Unfortunately, the well-posedness results for such problems are very scarce; hence, for the beginning we restrict our analysis to a 2D/1D FSI problem for which the existence of global-in-time strong solutions is available from [24]. Rigorous derivation of the nonlinear sixth-order thin-film equation, which is known in the engineering literature [26, 29, 34], as the reduced model for this FSI problem is still work in progress [7], and only main ideas are outlined in Sect. 8.3.3.

8.2 Heuristic Derivation of Reduced Models

We start with the heuristic derivation of reduced models, which is essentially based on the lubrication approximation in the fluid part and the assumption of the pressure balance on the interface.

Let us consider the Navier–Stokes system in a time-dependent domain $\Omega_\eta = \{(x,z) \in (0,L) \times (0,\eta(x,t))\}\mathbb{R}^2$, where $L > 0$ is a fixed length and $\eta(x,t) \geq \eta_0 > 0$ is a given uniformly positive, L-periodic smooth function that depicts the evolution of the upper boundary of Ω_η. The fluid velocity $\boldsymbol{v} = (v_1, v_2)$ and pressure p then satisfy

$$\varrho_f \partial_t \boldsymbol{v} + \varrho_f (\boldsymbol{v} \cdot \nabla)\boldsymbol{v} - \mu \Delta \boldsymbol{v} = -\nabla p, \quad \text{on } \Omega_\eta \times (0,\infty), \tag{8.1}$$

$$\operatorname{div} \boldsymbol{v} = 0, \quad \text{on } \Omega_\eta \times (0,\infty). \tag{8.2}$$

On the bottom boundary we consider fluid moving with a constant velocity in horizontal direction, i.e., $\boldsymbol{v}(\cdot,0,\cdot) = (v_D, 0)$, and on the top boundary the fluid moves with the velocity of the surface, i.e., $v_2 = \partial_t \eta$. Concerning the lateral boundaries, for the completeness of the problem we may take the periodic boundary conditions, but for the purpose of this exposition, they are not so important. Imposing initial velocity $\boldsymbol{v}(\cdot,\cdot,0) = \boldsymbol{v}_0$ completes the problem (8.1)–(8.2).

In many physical situations like spilled water on floor, fresh paint on a wall, or industrially more relevant microfluidics devices and lab-on-a-chip technologies [19, 33, 47], one typically has the length in one spatial direction (typically vertical) much smaller than the other one, i.e., $H := \sup_{x,t} \eta(x,t) \ll L$. Let us denote $\varepsilon := H/L$ and write down Eqs. (8.1) and (8.2) in a nondimensional form in terms of the small parameter ε. For that purpose, we introduce new nondimensional variables

$$\hat{x} = \frac{x}{L}, \quad \hat{z} = \frac{z}{H}, \quad \hat{v}_\alpha = \frac{v_\alpha}{V_\alpha}, \; \alpha = 1,2, \quad \hat{p} = \frac{p}{P}, \quad \hat{t} = \frac{t}{\mathsf{T}}, \quad \hat{\eta} = \frac{\eta}{H},$$

where V_α, P, and T denote nominal values of fluid velocities, pressure, and time scale, respectively. The following calculations are standard in many fluid mechanics textbooks or lecture notes, see, for instance, [46]. Performing the above change of variables in (8.1) and (8.2), and neglecting the hat notation in new variables, we find

$$\frac{\varrho_f V_1 L}{PT} \partial_t v_1 + \frac{\varrho_f V_1^2}{L}\left(v_1 \partial_x v_1 + \frac{V_2}{V_1} \varepsilon^{-1} v_2 \partial_z v_1 \right) - \frac{\mu V_1}{LP}\left(\partial_{xx} v_1 + \varepsilon^{-2} \partial_{zz} v_1 \right) + \partial_x p = 0,$$

$$\frac{\varrho_f V_2 L}{PT} \partial_t v_2 + \frac{\varrho_f V_2^2}{L}\left(\frac{V_1}{V_2} v_1 \partial_x v_2 + \varepsilon^{-1} v_2 \partial_z v_2 \right) - \frac{\mu V_2}{LP}\left(\partial_{xx} v_2 + \varepsilon^{-2} \partial_{zz} v_2 \right) + \varepsilon^{-1} \partial_z p = 0,$$

$$\frac{V_1}{V_2} \partial_x v_1 + \varepsilon^{-1} \partial_z v_2 = 0.$$

Notice that $\mu V_1/LP$ is a dimensionless quantity. It is customary in the lubrication approximation regime to assume that $\mu V_1/LP = \varepsilon^2$ and $V_2 = \varepsilon V_1$ [46]. Then the above system becomes

$$\varepsilon^2 \operatorname{Re}\left(\partial_t v_1 + v_1 \partial_x v_1 + v_2 \partial_z v_1 \right) - \varepsilon^2 \partial_{xx} v_1 - \partial_{zz} v_1 + \partial_x p = 0, \tag{8.3}$$

$$\varepsilon^4 \operatorname{Re}\left(\partial_t v_2 + v_2 \partial_x v_1 + v_2 \partial_z v_2 \right) - \varepsilon^4 \partial_{xx} v_2 - \varepsilon^2 \partial_{zz} v_2 + \partial_z p = 0, \tag{8.4}$$

$$\partial_x v_1 + \partial_z v_2 = 0, \tag{8.5}$$

where $\operatorname{Re} = \varrho_f L^2/(\mu \mathsf{T})$ denotes the Reynolds number. Under assumption of $\varepsilon \ll 1$ and $\varepsilon^2 \operatorname{Re} \ll 1$, and thus neglecting terms of those orders in (8.3)–(8.5), we arrive to the system

$$-\partial_{zz}v_1 + \partial_x p = 0\,, \tag{8.6}$$

$$\partial_z p = 0\,, \tag{8.7}$$

$$\partial_x v_1 + \partial_z v_2 = 0\,. \tag{8.8}$$

Integrating equation (8.8) with respect to z and employing the corresponding boundary conditions, we find

$$\partial_t \eta = -\partial_x \int_0^{\eta(x,t)} v_1(x,z)\,\mathrm{d}z\,. \tag{8.9}$$

Since Eq. (8.7) implies that the pressure is constant in the vertical direction, Eq. (8.6) is a Poisson equation with respect to z with Dirichlet boundary condition. Its explicit solution is then given by

$$v_1(x,z,t) = \frac{1}{2}z(z-\eta(x,t))\partial_x p(x,t) + \left(1 - \frac{z}{\eta}\right) v_D\,. \tag{8.10}$$

Integrating the last equation with respect to z and utilizing the obtained expression in (8.9), we arrive to the elliptic equation for the pressure p

$$-\,\partial_x\left(\frac{\eta^3}{12}\partial_x p\right) = -\partial_t \eta - \frac{1}{2}\partial_x(\eta v_D)\,. \tag{8.11}$$

Assuming that η is a stationary profile and horizontal velocity v_D is constant, Eq. (8.11) turns into the original Reynolds equation

$$-\,\partial_x\left(\eta^3 \partial_x p\right) = -6 v_D \partial_x \eta\,. \tag{8.12}$$

This is a fundamental equation of elastohydrodynamics, which has been derived by Reynolds himself in [45], and much later rigorously justified in [1, 13].

However, in most applications the upper boundary η is not a priori known, but it is coupled with pressure p through another equation. Physically, such equation describes the balance of forces on the interface between the two phases, and in the sequel we discuss three most common physical situations:

1. When the pressure is balanced by the gravity, then the pressure is proportional to η [26]. After appropriate time rescaling, Eq. (8.11) then becomes a porous medium-type equation

$$\partial_t \eta = \partial_x^2\left(\eta^4\right) - 6\partial_x(\eta v_D)\,. \tag{8.13}$$

   The mathematical theory of the porous medium-type equations like (8.13) is very well-developed and comprehended in the monograph [52].

2. In the presence of dominant surface tension force, the pressure is balanced by the linearized curvature, i.e., $p \sim -\partial_x^2 \eta$ [39]. Equation (8.11) then becomes the well-known thin-film equation [4, 39]

$$\partial_t \eta = -\partial_x \left(\eta^3 \partial_x^3 \eta\right) - 6\partial_x(\eta v_D)\,. \tag{8.14}$$

Fourth-order thin-film equations also gained huge attention in the applied mathematics and engineering community. We refer to [3, 14, 40] and references therein. Although they do not share some fundamental concepts of the second-order equations, like the maximum principle, which makes the analysis of fourth-order equations more difficult, they have rich mathematical structure that has been explored in numerous papers (cf., for instance, [2, 17, 23]).

3. When the fluid phase is covered by an elastic plate in dominantly bending regime, then the pressure satisfies $p \sim \partial_x^4 \eta$ and Eq. (8.11) reads

$$\partial_t \eta = \partial_x \left(\eta^3 \partial_x^5 \eta\right) - 6\partial_x(\eta v_D)\,. \tag{8.15}$$

Contrary to the fourth-order equations, the sixth-order thin-film equations that are physically as relevant as (8.14) did not gain a comparable attention in the literature. Equations of type (8.15) have been derived, for instance, in [29, 34].

Besides the gravity, other potential forces like capillarity, heating, Van der Waals forces, etc. with potential $\Phi(\eta)$ can be included into physical models (i)–(iii) leading to a general equation of type

$$\partial_t \eta = (-1)^{(\alpha-1)/2}\partial_x \left(\eta^3 \partial_x^\alpha \eta\right) + \partial_x \left(\eta^3 \partial_x \Phi'(\eta)\right) - 6\partial_x(\eta v_D)$$

with $\alpha = 1, 3, 5$. A review on the derivation of plethora of such fourth-order models can be found in [40].

8.3 Reduced Models for Fluid–Structure Interaction Problems: A Rigorous Approach

In this section we focus on FSI problems, formally the case (iii) of the previous section. We will consider fluid–structure interaction problems where both phases, fluid and structure, are relatively thin. Starting from an FSI problem, our aim is to rigorously justify the reduced model in terms of a sixth-order evolution equation of type (8.15). This means to prove that solutions of the original FSI problem converge in some sense to the solution of the sixth-order thin-film equation, and vice versa, by solving the sixth-order thin-film equation, one is able to construct an approximate solution to the original problem. Rigorous justification of reduced models is so far available only for linear problems due to the global-in-time well-posedness for the weak solutions, and such will be discussed here. Linear in this context means that equations of motion for both fluid and structure in FSI problem are linear, and moreover, the fluid domain is fixed and therefore the coupling is linear and is realized on the fixed fluid–structure interface.

As a model problem, we consider a three-dimensional channel of relative height $\varepsilon > 0$ that is filled with incompressible viscous fluid and the channel is covered by an elastic plate of relative height $h > 0$. We work in physical 3D space, although some results in the literature are available only in 2D, but we will emphasize when it comes to that point. Assume that the problem is properly nondimensionalized and denote by $\Omega = \Omega_\varepsilon \cup S_h \subset \mathbb{R}^3$ *the material domain*, where $\Omega_\varepsilon = (0,1)^2 \times (-\varepsilon, 0)$ denotes the fluid domain, and S_h denotes the structure domain, which depends on the structure model at hand. If we describe the structure dynamics by a lower-dimensional model, for instance, linear visco-elastic plate, then $S_h = (0,1)^2 \equiv \omega$ and we denote such problem as $\varepsilon{+}0$ *problem* (cf. Sect. 8.3.1). On the other hand, if the structure dynamics is fully described by linear elasticity equations, then $S_h = \omega \times (0, h)$ and the problem is denoted by $\varepsilon + h$ *problem* (cf. Sect. 8.3.2). The linear FSI problem is in general described by the system of partial differential equations

$$\varrho_f \partial_t \boldsymbol{v} - \operatorname{div} \sigma_f(\boldsymbol{v}, p) = f\,, \quad \Omega_\varepsilon \times (0, \infty)\,, \tag{8.16}$$

$$\operatorname{div} \boldsymbol{v} = 0\,, \quad \Omega_\varepsilon \times (0, \infty)\,, \tag{8.17}$$

$$\varrho_s \partial_{tt} \boldsymbol{u} + \mathbb{L}(\boldsymbol{u}, \partial_t \boldsymbol{u}) = g\,, \quad S_h \times (0, \infty)\,, \tag{8.18}$$

where Eqs. (8.16) and (8.17) denote the Stokes system for the fluid velocity $\boldsymbol{v}$ and the pressure p. The fluid Cauchy stress tensor is given by $\sigma_f(\boldsymbol{v}, p) = 2\nu \operatorname{sym} \nabla \boldsymbol{v} - pI_3$, where ν, $\varrho_f > 0$ denote the fluid viscosity and density, respectively, $\operatorname{sym}(\cdot)$ denotes symmetric part of the matrix, and f denotes the fluid external force. The structure displacement $\boldsymbol{u}$ is described by elasticity equation(s) (8.18), where $\mathbb{L}$ denotes the model-dependent elasticity operator, which can be decomposed as a sum $\mathbb{L}(\boldsymbol{u}, \partial_t \boldsymbol{u}) = \mathbb{B}\,\boldsymbol{u} + \mathbb{D}\,\partial_t \boldsymbol{u}$ of symmetric and positive operators $\mathbb{B}$ and $\mathbb{D}$ that correspond to elastic and viscoelastic energy of the structure, respectively, and are to be specified below. Coefficient ϱ_s is the structure density, and g denotes the force of fluid acting on the structure. The structure volume forces like, for instance, gravity are for simplicity excluded from our analysis.

The two subsystems need to be coupled through the interface conditions that we literary describe as: *continuity of velocities* (kinematic condition) and *balance of forces* (dynamic condition). Depending on the model at hand, they will be specified below. For simplicity of exposition, we assume periodic boundary conditions in horizontal variables for all unknowns. On the bottom of the channel, we assume the standard no-slip boundary condition for the fluid velocity, $\boldsymbol{v} = 0$, and the plate is free on the top boundary. The system is supplemented by trivial initial conditions

$$\boldsymbol{v}(0) = 0,\ \boldsymbol{u}(0) = 0,\ \partial_t \boldsymbol{u}(0) = 0\,, \tag{8.19}$$

although, all obtained results will also hold for nontrivial initial conditions under some additional smallness assumptions (cf., for instance, [6, Appendix]). A nontrivial volume force on the structure could also be involved, again under certain scaling assumptions (cf. again [6]), but the trivial one is in fact motivated by applications in microfluidics [47]. The previously settled framework also incorporates physically more relevant problem that involves prescribed pressure drop between inlet and outlet of the channel, instead of the periodic boundary conditions. As described in [41], this is a matter of the right choice of the fluid volume force f.

The simplified linear FSI problem (8.16)–(8.18) can be seen as a linearization of a truly nonlinear dynamics under the assumption of small displacements [53].

Let us now proceed with a formal analysis. Testing Eqs. (8.16) and (8.18) with assumed smooth solutions $\boldsymbol{v}$ and $\partial_t \boldsymbol{u}$, respectively, and integrating by parts and utilizing the divergence-free condition, we have: for every $t > 0$

$$\begin{aligned} &\frac{\varrho_f}{2}\int_{\Omega_\varepsilon} |\boldsymbol{v}(t)|^2 \mathrm{d}\boldsymbol{x} + 2\nu \int_0^t\int_{\Omega_\varepsilon} |\operatorname{sym}\nabla \boldsymbol{v}(s)|^2 \mathrm{d}\boldsymbol{x}\mathrm{d}s + \frac{\varrho_s}{2}\int_{S_h} |\partial_t \boldsymbol{u}(t)|^2 \mathrm{d}V \\ &+\frac{1}{2}\int_{S_h} \mathbb{B}\,\boldsymbol{u}(t)\cdot\boldsymbol{u}(t)\mathrm{d}V + \int_0^t\int_{S_h} \mathbb{D}\,\partial_t\boldsymbol{u}(s)\cdot\partial_t\boldsymbol{u}(s)\mathrm{d}V\mathrm{d}s = \int_0^t\int_{\Omega_\varepsilon} f\cdot\boldsymbol{v}\,\mathrm{d}\boldsymbol{x}\mathrm{d}s\,, \end{aligned} \tag{8.20}$$

where $\mathrm{d}V$ denotes the volume measure on the structure domain. Assuming that the fluid volume force satisfies $\|f\|_{L^\infty(0,\infty;L^\infty(\Omega_\varepsilon))} \le C$ and employing then the Poincaré and Korn inequalities on thin domains (cf. [6, Proposition A.2]), the right-hand side can be estimated and leads to the basic energy estimate: for every $t > 0$

$$\mathcal{E}_k(\boldsymbol{v}(t)) + \nu\int_0^t \mathcal{D}_f(\boldsymbol{v}(s))\mathrm{d}s + \mathcal{E}_k(\partial_t\boldsymbol{u}(t)) + \mathcal{E}_{el}(\boldsymbol{u}(t)) + \int_0^t \mathcal{D}_s(\partial_t\boldsymbol{u}(s))\mathrm{d}s \le Ct\varepsilon\,, \tag{8.21}$$

where $C > 0$ from now on denotes a generic positive constant independent of ε and t. Quantities $\mathcal{E}_k(\boldsymbol{v})$ and $\mathcal{E}_k(\partial_t\boldsymbol{u})$ denote kinetic energy of the fluid and the structure, $\mathcal{D}_f(\boldsymbol{v})$ and $\mathcal{D}_s(\partial_t\boldsymbol{u})$ denote the rate of the energy dissipation of the fluid and the structure, respectively, while $\mathcal{E}_{el}(\boldsymbol{u})$ denotes the elastic energy of the structure. All quantities are easily read off from (8.20).

Next, we briefly describe the concept of weak solutions on an abstract level, while details regarding the specific model are addressed in the respective subsections below. The choice of appropriate solution spaces is motivated by the above energy estimate. For the fluid velocity, this appears to be

$$\mathcal{V}_F(0,T;\Omega_\varepsilon) = L^\infty(0,T;L^2(\Omega_\varepsilon;\mathbb{R}^3)) \cap L^2(0,T;V_F(\Omega_\varepsilon))\,,$$

where $V_F(\Omega_\varepsilon) = \{\boldsymbol{v} \in H^1(\Omega_\varepsilon;\mathbb{R}^3) \ : \ \operatorname{div}\boldsymbol{v} = 0\,,\ \boldsymbol{v}|_{x_3=-\varepsilon} = 0\,,\ \boldsymbol{v} \text{ is } \omega\text{-periodic}\}$, and $T > 0$ is a given time horizon. Similarly, the structure function space will be

$$\mathcal{V}_S(0,T;S_h) = W^{1,\infty}(0,T;L^2(S_h;\mathbb{R}^3)) \cap L^\infty(0,T;\operatorname{Dom}(\mathbb{B}^{1/2})) \cap H^1(0,T;\operatorname{Dom}(\mathbb{D}^{1/2}))\,, \tag{8.22}$$

where $\operatorname{Dom}(\mathbb{B}^{1/2})$ and $\operatorname{Dom}(\mathbb{D}^{1/2})$ denote domains of respective operators. Finally, the solution space of the coupled problem (8.16)–(8.19) will be compound of previous spaces involving the kinematic interface condition (k. c.) as a constraint

$$\mathcal{V}(0,T;\Omega) = \{(\boldsymbol{v},\boldsymbol{u}) \in \mathcal{V}_F(0,T;\Omega_\varepsilon) \times \mathcal{V}_S(0,T;S_h) \ : \ \text{k. c. holds for a.e. } t \in (0,T)\}\,. \tag{8.23}$$

Now we can state the definition of weak solutions to our problem in the sense of Leray and Hopf.

Definition 8.3.1. We say that a pair $(\boldsymbol{v}^\varepsilon, \boldsymbol{u}^h) \in \mathcal{V}(0,T;\Omega)$ is a *weak solution* to the linear FSI problem (8.16)–(8.19), if the following variational equation holds in $\mathcal{D}'(0,T)$:

$$\begin{aligned}\varrho_f \frac{\mathrm{d}}{\mathrm{d}t}\int_{\Omega_\varepsilon} \boldsymbol{v}^\varepsilon \cdot \phi \,\mathrm{d}\boldsymbol{x} - \varrho_f \int_{\Omega_\varepsilon} \boldsymbol{v}^\varepsilon \cdot \partial_t\phi \,\mathrm{d}\boldsymbol{x} + 2\nu \int_{\Omega_\varepsilon} \operatorname{sym}\nabla\boldsymbol{v}^\varepsilon : \operatorname{sym}\nabla\phi \,\mathrm{d}\boldsymbol{x} + \varrho_s \frac{\mathrm{d}}{\mathrm{d}t}\int_{S_h} \boldsymbol{u}^h \cdot \partial_t\psi \,\mathrm{d}S \\ - \varrho_s \int_{S_h} \partial_t\boldsymbol{u}^h \cdot \partial_t\psi \,\mathrm{d}S + \int_{S_h} \mathbb{L}(\boldsymbol{u}^h, \partial_t\boldsymbol{u}^h)\cdot\psi \,\mathrm{d}S = \int_{\Omega_\varepsilon} f\cdot\phi\,\mathrm{d}\boldsymbol{x} + \int_{S_h} g\cdot\psi\,\mathrm{d}S\end{aligned} \tag{8.24}$$

for all $(\phi,\psi) \in \mathcal{W}(0,T;\Omega)$, where

$$\mathcal{W}(0,T;\Omega) = \{(\phi,\psi) \in C^1([0,T]; V_F(\Omega_\varepsilon) \times V_S(S_h)) \ : \ \phi(t)|_\omega = \psi(t)|_\omega \ \ \forall t \in [0,T]\}$$

denotes the space of test functions and $V_S(S_h) = \operatorname{Dom}(\mathbb{B}^{1/2}) \cap \operatorname{Dom}(\mathbb{D}^{1/2})$. Moreover, $(\boldsymbol{v}^\varepsilon, \boldsymbol{u}^h)$ verifies the energy inequality (8.21).

For the existence of weak solutions, one typically employs the Galerkin method, and formal estimate (8.21) provides crucial a priori estimates needed for the construction of a unique weak solution. The pressure in the system is treated in a standard manner, but unlike in the Stokes system solely, where the pressure is determined up to a function of time, here in the case of the full FSI problem the pressure is unique. This is a consequence of the fact that in the Stokes system the boundary (wall) is assumed to be rigid and therefore cannot "feel" the pressure, while in the present case elastic wall feels the pressure.

In the subsequent sections we address FSI problems (8.16)–(8.19) depending on the choice of the structure model, i.e., choice of the elasticity operator $\mathbb{L}$.

8.3.1 Linear $\varepsilon + 0$ Problem

First we discuss the FSI problem in which the elastic plate covering the fluid channel is already treated as a lower-dimensional object. One can see this approach as a two-step dimension reduction procedure, where first the structure model has been a priori reduced and afterward the dimension reduction for the fluid part is applied.

The structure domain will be $S_h = \omega \subset \mathbb{R}^2$ (independent of h), and we assume that the plate is linear visco-elastic and in the bending regime, i.e., the elasticity operator $\mathbb{L}$ from (8.18) is given by

$$\mathbb{L}(\boldsymbol{u}, \partial_t \boldsymbol{u}) = B\Delta'^4 \eta + \vartheta\, \Delta'^4 \partial_t \eta\,, \tag{8.25}$$

where the structure displacement is assumed to be of the form $\boldsymbol{u}(\boldsymbol{x},t)=(0,\eta(x',t)) \in \mathbb{R}^3$, and parameters B and ϑ describe material properties: rigidity and visco-elasticity, respectively. Since the structure model is considered as a boundary condition on the top boundary of the fluid domain, the dynamic interface condition becomes the balance of forces on the top boundary of the fluid domain, which is achieved by adding the right-hand side $g = -\sigma_f(\boldsymbol{v},p)e_3 \cdot e_3$ in (8.18). The kinematic condition reads $\partial_t \boldsymbol{u} = \boldsymbol{v}$. Analogous 2D model has been investigated in [41]. A similar model with

$$\mathbb{L}(\boldsymbol{u}, \partial_t \boldsymbol{u}) = B\Delta'^4 \eta - M\,\Delta'^2 \eta\,, \tag{8.26}$$

where $M > 0$ accounts for contribution of the horizontal tension to the vertical displacement and, also in 2D, was discussed in [16] and analogous results to those in [41] were obtained.

In the following we work with (8.25). Since horizontal components of the fluid velocity are zero on ω, we have the Korn equality for $\boldsymbol{v}$. Hence, testing Eqs. (8.16) and (8.18) with assumed smooth solutions $\boldsymbol{v}$ and $\partial_t \eta$, respectively, yields to the following energy estimate: for every $t > 0$

$$\begin{aligned} &\frac{\varrho_f}{2}\int_{\Omega_\varepsilon} |\boldsymbol{v}(t)|^2 \mathrm{d}\boldsymbol{x} + \frac{\nu}{2}\int_0^t \int_{\Omega_\varepsilon} |\nabla \boldsymbol{v}(s)|^2 \mathrm{d}\boldsymbol{x}\mathrm{d}s + \frac{\varrho_s}{2}\int_\omega |\partial_t \eta(t)|^2 \mathrm{d}x' \\ &\qquad + \frac{B}{2}\int_\omega |\Delta' \eta(t)|^2 \mathrm{d}x' + \vartheta \int_0^t \int_\omega |\Delta' \partial_t \eta(s)|^2 \mathrm{d}x' \mathrm{d}s \le Ct\varepsilon^3 . \end{aligned} \tag{8.27}$$

The obtained energy estimate motivates the following structure function space to be specified:

$$\mathcal{V}_S(0,T;\omega) = W^{1,\infty}(0,T;L^2(\omega)) \cap H^1(0,T;V_S(\omega))\,,$$

where $V_S(\omega) = \{\eta \in H^2(\omega) \ : \ \eta \text{ is } \omega\text{-periodic}\}$, and the solution space of the coupled problem (8.16)–(8.19) with (8.25) is then given by

$$\mathcal{V}(0,T;\Omega) = \{(\boldsymbol{v},\eta) \in \mathcal{V}_F(0,T;\Omega_\varepsilon) \times \mathcal{V}_S(0,T;\omega) \ : \qquad \boldsymbol{v}(t) = (0, \partial_t \eta(t)) \text{ on } \omega \text{ for a.e. } t \in (0,T)\}. \tag{8.28}$$

The well-posedness of the problem (8.16)–(8.19) in two space dimensions with the structure operator (8.25) was addressed in [41], and the following regularity of the weak solution $(\boldsymbol{v}^\varepsilon, \eta^\varepsilon)$ is obtained:

$$\partial_t \boldsymbol{v}^\varepsilon \in L^2(0,T;L^2(\Omega_\varepsilon;\mathbb{R}^2)) \quad \text{and} \quad \partial_{tt}\eta^\varepsilon \in L^2(\omega \times (0,T)), \quad \text{(time regularity)},$$
$$\boldsymbol{v}^\varepsilon \in L^\infty(0,T;H^1(\Omega_\varepsilon;\mathbb{R}^2)) \quad \text{and} \quad \eta^\varepsilon \in L^\infty(0,T;H^2(\omega)), \quad \text{(space regularity)}.$$

Moreover, there exists a unique pressure $p^\varepsilon \in L^2(0,T;H^1(\Omega_\varepsilon))$ such that $(\boldsymbol{v}^\varepsilon, p^\varepsilon, \eta^\varepsilon)$ solves (8.16)–(8.18) in the classical sense. Even though the result in [41] is stated and proved in 2D case, the Galerkin construction scheme and a priori estimates would provide the same results also in the 3D case. Here we work with the 3D case.

The aim is here to obtain a nontrivial limit behavior of the original system, as the small parameter ε tends to zero. The same problem has been analyzed in [41], but using the asymptotic expansion techniques, like also in [16]. Here we follow another concept developed in [6] for analysis of an $\varepsilon + h$ problem, which is based on careful quantitative energy estimates. For that purpose, the following scaling ansatz is assumed:

(S1) $B = \hat{B}\varepsilon^{-\kappa}$ and $\varrho_s = \hat{\varrho}_s \varepsilon^{-\kappa}$ for some $\kappa > 0$ and $\hat{B}, \hat{\varrho}_s > 0$ independent of ε;

(S2) $\mathsf{T} = \varepsilon^\tau$ for some $\tau \in \mathbb{R}$.

Scaling (S1) takes into account large rigidity of the structure where κ may be interpreted as a measure of the structure rigidity [11], and (S2) is the choice of the time scale T depending on ε. For now, the point of the above scalings is purely mathematical with the aim of finding a relation between free parameters κ and τ that will ensure the nontrivial coupled behavior of the system in the reduced model.

Next we perform the geometric change of variables from the thin fluid domain Ω_ε to the reference domain $\Omega_- = (0,1)^2 \times (-1,0)$ and obtain the uniform energy estimates on Ω_-. Let us denote by $\boldsymbol{v}(\varepsilon)$ and $\eta(\varepsilon)$ weak solutions to the rescaled system and denote the scaled differential operators by $\nabla_\varepsilon \boldsymbol{v} = (\nabla' \boldsymbol{v}, \varepsilon^{-1}\partial_3 \boldsymbol{v})$ (the scaled gradient) and $\operatorname{div}_\varepsilon v = \partial_1 v_1 + \partial_2 v_2 + \frac{1}{\varepsilon}\partial_3 v_3$ (the scaled divergence operator).

The energy estimate (8.27) on the reference domain and in the rescaled time then reads: for a.e. $t \in (0,T)$, it holds

$$\frac{\varrho_f}{2}\varepsilon \int_{\Omega_-} |\boldsymbol{v}(\varepsilon)(t)|^2 \mathrm{d}\boldsymbol{y} + \frac{\nu\varepsilon^{\tau+1}}{2} \int_0^t \int_{\Omega_-} |\nabla_\varepsilon \boldsymbol{v}(\varepsilon)|^2 \mathrm{d}\boldsymbol{y}\mathrm{d}s + \frac{\varrho_s}{2}\varepsilon^{-2\tau-\kappa} \int_\omega |\partial_t \eta(\varepsilon)(t)|^2 \mathrm{d}x'$$
$$+ \frac{B}{2}\varepsilon^{-\kappa} \int_\omega |\Delta' \eta(\varepsilon)(t)|^2 \mathrm{d}x' + \vartheta\varepsilon^{-\tau} \int_0^t \int_\omega |\Delta' \partial_t \eta(\varepsilon)(s)|^2 \mathrm{d}x'\mathrm{d}s \le C\varepsilon^{\tau+3}, \tag{8.29}$$

where $'$ denotes horizontal variables and respective operators. Estimate (8.29) provides the uniform bound for the fluid velocity

$$\int_0^T \int_{\Omega_-} |\partial_3 \boldsymbol{v}(\varepsilon)|^2 \mathrm{d}\boldsymbol{y} \mathrm{d}s \le C\varepsilon^4 ,$$

which, using the no-slip boundary condition on the bottom of the channel, directly implies

$$\|\boldsymbol{v}(\varepsilon)\|_{L^2(0,T;L^2(\Omega_-)} \le C\varepsilon^2 .$$

This motivates to rescale the fluid velocity according to $\bar{\boldsymbol{v}}(\varepsilon) = \varepsilon^{-2}\boldsymbol{v}(\varepsilon)$ and neglecting the bar notation, uniform a priori estimates imply the following weak convergence results (on a subsequence as $\varepsilon \downarrow 0$):

$$\boldsymbol{v}(\varepsilon) \rightharpoonup \boldsymbol{v} \quad \text{and} \quad \partial_3 \boldsymbol{v}(\varepsilon) \rightharpoonup \partial_3 \boldsymbol{v} \quad \text{weakly in } L^2(0,T;L^2(\Omega_-;\mathbb{R}^3)) . \tag{8.30}$$

The pressure in the system is treated in a standard manner. Define $\pi_\varepsilon(t) = \int_{\Omega_-} p(\varepsilon)(\boldsymbol{y},t)\mathrm{d}\boldsymbol{y}$ to be the mean value of the pressure at time $t \in (0,T)$. First, the zero mean value part of the pressure $p(\varepsilon) - \pi_\varepsilon$ is estimated in a classical way by utilizing the Bogovski operator for the construction of an appropriate test function (cf. [6, Section 3.1]), which provides

$$\|p(\varepsilon) - \pi_\varepsilon\|_{L^2(0,T;L^2(\Omega_-))} \le C .$$

Estimating the mean value as $\|\pi_\varepsilon\|_{L^2(0,T)} \le C$ (cf. again [6, Section 3.1]) yields the uniform bound for the pressure $\|p(\varepsilon)\|_{L^2(0,T;L^2(\Omega_-))} \le C$, which implies the existence of $p \in L^2(0,T;L^2(\Omega_-))$ such that (on a subsequence as $\varepsilon \downarrow 0$)

$$p(\varepsilon) \rightharpoonup p \quad \text{weakly in } L^2(0,T;L^2(\Omega_-)) . \tag{8.31}$$

According to (8.29), for the structure displacement we have the bound

$$\operatorname*{ess\,sup}_{t\in(0,T)} \int_\omega |\Delta'\eta(\varepsilon)(t)|^2 \mathrm{d}x' \le C\varepsilon^{\tau+\kappa+3} ,$$

which due to the Poincaré inequality, since $\int_\omega \eta(\varepsilon)(t)\mathrm{d}x' = 0$ for a.e. $t \in (0,T)$, yields

$$\|\eta(\varepsilon)\|_{L^\infty(0,T;H^2(\omega))} \le C\varepsilon^{(\tau+\kappa+3)/2} . \tag{8.32}$$

Rescaling $\eta(\varepsilon)$ according to $\bar{\eta}(\varepsilon) = \varepsilon^{-(\tau+\kappa+3)/2}\eta(\varepsilon)$ and taking all previous rescalings into account yield the weak form on the reference domain that includes the pressure

$$
\begin{aligned}
-\varrho_f \varepsilon^{3-\tau} \int_0^T \int_{\Omega_-} \boldsymbol{v}(\varepsilon) \cdot \partial_t \phi \, \mathrm{d}\boldsymbol{y}\mathrm{d}t + 2\nu\varepsilon^3 \int_0^T \int_{\Omega_-} \operatorname{sym} \nabla_\varepsilon \boldsymbol{v}(\varepsilon) : \operatorname{sym} \nabla_\varepsilon \phi \, \mathrm{d}\boldsymbol{y}\mathrm{d}t \\
-\varepsilon \int_0^T \int_{\Omega_-} p(\varepsilon) \operatorname{div}_\varepsilon \phi \, \mathrm{d}\boldsymbol{y}\mathrm{d}t + \varrho_s \varepsilon^{\delta-2\tau} \int_0^T \int_\omega \eta(\varepsilon) \partial_{tt}\psi \, \mathrm{d}x'\mathrm{d}t \\
+\varepsilon^\delta \int_0^T \int_\omega \left(B\Delta'\eta(\varepsilon)\Delta'\psi - \vartheta\varepsilon^{\kappa-\tau}\Delta'\eta(\varepsilon)\Delta'\partial_t\psi \right) \mathrm{d}x'\mathrm{d}t \\
= \varepsilon \int_0^T \int_{\Omega_-} f(\varepsilon) \cdot \phi \, \mathrm{d}\boldsymbol{y}\mathrm{d}t \,,
\end{aligned}
\tag{8.33}
$$

for all $(\phi, \psi) \in C_c^2 \left([0,T); V(\Omega_-) \times V_S(\omega)\right)$ such that $\phi(t) = (0, \psi(t))$ on ω for all

$t \in [0,T)$, and where $\delta = (\tau+\kappa+3)/2 - \kappa$. Omitting the divergence-free condition in fluid test functions, the above fluid space is $V(\Omega_-) = \big\{\boldsymbol{v} \in H^1(\Omega_-;\mathbb{R}^3) \ : \ \boldsymbol{v}|_{\{y_3=-1\}} = 0\,,\ \boldsymbol{v} \text{ is } \omega\text{-periodic}\big\}$.

In order to realize a nontrivial coupling between the fluid and the structure part in the reduced model, we need to adjust $\delta = 0$. Namely, in this case the fluid pressure will balance the structure bending. This condition then yields the choice of the right time scale $\mathsf{T} = \varepsilon^\tau$ with

$$
\tau = \kappa - 3\,. \tag{8.34}
$$

Remark 8.3.1. The obtained relation (8.34) relates the rigidity of the structure with the corresponding time scale that "sees" the interaction between the subsystems in the reduced model. (8.34) is consistent with assumptions and results obtained in [41] and [16]. In [41], assumed time scale corresponds to $\tau = 0$ and the nontrivial coupling between the subsystems in the reduced model is realized for the structure rigidity that corresponds to $\kappa = 3$. On the other hand, in [16] the assumed time scale is given by $\tau = -2$ and the nontrivial coupling is realized when the structure rigidity corresponds to $\kappa = 1$.

The leading order terms in (8.33) ($O(1)$ with respect to ε) are the pressure term in the fluid part and the bending term in the structure. Hence, under additional assumption $\tau < 0$, which ensures that the inertial term of the structure vanishes, the limit form of (8.33) (on a subsequence as $\varepsilon \downarrow 0$) reads

$$
-\int_0^T \int_{\Omega_-} p\partial_3\phi_3 \, \mathrm{d}\boldsymbol{y}\mathrm{d}t + B \int_0^T \int_\omega \Delta'\eta\Delta'\psi \, \mathrm{d}x'\mathrm{d}t = 0\,. \tag{8.35}
$$

Taking a test function ϕ in (8.35) that is compactly supported in space and taking $\psi = 0$, it follows from (8.35) that limit pressure is independent of the vertical variable z_3, and therefore p (although L^2-function) has the trace on ω. Since $\phi_3 = \psi_3$ on $\omega \times (0,T)$, after integrating by parts in the pressure term, the limit form (8.35) then becomes

$$-\int_0^T \int_\omega p\psi \,\mathrm{d}x'\mathrm{d}t + B\int_0^T \int_\omega \Delta'\eta \Delta'\psi \,\mathrm{d}x'\mathrm{d}t = 0 \tag{8.36}$$

for arbitrary $\psi \in C_c^2([0,T); H^2_{\mathrm{per}}(\omega))$.

In order to close the limit model, we need to further explore on the fluid part. First, multiplying $\mathrm{div}_\varepsilon\, \boldsymbol{v}(\varepsilon) = 0$ by a test function $\varphi \in C_c^1([0,T); H^1_{\mathrm{per}}(\omega))$, integrating over space and time, integrating by parts, and employing the rescaled kinematic condition for the vertical component $v_3(\varepsilon)/\varepsilon = \partial_t \eta(\varepsilon)$ a.e. on $\omega \times (0,T)$, we find

$$-\int_0^T \int_{\Omega_-} (v_1(\varepsilon)\partial_1\varphi + v_2(\varepsilon)\partial_2\varphi)\,\mathrm{d}\boldsymbol{y}\mathrm{d}t - \int_0^T \int_\omega \eta(\varepsilon)\partial_t\varphi \,\mathrm{d}y'\mathrm{d}t = 0\,, \tag{8.37}$$

which (on a subsequence as $\varepsilon \downarrow 0$) implies

$$-\int_0^T \int_{\Omega_-} (v_1\partial_1\varphi + v_2\partial_2\varphi)\,\mathrm{d}\boldsymbol{y}\mathrm{d}t - \int_0^T \int_\omega \eta\partial_t\varphi \,\mathrm{d}y'\mathrm{d}t = 0\,, \tag{8.38}$$

for all $\varphi \in C_c^1([0,T); H^1_{\mathrm{per}}(\omega))$. This relates the limit vertical displacement of the structure with limit horizontal fluid velocities. For the vertical fluid velocity, the divergence-free equation and the no-slip boundary condition imply $v_3 = 0$. Finally, to close the reduced model, relation between horizontal fluid velocities v_α and the pressure p is obtained from (8.65) by appropriate choice of test functions: $\phi = (\phi_1/\varepsilon, \phi_2/\varepsilon, 0)$ with $\phi_\alpha \in C_c^1([0,T); C_c^\infty(\Omega_-))$ and $\psi = 0$. Convergence results (8.30) and (8.31) then yield the limit equation

$$\nu\int_0^T \int_{\Omega_-} (\partial_3 v_1\partial_3\phi_1 + \partial_3 v_2\partial_3\phi_2)\,\mathrm{d}\boldsymbol{y}\mathrm{d}t - \int_0^T \int_{\Omega_-} p(\partial_1\phi_1 + \partial_2\phi_2)\,\mathrm{d}\boldsymbol{y}\mathrm{d}t \tag{8.39}$$
$$= \int_0^T \int_{\Omega_-} (f_1\phi_1 + f_2\phi_2)\,\mathrm{d}\boldsymbol{y}\mathrm{d}t\,.$$

Since the pressure p is independent of the vertical variable y_3, Eq. (8.39) can be solved for v_α explicitly in terms of y_3 and p. The boundary conditions are inherited from the original no-slip conditions, i.e., $v_\alpha(\cdot,-1,\cdot) = v_\alpha(\cdot,0,\cdot) = 0$. Explicit solution of v_α from (8.39) is then given by

$$v_\alpha(\boldsymbol{y},t) = \frac{1}{2\nu} y_3(y_3+1)\partial_\alpha p(y',t) + F_\alpha(\boldsymbol{y},t)\,, \quad (\boldsymbol{y},t) \in \Omega_- \times (0,T)\,,\ \alpha = 1,2\,, \tag{8.40}$$

where $F_\alpha(\cdot, y_3, \cdot) = \dfrac{y_3+1}{\nu}\displaystyle\int_{-1}^{0} \zeta_3 f_\alpha(\cdot, \zeta_3, \cdot)\,\mathrm{d}\zeta_3 + \frac{1}{\nu}\int_{-1}^{y_3} (y_3-\zeta_3) f_\alpha(\cdot, \zeta_3, \cdot)\,\mathrm{d}\zeta_3$. Replacing v_α from (8.40) into Eq. (8.38), we obtain a Reynolds-type equation

$$\int_0^T \int_\omega \left(-\frac{1}{12\nu}\Delta' p - F + \partial_t \eta \right) \varphi \, \mathrm{d}y' \mathrm{d}t = 0\,, \tag{8.41}$$

where $F(y', t) = -\displaystyle\int_{-1}^{0} (\partial_1 F_1 + \partial_2 F_2)\,\mathrm{d}y_3$. Combining the latter with Eq. (8.36), we finally obtain the reduced model in terms of the sixth-order evolution equation for the vertical displacement

$$\partial_t \eta - \frac{B}{12\nu}(\Delta')^3 \eta = F\,. \tag{8.42}$$

Equation (8.42) is accompanied by trivial initial data $\eta(0) = 0$ and periodic boundary conditions.

Based on the reduced model, i.e., knowing η solely, we are able to recover approximate solutions to the original FSI problem. The limit pressure p and horizontal fluid velocities v_α are calculated according to (8.36) and (8.40), respectively. The approximate fluid velocity is then defined by

$$\mathsf{v}^\varepsilon(\boldsymbol{x}, t) = \varepsilon^2 \Big(v_1(x', \frac{x_3}{\varepsilon}, t), v_2(x', \frac{x_3}{\varepsilon}, t), \mathsf{v}_3^\varepsilon \Big)\,, \quad (\boldsymbol{x}, t) \in \Omega_\varepsilon \times (0, T)\,, \tag{8.43}$$

where $\mathsf{v}_3^\varepsilon(\boldsymbol{x}, t) = -\varepsilon \displaystyle\int_{-1}^{x_3/\varepsilon} (\partial_1 v_1 + \partial_2 v_2)(x', \xi, t)\,\mathrm{d}\xi$, and the approximate pressure by $\mathsf{p}^\varepsilon(\boldsymbol{x}, t) = p(x', t)$ for all $(\boldsymbol{x}, t) \in \Omega_\varepsilon \times (0, T)$. Moreover, the approximate vertical displacement of the structure is defined by

$$\mathsf{\eta}^\varepsilon(x', t) = \varepsilon^\kappa \eta(x', t)\,, \quad (x', t) \in \omega \times (0, T)\,. \tag{8.44}$$

The following theorem is the key result of this section.

Theorem 8.3.2. Let $(\boldsymbol{v}^\varepsilon, p^\varepsilon, \eta^\varepsilon)$ be the classical solution to the FSI problem (8.16)–(8.19) with (8.25) in rescaled time, let $(\mathsf{v}^\varepsilon, \mathsf{p}^\varepsilon, \mathsf{\eta}^\varepsilon)$ be approximate solution constructed from the reduced model as above, and assume that $0 < \kappa \leq 5/2$, then

$$\|\boldsymbol{v}^\varepsilon - \mathsf{v}^\varepsilon\|_{L^2(0,T;L^2(\Omega_\varepsilon))} \leq C\varepsilon^3\,,$$
$$\|p^\varepsilon - \mathsf{p}^\varepsilon\|_{L^2(0,T;L^2(\Omega_\varepsilon))} \leq C\varepsilon\,,$$
$$\|\eta^\varepsilon - \mathsf{\eta}^\varepsilon\|_{L^\infty(0,T;H^2(\omega))} \leq C\varepsilon^{\kappa+1/2}\,,$$

where $C > 0$ denotes a generic positive constant independent of ε.

Remark 8.3.2. Error estimates for approximate solutions of the same problem have been derived in [41, Theorem 6.1]. However, employing the asymptotic expansion techniques will provide good error estimates only for higher-order approximations, while here we present optimal estimates for the zero-order approximation.

Proof. Based on the limit Eq. (8.39), the approximate fluid velocity v^ε and the pressure p^ε satisfy the modified Stokes system

$$\varrho_f \mathsf{T}^{-1}\partial_t \mathsf{v}^\varepsilon - \operatorname{div}\sigma_f(\mathsf{v}^\varepsilon,\mathsf{p}^\varepsilon) = f^\varepsilon - f_3^\varepsilon e_3 + \mathsf{r}_f^\varepsilon\,, \tag{8.45}$$

where the residual term r_f^ε is given by $\mathsf{r}_f^\varepsilon = \varrho_f \mathsf{T}^{-1}\partial_t \mathsf{v}^\varepsilon - \nu\Delta'\mathsf{v}^\varepsilon - \nu\partial_{33}\mathsf{v}_3^\varepsilon\, e_3$ and

enjoys the uniform bound $\|\mathsf{r}_f^\varepsilon\|_{L^2(0,T;L^2(\Omega_\varepsilon))} \le C\varepsilon^{3/2}$. Multiplying Eq. (8.45) by a test function $\phi \in C_c^1([0,T);V_F(\Omega_\varepsilon))$, and then integrating over $\Omega_\varepsilon \times (0,T)$, we find

$$\begin{aligned} -\varrho_f \int_0^T\!\!\int_{\Omega_\varepsilon} \mathsf{v}^\varepsilon\cdot\partial_t\phi\, d\boldsymbol{x}dt + 2\nu\mathsf{T}\int_0^T\!\!\int_{\Omega_\varepsilon} \operatorname{sym}\nabla\mathsf{v}^\varepsilon : \operatorname{sym}\nabla\phi\, d\boldsymbol{x}dt \\ -\mathsf{T}\int_0^T\!\!\int_\omega \sigma_f(\mathsf{v}^\varepsilon,\mathsf{p}^\varepsilon)\phi\cdot e_3\, d\boldsymbol{x}dt = \mathsf{T}\int_0^T\!\!\int_{\Omega_\varepsilon}(f_1^\varepsilon\phi_1 + f_2^\varepsilon\phi_2)\, d\boldsymbol{x}dt + \mathsf{T}\int_0^T\!\!\int_{\Omega_\varepsilon}\mathsf{r}_f^\varepsilon\cdot\phi\, d\boldsymbol{x}dt\,. \end{aligned} \tag{8.46}$$

Expanding the boundary term, employing the pressure relation (8.36), and utilizing the definition of the approximate displacement $\mathsf{\eta}^\varepsilon$, we find

$$\begin{aligned} -\varrho_f \int_0^T\!\!\int_{\Omega_\varepsilon} \mathsf{v}^\varepsilon\cdot\partial_t\phi\, d\boldsymbol{x}dt + 2\nu\mathsf{T}\int_0^T\!\!\int_{\Omega_\varepsilon} \operatorname{sym}\nabla\mathsf{v}^\varepsilon : \operatorname{sym}\nabla\phi\, d\boldsymbol{x}dt \\ -\varrho_s^\varepsilon\mathsf{T}^{-1}\int_0^T\!\!\int_\omega \partial_t\mathsf{\eta}^\varepsilon\partial_t\psi\, dx'dt + B^\varepsilon\mathsf{T}\int_0^T\!\!\int_\omega \Delta'\mathsf{\eta}^\varepsilon\Delta'\psi\, dx'dt + \vartheta\int_0^T\!\!\int_\omega \Delta'\partial_t\mathsf{\eta}^\varepsilon\Delta'\psi\, dx'dt \\ = \mathsf{T}\int_0^T\!\!\int_{\Omega_\varepsilon}(f_1^\varepsilon\phi_1 + f_2^\varepsilon\phi_2)\, d\boldsymbol{x}dt + \mathsf{T}\int_0^T\!\!\int_{\Omega_\varepsilon}\mathsf{r}_f^\varepsilon\cdot\phi\, d\boldsymbol{x}dt + \langle \mathsf{r}_s^\varepsilon,\psi\rangle\,, \end{aligned}$$

where the structure residual term is given by

$$\langle \mathsf{r}_s^\varepsilon,\psi\rangle = -\varrho_s^\varepsilon\mathsf{T}^{-1}\int_0^T\!\!\int_\omega \partial_t\mathsf{\eta}^\varepsilon\partial_t\psi\, dx'dt + \vartheta\int_0^T\!\!\int_\omega \Delta'\partial_t\mathsf{\eta}^\varepsilon\Delta'\psi\, dx'dt\,,$$

while $B^\varepsilon := B\varepsilon^{-\kappa}$ and $\varrho^\varepsilon = \varrho\varepsilon^{-\kappa}$ as in the scaling ansatz (S1). Introducing the error functions $e_f^\varepsilon := \boldsymbol{v}^\varepsilon - \mathsf{v}^\varepsilon$ and $e_\eta^\varepsilon := \eta^\varepsilon - \mathsf{\eta}^\varepsilon$, we arrive to the error equation in the weak form

$$-\varrho_f \int_0^T \int_{\Omega_\varepsilon} e_f^\varepsilon \cdot \partial_t \phi \, \mathrm{d}\boldsymbol{x}\mathrm{d}t + 2\nu \mathsf{T} \int_0^T \int_{\Omega_\varepsilon} \operatorname{sym} \nabla e_f^\varepsilon : \operatorname{sym} \nabla \phi \, \mathrm{d}\boldsymbol{x}\mathrm{d}t$$

$$-\varrho_s^\varepsilon \mathsf{T}^{-1} \int_0^T \int_\omega \partial_t e_\eta^\varepsilon \partial_t \psi \, \mathrm{d}x'\mathrm{d}t + B^\varepsilon \mathsf{T} \int_0^T \int_\omega \Delta' e_\eta^\varepsilon \Delta' \psi \, \mathrm{d}x'\mathrm{d}t + \vartheta \int_0^T \int_\omega \Delta' \partial_t e_\eta^\varepsilon \Delta' \psi \, \mathrm{d}x'\mathrm{d}t \tag{8.47}$$

$$= \mathsf{T} \int_0^T \int_{\Omega_\varepsilon} f_3^\varepsilon \phi_3 \, \mathrm{d}\boldsymbol{x}\mathrm{d}t - \mathsf{T} \int_0^T \int_{\Omega_\varepsilon} \mathsf{r}_f^\varepsilon \cdot \phi \, \mathrm{d}\boldsymbol{x}\mathrm{d}t - \langle \mathsf{r}_s^\varepsilon, \psi \rangle .$$

Observe that by construction, approximate solutions satisfy the same boundary conditions as the true solutions to the original problem. Namely, $\mathsf{v}_\alpha^\varepsilon|_{\{y_3=-1\}} = \mathsf{v}_\alpha^\varepsilon|_\omega = 0$, while for the vertical component we have

$$\mathsf{v}_3^\varepsilon|_\omega = -\varepsilon^3 \int_{-1}^0 (\partial_1 v_1 + \partial_2 v_2) \mathrm{d}y_3 = \varepsilon^3 \partial_t \eta = \mathsf{T}^{-1} \partial_t \mathsf{\eta}^\varepsilon .$$

Thus, utilizing the Korn equality and $(\phi, \psi) = (e_f^\varepsilon, \mathsf{T}^{-1} \partial_t e_\eta^\varepsilon)$ as test functions in

(8.47), we find: for a.e. $t \in (0, T)$

$$\frac{\varrho_f}{2} \int_{\Omega_\varepsilon} |e_f^\varepsilon(t)|^2 \mathrm{d}\boldsymbol{x} + \nu \mathsf{T} \int_0^t \int_{\Omega_\varepsilon} |\nabla e_f^\varepsilon|^2 \, \mathrm{d}\boldsymbol{x}\mathrm{d}s + \frac{\varrho_s^\varepsilon \mathsf{T}^{-2}}{2} \int_\omega |\partial_t e_\eta^\varepsilon(t)|^2 \mathrm{d}x'$$
$$+ \frac{B^\varepsilon}{2} \int_\omega |\Delta' e_\eta^\varepsilon(t)|^2 \mathrm{d}x' + \vartheta \mathsf{T}^{-1} \int_0^t \int_\omega |\Delta' \partial_t e_\eta^\varepsilon(s)|^2 \mathrm{d}x'\mathrm{d}s$$
$$= \mathsf{T} \int_0^t \int_{\Omega_\varepsilon} f_3^\varepsilon e_{f,3}^\varepsilon \, \mathrm{d}\boldsymbol{x}\mathrm{d}s - \mathsf{T} \int_0^t \int_{\Omega_\varepsilon} \mathsf{r}_f^\varepsilon \cdot e_f^\varepsilon \, \mathrm{d}\boldsymbol{x}\mathrm{d}s - \mathsf{T}^{-1} \langle \mathsf{r}_s^\varepsilon, \partial_t e_\eta^\varepsilon \rangle .$$

Let us now estimate the right-hand side. Employing a higher-order energy estimate (cf. [6, Section 2.4]), one can conclude $\|\partial_\alpha \boldsymbol{v}^\varepsilon\|_{L^2(0,T;L^2(\Omega_\varepsilon))} \leq C\varepsilon^{5/2}$, which combined with the diver gence-free condition on the thin domain provides $\|v_3^\varepsilon\|_{L^2(0,T;L^2(\Omega_\varepsilon))} \leq C\varepsilon^{7/2}$. As a consequence, we easily conclude $\|e_{f,3}^\varepsilon\|_{L^2(0,T;L^2(\Omega_\varepsilon))} \leq C\varepsilon^{7/2}$, which in further gives

$$\mathsf{T} \left| \int_0^t \int_{\Omega_\varepsilon} f_3^\varepsilon e_{f,3}^\varepsilon \, \mathrm{d}\boldsymbol{x}\mathrm{d}s \right| \leq C\mathsf{T}\varepsilon^4 .$$

For the fluid residual, we employ Cauchy–Schwarz, Poincaré, and Young inequality, respectively, and obtain

$$\mathsf{T} \left| \int_0^t \int_{\Omega_\varepsilon} \mathsf{r}_f^\varepsilon \cdot e_f^\varepsilon \, \mathrm{d}\boldsymbol{x}\mathrm{d}s \right| \leq C\mathsf{T} \int_0^t \varepsilon \|\mathsf{r}_f^\varepsilon\|_{L^2(\Omega_\varepsilon)} \|\nabla e_f^\varepsilon\|_{L^2(\Omega_\varepsilon)} \mathrm{d}s \leq C\mathsf{T}\varepsilon^5 + \frac{\nu \mathsf{T}}{2} \int_0^t \int_{\Omega_\varepsilon} |\nabla e_f^\varepsilon|^2 \, \mathrm{d}\boldsymbol{x}\mathrm{d}s .$$

Finally, we estimate the structure residual

$$\mathsf{T}^{-1}\left|\langle \mathsf{r}_s^\varepsilon, \partial_t e_\eta^\varepsilon\rangle\right| \le \varrho_s^\varepsilon \mathsf{T}^{-2}\left|\int_0^t\int_\omega \partial_t \eta^\varepsilon \partial_{tt} e_\eta^\varepsilon \,\mathrm{d}x'\mathrm{d}s\right| + \vartheta \mathsf{T}^{-1}\left|\int_0^t\int_\omega \Delta'\partial_t\eta^\varepsilon \Delta'\partial_t e_\eta^\varepsilon \,\mathrm{d}x'\mathrm{d}s\right|.$$

Integrating by parts in time, the first term can be estimated as

$$\varrho_s^\varepsilon \mathsf{T}^{-2}\left|\int_0^t\int_\omega \partial_t\eta^\varepsilon \partial_{tt} e_\eta^\varepsilon \,\mathrm{d}x'\mathrm{d}s\right| \le C\mathsf{T}\varepsilon^{3-2\tau} + \frac{\varrho_s^\varepsilon \mathsf{T}^{-2}}{4}\int_\omega |\partial_t e_\eta^\varepsilon(t)|^2 \mathrm{d}x' + \frac{\varrho_s^\varepsilon \mathsf{T}^{-2}}{4}\int_0^t\int_\omega |\partial_t e_\eta^\varepsilon(s)|^2 \mathrm{d}x'\mathrm{d}s,$$

while for the second term we have

$$\vartheta \mathsf{T}^{-1}\left|\int_0^t\int_\omega \Delta'\partial_t\eta^\varepsilon \Delta'\partial_t e_\eta^\varepsilon \,\mathrm{d}x'\mathrm{d}s\right| \le C\mathsf{T}\varepsilon^6 + \frac{\vartheta \mathsf{T}^{-1}}{2}\int_0^t\int_\omega |\Delta'\partial_t e_\eta^\varepsilon(s)|^2 \mathrm{d}x'\mathrm{d}s.$$

Summing all up and employing the Gronwall inequality, we have: for a.e. $t \in (0, T)$

$$\frac{\varrho_f}{2}\int_{\Omega_\varepsilon} |e_f^\varepsilon(t)|^2 \mathrm{d}\boldsymbol{x} + \frac{\nu \mathsf{T}}{2}\int_0^t\int_{\Omega_\varepsilon} |\nabla e_f^\varepsilon|^2 \,\mathrm{d}\boldsymbol{x}\mathrm{d}s + \frac{\varrho_s^\varepsilon \mathsf{T}^{-2}}{4}\int_\omega |\partial_t e_\eta^\varepsilon(t)|^2 \mathrm{d}x'$$
$$+ \frac{B^\varepsilon}{2}\int_\omega |\Delta' e_\eta^\varepsilon(t)|^2 \mathrm{d}x' + \frac{\vartheta \mathsf{T}^{-1}}{2}\int_0^t\int_\omega |\Delta'\partial_t e_\eta^\varepsilon(s)|^2 \mathrm{d}x'\mathrm{d}s \le C\mathsf{T}\varepsilon^4 \tag{8.48}$$

provided $\tau \le -1/2$, i.e., $\kappa \le 5/2$.

Combining estimate (8.48) and the Poincaré inequality, we find

$$\|e_f^\varepsilon\|_{L^2(0,T;L^2(\Omega_\varepsilon))} \le C\varepsilon^3, \tag{8.49}$$

which is the desired error estimate for the fluid velocity. Due to the zero mean value on ω, according to (8.48), the displacement error can be controlled as

$$\|e_\eta^\varepsilon\|^2_{L^\infty(0,T;H^2(\omega))} \le C\varepsilon^{2\kappa+1}, \tag{8.50}$$

which is the required estimate for structure displacement. The error estimate for the pressure follows somewhat different approach, which we omit here and refer to [6, Section 4.5] for details.

□

Remark 8.3.3. Observe that the error estimate of horizontal fluid velocities relative to the norm of velocities is $O(\sqrt{\varepsilon})$. The same holds true for the relative error estimate of the pressure and displacement errors.

8.3.2 Linear $\varepsilon + h$ Problem

In practice we often have that both layers are relatively thin and of equivalent size. Moreover, we expect then that the interplay between the two thicknesses: ε of the fluid and h od the structure plays the role in the derivation of reduced models. Unlike in the previous section, here we take a full structure model and aim to perform a simultaneous dimension reduction.

In the FSI problem (8.16)–(8.19), the channel is now covered by an elastic plate of relative height $h > 0$ with material configuration $S_h = (0,1)^2 \times (0,h)$ and the plate is described by the linear 3D elasticity equations, i.e., the elasticity operator $\mathbb{L}$ in (8.18) is given by

$$\mathbb{L}(\boldsymbol{u}, \partial_t \boldsymbol{u}) = -\operatorname{div} \sigma_s(\boldsymbol{u}), \tag{8.51}$$

where $\sigma_s(\boldsymbol{u}) = 2\mu \operatorname{sym} \nabla \boldsymbol{u} + \lambda(\operatorname{div} \boldsymbol{u}) I_3$ denotes the Cauchy stress tensor, μ and λ are Lamé constants, and I_3 is 3×3 identity matrix. The two subsystems are coupled through the interface conditions on the fixed interface ω

$$\partial_t \boldsymbol{u} = \boldsymbol{v}, \quad \omega \times (0,\infty), \quad \text{(kinematic—continuity of velocities)}, \tag{8.52}$$

$$(\sigma_f(\boldsymbol{v}, p) - \sigma_s(\boldsymbol{u})) e_3 = 0, \quad \omega \times (0,\infty), \quad \text{(dynamic—balance of forces)}. \tag{8.53}$$

The action of the fluid on the structure and vice versa is considered through the dynamic coupling condition (8.53), and we assume the absence of the structure volume forces, hence, $g = 0$ in (8.18). Following [6] the basic energy inequality

(8.21), in this case reads: for every $t > 0$

$$\frac{\varrho_f}{2} \int_{\Omega_\varepsilon} |\boldsymbol{v}(t)|^2 \mathrm{d}\boldsymbol{x} + \nu \int_0^t \int_{\Omega_\varepsilon} |\operatorname{sym} \nabla \boldsymbol{v}(s)|^2 \mathrm{d}\boldsymbol{x} \mathrm{d}s + \frac{\varrho_s}{2} \int_{S_h} |\partial_t \boldsymbol{u}(t)|^2 \mathrm{d}\boldsymbol{x}$$
$$+ \mu \int_{S_h} |\operatorname{sym} \nabla \boldsymbol{u}(t)|^2 \mathrm{d}\boldsymbol{x} + \frac{\lambda}{2} \int_{S_h} |\operatorname{div} \boldsymbol{u}(t)|^2 \mathrm{d}\boldsymbol{x} \leq C t \varepsilon, \tag{8.54}$$

and the structure function space (8.22) is now specified to be

$$\mathcal{V}_S(0,T; S_h) = W^{1,\infty}(0,T; L^2(S_h; \mathbb{R}^3)) \cap L^\infty(0,T; V_S(S_h)),$$

where $V_S(S_h) = \{\boldsymbol{u} \in H^1(S_h; \mathbb{R}^3) \;:\; \boldsymbol{u} \text{ is } \omega\text{-periodic}\}$. The solution space of the coupled problem (8.16)–(8.19) with (8.51)–(8.53) is now given by

$$\mathcal{V}(0,T;\Omega) = \{(\boldsymbol{v}, \boldsymbol{u}) \in \mathcal{V}_F(0,T;\Omega_\varepsilon) \times \mathcal{V}_S(0,T;S_h) \;: \tag{8.55}$$
$$\boldsymbol{v}(t) = \partial_t \boldsymbol{u}(t) \text{ on } \omega \text{ for a.e. } t \in (0,T)\}.$$

The well-posedness of the problem (8.16)–(8.19) with (8.51)–(8.53) in the sense of Definition 8.3.1 is addressed in [6], and the regularity of the weak solution $(\boldsymbol{v}^\varepsilon, \boldsymbol{u}^h)$ required for the subsequent analysis is obtained as follows:

$$\partial_t \boldsymbol{v}^\varepsilon \in L^\infty(0,T; L^2(\Omega_\varepsilon; \mathbb{R}^3)) \quad \text{and} \quad \partial_{tt} \boldsymbol{u}^h \in L^\infty(0,T; L^2(S_h; \mathbb{R}^3)), \quad \text{(time regularity)},$$
$$\boldsymbol{v}^\varepsilon \in L^\infty(0,T; H^2(\Omega_\varepsilon; \mathbb{R}^3)) \quad \text{and} \quad \boldsymbol{u}^h \in L^\infty(0,T; H^2(S_h; \mathbb{R}^3)), \quad \text{(space regularity)}.$$

Moreover, a refined energy estimate is obtained (cf. [6, Section 2.4]): for a.e. $t \in [0,T)$

$$\frac{\varrho_f}{2}\int_{\Omega_\varepsilon}|\boldsymbol{v}^\varepsilon(t)|^2\mathrm{d}\boldsymbol{x}+\frac{\nu}{2}\int_0^t\int_{\Omega_\varepsilon}|\nabla\boldsymbol{v}^\varepsilon|^2\mathrm{d}\boldsymbol{x}\mathrm{d}s+\frac{\varrho_s}{2}\int_{S_h}|\partial_t\boldsymbol{u}^h(t)|^2\mathrm{d}\boldsymbol{x} \\ +\int_{S_h}\Big(\mu|\operatorname{sym}\nabla\boldsymbol{u}^h(t)|^2+\frac{\lambda}{2}|\operatorname{div}\boldsymbol{u}^h(t)|^2\Big)\mathrm{d}\boldsymbol{x}\le Ct\varepsilon^3\,, \tag{8.56}$$

which will be the cornerstone for the derivation of the reduced model. Furthermore, there exists a unique pressure $p^\varepsilon \in L^2(0,T;H^1(\Omega_\varepsilon))$ such that $(\boldsymbol{v}^\varepsilon, p^\varepsilon, \boldsymbol{u}^h)$ solves (8.16)–(8.18) in the classical sense.

A similar problem has been analyzed in [42], where starting from a 2D/2D linear FSI problem of type $1+h$, the FSI problem analyzed in [41] has been justified. Our main aim here is to obtain a nontrivial limit behavior of the original system as both small parameters ε and h simultaneously vanish. In order to achieve that, we need to assume some scaling ansatz:

(S1) $\varepsilon = h^\gamma$ for some $\gamma > 0$ independent of h;

(S2) $\mu = \hat{\mu}h^{-\kappa}$, $\lambda = \hat{\lambda}h^{-\kappa}$ and $\varrho_s = \hat{\varrho}_s h^{-\kappa}$ for some $\kappa > 0$ and $\hat{\mu}$, $\hat{\lambda}$, $\hat{\varrho}_s$ independent of h;

(S3) $\mathsf{T} = h^\tau$ for some $\tau \in \mathbb{R}$.

Scaling (S1) is a geometric relation between small parameters, (S2), like in the previous section, takes into account the large rigidity of the structure where κ may be interpreted as a measure of the structure rigidity [11], and (S3) simply means the choice of the time scale T depending on h. Again, the point of the above scalings is for now purely mathematical and we aim to find a relation between free parameters γ, κ, and τ, similar to (8.34), which will ensure the nontrivial limit behavior of the full system.

Performing the standard change of variables, we move to the reference domain $\Omega_- \cup \omega \cup S_+$ and obtain the uniform energy estimates there. Let us denote by $\boldsymbol{v}(\varepsilon)$ and $\boldsymbol{u}(h)$ weak solutions to the rescaled system and by ∇_ε and ∇_h the corresponding scaled gradients, then the energy estimate (8.56) on the reference domain and in rescaled time reads: for a.e. $t \in (0,T)$ it holds

$$\frac{\varrho_f}{2}\varepsilon\int_{\Omega_-}|\boldsymbol{v}(\varepsilon)(t)|^2\mathrm{d}\boldsymbol{y}+\frac{\nu h^\tau}{2}\varepsilon\int_0^t\int_{\Omega_-}|\nabla_\varepsilon\boldsymbol{v}(\varepsilon)|^2\mathrm{d}\boldsymbol{y}\mathrm{d}s+\frac{\varrho_s}{2}h^{-\kappa-2\tau+1}\int_{S_+}|\partial_t\boldsymbol{u}(h)(t)|^2\mathrm{d}\boldsymbol{z} \\ +h^{-\kappa+1}\int_{S_+}\Big(\mu|\operatorname{sym}\nabla_h\boldsymbol{u}(h)(t)|^2+\frac{\lambda}{2}|\operatorname{div}_h\boldsymbol{u}(h)(t)|^2\Big)\mathrm{d}\boldsymbol{z}\le Ch^\tau\varepsilon^3\,. \tag{8.57}$$

The rescaled energy estimate (8.57) gives us for the fluid part same uniform bounds and the same convergence results as in (8.30) and (8.31). For the structure part, the energy estimate (8.57) provides an L^∞-L^2 estimate of the symmetrized scaled gradient of the displacement

$$\operatorname*{ess\,sup}_{t\in(0,T)} \int_{S_+} |\operatorname{sym}\nabla_h \boldsymbol{u}(h)|^2 \mathrm{d}\boldsymbol{z} \le C h^{3\gamma-1+\tau+\kappa}, \tag{8.58}$$

which motivates rescaling of the structure displacements according to $\bar{\boldsymbol{u}}(h) = h^{-(3\gamma-1+\tau+\kappa)/2}$
$\boldsymbol{u}(h)$.

The uniform bound on the symmetrized scaled gradient of displacements motivates to invoke the framework of the Griso decomposition [25]—for every $h>0$, scaled structure displacement $\boldsymbol{u}(h)$ is, at almost every time instance $t\in(0,T)$, decomposed into a sum of the so-called elementary plate displacement and warping:

$$\boldsymbol{u}(h)(\boldsymbol{z}) = w(h)(z') + r(h)(z') \times (z_3 - \frac{1}{2})e_3 + \tilde{\boldsymbol{u}}(h), \tag{8.59}$$

where

$$w(h)(t,z') = \int_0^1 \boldsymbol{u}(h)(t,\boldsymbol{z})\mathrm{d}z_3\,, \quad r(h)(t,z') = \frac{3}{h}\int_0^1 (z_3-\frac{1}{2})e_3 \times \boldsymbol{u}(h)(t,\boldsymbol{z})\mathrm{d}z_3\,,$$

$\tilde{\boldsymbol{u}}(h) \in L^\infty(0,T;H^1(S_+))$ is the warping term, and $\times$ denotes the cross product in $\mathbb{R}^3$. Moreover, the following uniform estimate holds

$$\begin{aligned}&\|\operatorname{sym}\nabla_h\left(w(h)(z')+r(h)(z')\times(z_3-1/2)e_3\right)\|^2_{L^\infty(0,T;L^2(S_+))}\\ &\quad+\|\nabla_h\tilde{\boldsymbol{u}}(h)\|^2_{L^\infty(0,T;L^2(S_+))}+\frac{1}{h^2}\|\tilde{\boldsymbol{u}}(h)\|^2_{L^\infty(0,T;L^2(S_+))} \le C,\end{aligned}$$

with $C>0$ independent of h and $\boldsymbol{u}(h)$. Following [25, Theorem 2.6], the above uniform estimate implies the existence of a sequence of in-plane translations $a(h) = (a_1(h), a_2(h))(L^\infty(0,T))^2$, as well as limit displacements $\eta_1, \eta_2 \in L^\infty(0,T;$

$H^1_{\mathrm{per}}(\omega))$, $\eta_3 \in L^\infty(0,T;H^2_{\mathrm{per}}(\omega))$ and $\bar{\boldsymbol{u}} \in L^2(\omega;H^1((0,1);\mathbb{R}^3))$ such that the following weak-$\star$ convergence results hold:

$$w_\alpha(h) - a_\alpha(h) \overset{*}{\rightharpoonup} \eta_\alpha \quad \text{in } L^\infty(0,T;H^1_{\mathrm{per}}(\omega))\,, \quad \alpha=1,2\,, \tag{8.60}$$

$$hw_3(h) \overset{*}{\rightharpoonup} \eta_3 \quad \text{in } L^\infty(0,T;H^1_{\mathrm{per}}(\omega))\,, \tag{8.61}$$

$$u_\alpha(h) - a_\alpha(h) \overset{*}{\rightharpoonup} \eta_\alpha - (z_3-\frac{1}{2})\partial_\alpha\eta_3 \quad \text{in } L^\infty(0,T;H^1(S_+))\,, \quad \alpha=1,2\,, \tag{8.62}$$

$$hu_3(h) \overset{*}{\rightharpoonup} \eta_3 \quad \text{in } L^\infty(0,T;H^1(S_+))\,, \tag{8.63}$$

$$\operatorname{sym}\nabla_h \boldsymbol{u}(h) \overset{*}{\rightharpoonup} \imath\Big(\operatorname{sym}\nabla'(\eta_1,\eta_2) - (z_3-\frac{1}{2})\nabla'^2\eta_3\Big) + \operatorname{sym}(e_3\otimes(\partial_3\bar{\boldsymbol{u}}))\,. \tag{8.64}$$

Taking all the above rescalings into account, the weak form, which now includes the pressure, on the reference domain reads

$$\begin{aligned}
-\varrho_f h^{-\tau}\varepsilon^3 \int_0^T\int_{\Omega_-} \boldsymbol{v}(\varepsilon)\cdot\partial_t\phi\,\mathrm{d}\boldsymbol{y}\mathrm{d}t + 2\nu\varepsilon^3\int_0^T\int_{\Omega_-} \operatorname{sym}\nabla_\varepsilon \boldsymbol{v}(\varepsilon) : \operatorname{sym}\nabla_\varepsilon\phi\,\mathrm{d}\boldsymbol{y}\mathrm{d}t \\
-\varepsilon\int_0^T\int_{\Omega_-} p(\varepsilon)\operatorname{div}_\varepsilon\phi\,\mathrm{d}\boldsymbol{y}\mathrm{d}t + \varrho_s h^{\delta-2\tau}\int_0^T\int_{S_+} \boldsymbol{u}(h)\cdot\partial_{tt}\psi\,\mathrm{d}\boldsymbol{z}\mathrm{d}t \\
+h^\delta\int_0^T\int_{S_+} (2\mu\operatorname{sym}\nabla_h\boldsymbol{u}(h) : \operatorname{sym}\nabla_h\psi + \lambda\operatorname{div}_h\boldsymbol{u}(h)\operatorname{div}_h\psi)\,\mathrm{d}\boldsymbol{z}\mathrm{d}t \\
= \varepsilon\int_0^T\int_{\Omega_-} f(\varepsilon)\cdot\phi\,\mathrm{d}\boldsymbol{y}\mathrm{d}t\,,
\end{aligned} \tag{8.65}$$

for all $(\phi,\psi)\in C_c^2([0,T);V(\Omega_-)\times V_S(S_+))$ such that $\phi(t)=\psi(t)$ on ω for all $t\in[0,T)$, and where $\delta=(3\gamma-1+\tau+\kappa)/2+1-\kappa$.

In order to realize a nontrivial coupling in the reduced model on the limit as $h\downarrow 0$, like in the previous section we need to adjust the parameter δ. The linear theory of plates (cf. [10, Section 1.10]) suggests $\delta=-1$. Namely, the fluid pressure that is here $O(1)$ is acting as a normal force on the structure and therefore has to balance the structure stress terms in the right way. This condition then yields the choice of the right time scale $\mathsf{T}=h^\tau$ with

$$\tau=\kappa-3\gamma-3\,. \tag{8.66}$$

Remark 8.3.4. Relation (8.66) is analogous to (8.34) and mathematically has two free parameters: measure of the structure rigidity κ and geometric relation between thicknesses γ, which fix the choice of the right time scale that "sees" the interaction between the two subsystems in the reduced model.

Exploring the structure of elasticity equations and taking appropriate test functions that imitate the shape of the limit of scaled displacements (8.62)–(8.63), i.e., $\psi=(h\psi_1,h\psi_2,\psi_3)$ satisfying $\partial_1\psi_3+\partial_3\psi_1=\partial_2\psi_3+\partial_3\psi_2=\partial_3\psi_3=0$ (cf. [10, Theorem 1.4-1]) and $\phi=(h\phi_1,h\phi_2,\phi_3)$, under additional assumption $\tau<-1$, the weak limit form of (8.65) (on a subsequence as $h\downarrow 0$, as well as $\varepsilon\downarrow 0$) reads

$$\begin{aligned}
-\int_0^T\int_{\Omega_-} p\partial_3\phi_3\,\mathrm{d}\boldsymbol{y}\mathrm{d}t + \int_0^T\int_{S_+}\Big(2\mu\big(\operatorname{sym}\nabla'(\eta_1,\eta_2)-(z_3-\frac{1}{2})\nabla'^2\eta_3\big) : \operatorname{sym}\nabla'(\psi_1,\psi_2) \\
+\frac{2\mu\lambda}{2\mu+\lambda}\operatorname{div}\big((\eta_1,\eta_2)-(z_3-\frac{1}{2})\nabla'\eta_3\big)\operatorname{div}(\psi_1,\psi_2)\Big)\,\mathrm{d}\boldsymbol{z}\mathrm{d}t = 0\,.
\end{aligned} \tag{8.67}$$

The obtained limit equation can be interpreted as a linear plate model [10] coupled with the limit fluid pressure acting as a normal force on the structure interface ω. Exploring in further the structure of the space of test functions for the elasticity part [10, Theorem 1.4-1 (c)] leads to an equivalent decoupled system for horizontal and vertical displacements

$$-\int_0^T\int_\omega p\zeta_3\,\mathrm{d}z'\mathrm{d}t+\int_0^T\int_\omega\left(\frac{4\mu}{3}\nabla'^2\eta_3:\nabla'^2\zeta_3+\frac{4\mu\lambda}{3(2\mu+\lambda)}\Delta'\eta_3\Delta'\zeta_3\right)\mathrm{d}z'\mathrm{d}t=0 \tag{8.68}$$

for arbitrary $\zeta_3\in C_c^2([0,T);H^2_{\mathrm{per}}(\omega))$ and

$$\int_0^T\int_\omega\left(4\mu\,\mathrm{sym}\,\nabla'(\eta_1,\eta_2):\nabla'(\zeta_1,\zeta_2)+\frac{4\mu\lambda}{2\mu+\lambda}\,\mathrm{div}'\,(\eta_1,\eta_2)\,\mathrm{div}'(\zeta_1,\zeta_2)\right)\mathrm{d}z'\mathrm{d}t=0 \tag{8.69}$$

for arbitrary $\zeta_\alpha\in C_c^1([0,T);H^1_{\mathrm{per}}(\omega))$. Equation (8.69) implies that horizontal displacements (η_1,η_2) are spatially constant functions, and as such they will not affect the reduced model. Moreover, they are dominated by potentially large horizontal translations as discussed in [6, Section 2.5]; hence, we omit them in further analysis. Thus, the limit system (8.67) is now essentially described with (8.68), which relates the limit fluid pressure p with the limit vertical displacement of the structure η_3.

Analysis of the fluid part completely follows the lines of the previous section and results in limit equations (8.38) and (8.39). The limit horizontal velocities v_α can again be explicitly calculated in terms of y_3 and p, but the top boundary condition for v_α is no longer trivial. It follows from the interface kinematic condition that $v_\alpha(\cdot,0,\cdot)=\partial_t a_\alpha$, $\alpha=1,2$, where $\partial_t a_\alpha$ are translational limit velocities of the structure defined below and discussed in [6, Section 3.3]. Explicit solution of v_α from (8.39) is then given by

$$v_\alpha(\boldsymbol{y},t)=\frac{1}{2\nu}y_3(y_3+1)\partial_\alpha p(y',t)+F_\alpha(\boldsymbol{y},t)+(1+y_3)\partial_t a_\alpha\,,\quad(\boldsymbol{y},t)\in\Omega_-\times(0,T)\,, \tag{8.70}$$

where $F_\alpha(\cdot,y_3,\cdot)=\dfrac{y_3+1}{\nu}\displaystyle\int_{-1}^0\zeta_3 f_\alpha(\cdot,\zeta_3,\cdot)\,\mathrm{d}\zeta_3+\frac{1}{\nu}\int_{-1}^{y_3}(y_3-\zeta_3)f_\alpha(\cdot,\zeta_3,\cdot)\,\mathrm{d}\zeta_3$. Replacing v_α from (8.70) into Eq. (8.38), we obtain the same Reynolds equation (8.41), which combined with Eq. (8.68) results in the reduced model in terms of the vertical displacement only. Substituting $\eta\equiv\eta_3$, we arrive to the sixth-order evolution equation

$$\partial_t\eta-\frac{2\mu(\mu+\lambda)}{9\nu(2\mu+\lambda)}(\Delta')^3\eta=F\,, \tag{8.71}$$

which is (up to the coefficient in front of the spatial operator) the same as (8.42).

Knowing η solely, the pressure p and horizontal fluid velocities v_α are then calculated according to (8.41) and (8.70), respectively. Based on that, we can construct approximate fluid velocity

$$\mathsf{v}^\varepsilon(\boldsymbol{x},t) = \varepsilon^2\Big(v_1(x',\frac{x_3}{\varepsilon},t), v_2(x',\frac{x_3}{\varepsilon},t), \mathsf{v}_3^\varepsilon\Big), \quad (\boldsymbol{x},t)\in\Omega_\varepsilon\times(0,T), \tag{8.72}$$

where $\mathsf{v}_3^\varepsilon(\boldsymbol{x},t) = -\varepsilon\int_{-1}^{x_3/\varepsilon}(\partial_1 v_1 + \partial_2 v_2)(x',\xi,t)\,\mathrm{d}\xi$, and the approximate pressure by $\mathsf{p}^\varepsilon(\boldsymbol{x},t) = p(x',t)$ for all $(\boldsymbol{x},t)\in\Omega_\varepsilon\times(0,T)$. Moreover, the approximate displacement is defined by

$$\mathsf{u}^h(\boldsymbol{x},t) = h^{\kappa-3}\left(h^{-\gamma}a_1 - \Big(x_3-\frac{h}{2}\Big)\partial_1\eta(x',t), h^{-\gamma}a_2 - \Big(x_3-\frac{h}{2}\Big)\partial_2\eta(x',t), \eta(x',t)\right), \tag{8.73}$$

for all $(\boldsymbol{x},t)\in\Omega_h\times(0,T)$, where a_α are horizontal time-dependent translations calculated by $a_\alpha(t) = -\int_0^t\int_\omega \partial_3 F_\alpha(y',0,t)\mathrm{d}y'\mathrm{d}s$, $\alpha = 1,2$. The virtue of the reduced model is then revealed by the following convergence results of approximate solutions.

Theorem 8.3.3 ([6]). Let $(\boldsymbol{v}^\varepsilon, p^\varepsilon, \boldsymbol{u}^h)$ be the classical solution to the FSI problem (8.16)–(8.19) under (8.51)–(8.53) in rescaled time, and let $(\mathsf{v}^\varepsilon, \mathsf{p}^\varepsilon, \mathsf{u}^h)$ be approximate solution constructed from the reduced model as above. Let us additionally assume that $\max\{2\gamma+1, \frac{7}{4}\gamma+\frac{3}{2}\}\le\kappa<2+2\gamma$, then

$$\begin{aligned}
\|\boldsymbol{v}^\varepsilon - \mathsf{v}^\varepsilon\|_{L^2(0,T;L^2(\Omega_\varepsilon))} &\le C\varepsilon^{5/2}h^{\min\{\gamma/2,\,2\gamma-\kappa+2\}},\\
\|p^\varepsilon - \mathsf{p}^\varepsilon\|_{L^2(0,T;L^2(\Omega_\varepsilon))} &\le C\varepsilon^{1/2}h^{\min\{\gamma/2,2\gamma-\kappa+2\}},\\
\|u_\alpha^h - \mathsf{u}_\alpha^h\|_{L^\infty(0,T;L^2(\Omega_h))} &\le Ch^{\kappa-3/2}h^{\min\{1,\gamma/2,2\gamma+2-\kappa\}} + C\sqrt{h}\|a_\alpha^h - h^{\kappa-3-\gamma}a_\alpha\|_{L^\infty(0,T)},\\
\|u_3^h - \mathsf{u}_3^h\|_{L^\infty(0,T;L^2(\Omega_h))} &\le Ch^{\kappa-5/2}h^{\min\{1/2,\gamma/2,2\gamma+2-\kappa\}},
\end{aligned}$$

where $C>0$ denote generic positive constants independent of ε and h.

Remark 8.3.5. Observe that, like in the previous section, the error estimate of horizontal fluid velocities relative to the norm of velocities is $O(\sqrt{\varepsilon})$ for $\kappa\le\frac{3}{2}\gamma+2$. The same holdes true for the relative error estimate of the pressure. For the vertical fluid velocity, which is of lower order, there is a lack of the error estimate. In the leading order of the structure displacement, namely in the vertical component, for $\kappa\le\frac{3}{2}\gamma+2$ we have the relative convergence rate $O(h^{\min\{1/2,\gamma/2\}})$, which means $O(\sqrt{\varepsilon})$ for $\gamma\le 1$ and $O(\sqrt{h})$ for $\gamma>1$. In horizontal structure displacements, dominant part of the error estimates are errors in horizontal translations, which are actually artifact of periodic boundary conditions (cf. [6, Section 2.5]). Neglecting these errors, which cannot be controlled in a better way, the relative error estimate of horizontal displacements for $\kappa \le \frac{3}{2}\gamma + 2$ is $O(h^{\min\{1,\gamma/2\}})$, which means $O(\sqrt{\varepsilon})$ for $\gamma \le 2$ and $O(h)$ for $\gamma > 2$. Let us point out that one cannot expect better convergence rates for such first-order approximation without dealing with boundary layers, which arise around the interface ω due to mismatch of the interface conditions for approximate solutions. Moreover, in [35] the obtained convergence rate for the Poiseuille flow in the case of rigid walls of the fluid channel is $O(\sqrt{\varepsilon})$. On the other hand, convergence rate for the clamped Kirchhoff–Love plate is found to be $O(\sqrt{h})$ [20].

Outline of the Proof of Theorem 8.3.3. The proof follows the idea of the proof of Theorem 8.3.2, but mostly due to mismatch of the interface conditions for approximate solutions, the analysis is much more involved.

Starting from Eq. (8.45) satisfied by the approximate fluid velocity v^ε, expanding the boundary terms, employing the pressure relation (8.67), and utilizing the definition of the approximate displacement u^h, we find the weak form for approximate solutions to be

$$\begin{aligned} -\varrho_f \int_0^T\int_{\Omega_\varepsilon} \mathsf{v}^\varepsilon \cdot \partial_t\phi \,\mathrm{d}\boldsymbol{x}\mathrm{d}t + 2\nu\mathsf{T}\int_0^T\int_{\Omega_\varepsilon} \operatorname{sym}\nabla\mathsf{v}^\varepsilon : \operatorname{sym}\nabla\phi \,\mathrm{d}\boldsymbol{x}\mathrm{d}t \\ -\varrho_s^h\mathsf{T}^{-1}\int_0^T\int_{S_h} \partial_t\mathsf{u}^h \cdot \partial_t\psi \,\mathrm{d}\boldsymbol{x}\mathrm{d}t + \mathsf{T}\int_0^T\int_{S_h} \left(2\mu^h \operatorname{sym}\nabla\mathsf{u}^h : \operatorname{sym}\nabla\psi + \lambda^h \operatorname{div}\mathsf{u}^h \operatorname{div}\psi\right)\mathrm{d}\boldsymbol{x}\mathrm{d}t \\ = \mathsf{T}\int_0^T\int_{\Omega_\varepsilon} (f_1^\varepsilon\phi_1 + f_2^\varepsilon\phi_2)\,\mathrm{d}\boldsymbol{x}\mathrm{d}t + \mathsf{T}\int_0^T\int_{\Omega_\varepsilon} \mathsf{r}_f^\varepsilon\cdot\phi \,\mathrm{d}\boldsymbol{x}\mathrm{d}t + \mathsf{T}\int_0^T\int_\omega \mathsf{r}_b^\varepsilon\cdot\phi\,\mathrm{d}x'\mathrm{d}t + \langle \mathsf{r}_s^h, \psi\rangle, \end{aligned} \tag{8.74}$$

where r_b^ε denotes the boundary residual term given by

$$\mathsf{r}_b^\varepsilon = \nu\left(\varepsilon\partial_3 v_1 - \varepsilon^3\int_{-1}^0 (\partial_{11}v_1 + \partial_{12}v_2), \varepsilon\partial_3 v_2 - \varepsilon^3\int_{-1}^0 (\partial_{21}v_1 + \partial_{22}v_2), -2\varepsilon^2(\partial_1 v_1 + \partial_2 v_2)\right),$$

$\langle \mathsf{r}_s^h, \psi\rangle$ denotes the structure residual term r_s^h acting on a test function ψ as

$$\begin{aligned} \langle \mathsf{r}_s^h, \psi\rangle = -\varrho_s^h\mathsf{T}^{-1}\int_0^T\int_{S_h} \partial_t\mathsf{u}^h\cdot\partial_t\psi\,\mathrm{d}\boldsymbol{x}\mathrm{d}t \\ + \mathsf{T}\int_0^T\int_{S_h}\left(\frac{(\lambda^h)^2}{2\mu^h+\lambda^h}\operatorname{div}\mathsf{u}^h\operatorname{div}(\psi_1,\psi_2) + \lambda^h\operatorname{div}\mathsf{u}^h\partial_3\psi_3\right)\mathrm{d}\boldsymbol{x}\mathrm{d}t, \end{aligned}$$

and coefficients $\mu^h = \mu h^{-\kappa}$, $\lambda^h = \lambda h^{-\kappa}$, and $\varrho_s^h = \varrho_s h^{-\kappa}$ are according to the scaling ansatz (S2). Defining the fluid error $e_f^\varepsilon := \boldsymbol{v}^\varepsilon - \mathsf{v}^\varepsilon$ and the structure error $e_s^h := \boldsymbol{u}^h - \mathsf{u}^h$, and subtracting (8.74) from the weak form of the original weak form, we find the variational equation for the errors

$$
\begin{aligned}
-\varrho_f \int_0^T \int_{\Omega_\varepsilon} e_f^\varepsilon \cdot \partial_t \phi \, d\boldsymbol{x}dt + 2\nu \mathsf{T} \int_0^T \int_{\Omega_\varepsilon} \operatorname{sym} \nabla e_f^\varepsilon : \operatorname{sym} \nabla \phi \, d\boldsymbol{x}dt \\
-\varrho_s^h \mathsf{T}^{-1} \int_0^T \int_{S_h} \partial_t e_s^h \cdot \partial_t \psi \, d\boldsymbol{x}dt + \mathsf{T} \int_0^T \int_{S_h} \left(2\mu^h \operatorname{sym} \nabla e_s^h : \operatorname{sym} \nabla \psi + \lambda^h \operatorname{div} e_s^h \operatorname{div} \psi\right) d\boldsymbol{x}dt \\
= \mathsf{T} \int_0^T \int_{\Omega_\varepsilon} f_3^\varepsilon \phi_3 \, d\boldsymbol{x}dt - \mathsf{T} \int_0^T \int_{\Omega_\varepsilon} \mathsf{r}_f^\varepsilon \cdot \phi \, d\boldsymbol{x}dt - \mathsf{T} \int_0^T \int_\omega \mathsf{r}_b^\varepsilon \cdot \phi \, dx'dt - \langle \mathsf{r}_s^h, \psi \rangle
\end{aligned}
\tag{8.75}
$$

for all test functions $(\phi, \psi) \in \mathcal{W}(0,T;\Omega)$.

Next step is a careful selection of test functions in (8.75). First we choose

$$
\psi = \mathsf{T}^{-1}(\partial_t e_{s,1}^{\mathrm{e}}, \partial_t e_{s,2}^{\mathrm{e}}, \partial_t e_{s,3}^{\mathrm{o}}), \tag{8.76}
$$

where superscripts e and o denote even and odd components of the orthogonal decomposition of respective functions with respect to the variable $(x_3 - h/2)$. Observe in (8.73) that, up to time dependent constants, components of the approximate displacement u^h are, respectively, odd, odd, and even with respect to $(x_3 - h/2)$. The idea of using this particular test function comes from the fact that such ψ annihilates large part of the structure residual term r_s^h on the right-hand side in (8.75), and the rest can be controlled (cf. [6, Section 4.2] for details). Concerning the fluid part, observe that approximate solutions do not satisfy the kinematic interface condition in the horizontal components, i.e., $\mathsf{v}_\alpha^\varepsilon \neq \partial_t \mathsf{u}_\alpha^h$ on $\omega \times (0,T)$ and therefore $(\mathsf{v}^\varepsilon, \mathsf{u}^h)$ does not belong to the space $\mathcal{V}(0,T;\Omega)$. For the third component however, the interface condition is satisfied. In order to match interface values of ψ, the fluid test function ϕ has to be accordingly corrected fluid error, i.e., we take

$$
\phi = e_f^\varepsilon + \varphi, \tag{8.77}
$$

where the correction φ satisfies

$$
\begin{aligned}
\operatorname{div}\varphi &= 0 \quad \text{on } \Omega_\varepsilon \times (0,T)\,,\\
\varphi_\alpha\big|_{\omega\times(0,T)} &= -\mathsf{T}^{-1}\,\partial_t u^{\mathrm{o}}_\alpha\big|_{\omega\times(0,T)}\,,\\
\varphi_3\big|_{\omega\times(0,T)} &= -\mathsf{T}^{-1}\,\partial_t e^{\mathrm{e}}_{s,3}\big|_{\omega\times(0,T)}\,,\\
\varphi\big|_{\{x_3=-\varepsilon\}\times(0,T)} &= 0\,,
\end{aligned}
$$

and $\varphi(\cdot,t)$ is ω-periodic for every $t\in(0,T)$. This choice of φ ensures the kinematic boundary condition $\phi=\psi$ a.e. on $\omega\times(0,T)$. Following [6], it can be proved that the corrector φ satisfies the uniform bound

$$
\|\nabla\varphi\|_{L^\infty(0,T;L^2(\Omega_\varepsilon))}\le C\varepsilon^{5/2}
$$

with $C>0$ independent of φ and ε. Moreover, careful estimation of other residual terms in (8.75) provides the basic error estimate: for a.e. $t\in(0,T)$ we have

$$
\begin{aligned}
&\frac{\varrho_f}{4}\int_{\Omega_\varepsilon}|e^\varepsilon_f(t)|^2\mathrm{d}\boldsymbol{x}+\frac{\nu\mathsf{T}}{2}\int_0^t\int_{\Omega_\varepsilon}|\nabla e^\varepsilon_f|^2\,\mathrm{d}\boldsymbol{x}\mathrm{d}s+\frac{\varrho^h_s\mathsf{T}^{-2}}{4}\int_{S_h}\left((\partial_t e^{\mathrm{e}}_{s,\alpha}(t))^2+(\partial_t e^{\mathrm{o}}_{s,3}(t))^2\right)\mathrm{d}\boldsymbol{x}\\
&\quad+\int_{S_h}\left(\mu^h\left|\operatorname{sym}\nabla(e^{\mathrm{e}}_{s,1},e^{\mathrm{e}}_{s,2},e^{\mathrm{o}}_{s,3})(t)\right|^2+\frac{\lambda^h}{2}\left|\operatorname{div}(e^{\mathrm{e}}_{s,1},e^{\mathrm{e}}_{s,2},e^{\mathrm{o}}_{s,3})(t)\right|^2\right)\mathrm{d}\boldsymbol{x} \qquad (8.78)\\
&\quad\le C\mathsf{T}\varepsilon^3(h^\gamma+h^{4\gamma-2\kappa+4})\,.
\end{aligned}
$$

Estimate (8.78) is now sufficient to conclude the error estimates for the fluid part: velocities and the pressure, while for the structure part, the Griso decomposition of the structure error needs to be examined and employing another pair of test functions $(\psi,\phi)=(\mathsf{T}^{-1}\partial_t e^{\mathrm{o}}_s, e^\varepsilon_f+\varphi)$ with appropriate corrector φ will provide sufficient conditions to conclude the error estimates also for the structure part (cf. [6, Section 4.4]).

□

8.3.3 Nonlinear $\varepsilon + 0$ Problem

In this section we discuss a nonlinear FSI problem in which nonlinearities appear both in equations for the fluid motion and in geometry of the fluid domain. More precisely, the coupling conditions are also nonlinear, and the coupling is realized on the moving interface. Unlike in the previous section, the structure is here modeled by a lower-dimensional elasticity model, and additionally we decrease dimensionality of the original problem to two space dimensions for the fluid and one space dimension for the structure. The main reason for this ad hoc dimension reduction in the FSI problem is availability of the well-posedness results.

Let us now describe our setting. The fluid domain at time t is assumed to be of the form

$$\Omega_\eta(t) = \{(x,z) : x \in \omega,\ z \in (0, \eta(t,x))\}\mathbb{R}^2,$$

where $\omega = (0,1)$, and function η describes the dynamics of the vertical displacement of the top boundary. Let us further denote the space–time cylinder

$$\Omega_\eta(t) \times (0,T) := \bigcup_{t\in(0,T)} \Omega_\eta(t) \times \{t\} \subset \mathbb{R}^2 \times (0,\infty), \quad T \in (0,\infty],$$

to be domain of our problem. The FSI problem is described by the system of partial differential equations

$$\varrho_f(\partial_t \boldsymbol{v} + (\boldsymbol{v}\cdot\nabla)\boldsymbol{v}) - \operatorname{div}\sigma_f(\boldsymbol{v},p) = f\,, \quad \Omega_\eta(t)\times(0,\infty)\,, \tag{8.79}$$

$$\operatorname{div}\boldsymbol{v} = 0\,, \quad \Omega_\eta(t)\times(0,\infty)\,, \tag{8.80}$$

$$\varrho_s\partial_{tt}\eta - D\partial_x^2\partial_t\eta + B\partial_x^4\eta = -J^\eta\big(\sigma_f(\boldsymbol{v},p)\mathbf{n}^\eta\big)(t,x,\eta(t,x))\cdot\mathbf{e}_z\,, \quad \omega\times(0,\infty)\,, \tag{8.81}$$

$$\boldsymbol{v}(t,x,\eta(t,x)) = (0,\partial_t\eta(t,x))\,, \quad \omega\times(0,\infty)\,, \tag{8.82}$$

where $\sigma_f(\boldsymbol{v},p) = 2\nu\,\mathrm{sym}\,\nabla\boldsymbol{v} - pI$ denotes the Cauchy stress tensor of the viscous fluid, ν and $\varrho_f > 0$ are the fluid viscosity and density, respectively, and f denotes the fluid external force. Furthermore, ϱ_s is the structure density, n^η is the unit outer normal to the deformed configuration Ω_η, $J^\eta(t,x) = \sqrt{1+\partial_x\eta(t,x)^2}$ is Jacobian of the transformation from Eulerian to Lagrangian coordinates, and constants $D, B > 0$ describe visco-elasticity and elasticity properties of the structure, respectively.

Equations (8.79) and (8.80) are standard incompressible Navier–Stokes equations describing the flow of the Newtonian fluid, while the structure is described by a linear equation of visco-elastic plate (8.81). The fluid and the structure are coupled via dynamic and kinematic coupling conditions (8.81) and (8.82) representing the balance of forces in $\mathbf{e}_z$ direction and continuity of the velocity, respectively. Additional simplifying assumption is that the structure moves only in the vertical direction. This is not fully justified from the physical grounds, but it is reasonable in the view of results of the previous section, where it is shown that, up to time-dependent translations, the bending regime is the dominant one and the displacement in the horizontal direction is of the lower order. For more details about physical background of system (8.79)–(8.82) and corresponding lower-dimensional elasticity models, we refer to [10, 38] and references therein. The bottom boundary is rigid, and we prescribe the standard no-slip boundary condition for the fluid velocity $\boldsymbol{v}(t,x',0)=0$ for all $(x',t)\in\omega\times(0,\infty)$. On the lateral boundaries, we prescribe the periodic boundary conditions in the horizontal direction, which is taken for technical simplicity and because of availability of the global existence results [24]. In such a case the flow is driven by the right-hand side f. Finally, for simplicity of exposition, we impose some trivial initial conditions: $\boldsymbol{v}(0,.)=0$, $\eta(0,.)=\eta_0$, and $\partial_t\eta(0,.)=0$.

Since our aim is to derive the reduced model in the regime of relatively thin domain, we assume that the initial thickness of the domain is $O(\varepsilon)$, i.e., $\eta_0^\varepsilon(x)=\varepsilon\eta_0(x)$ for $x\in(0,L)$. Moreover, like in the previous section, we assume that $\|f\|_{L^\infty(0,\infty;L^\infty(\Omega_\eta(t);\mathbb{R}^2))}\le C$, which is satisfied by physically relevant volume forces. Testing formally Eqs. (8.79) and (8.81) with classical solutions $\boldsymbol{v}$ and $\partial_t\eta$, respectively, and integrating by parts yield the basic energy inequality: for every $t\in(0,T)$

$$
\begin{aligned}
&\frac{\varrho_f}{2}\|\boldsymbol{v}(t)\|^2_{L^2(\Omega_\eta(t))}+2\nu\int_0^t\int_{\Omega_\eta(s)}|\operatorname{sym}\nabla\boldsymbol{v}|^2\mathrm{d}\boldsymbol{x}\mathrm{d}s\\
&+\frac{\varrho_s}{2}\|\partial_t\eta(t)\|^2_{L^2(\omega)}+D\int_0^t\|\partial_t\partial_x\eta\|^2_{L^2(\omega)}\mathrm{d}s+\frac{B}{2}\|\partial_x^2\eta(t)\|^2_{L^2(\omega)}\\
&\qquad\le\frac{B}{2}\|\partial_x^2\eta_0\|^2_{L^2(\omega)}+\int_0^t\int_{\Omega_\eta(s)}f\cdot\boldsymbol{v}\,\mathrm{d}\boldsymbol{x}\mathrm{d}s\,.
\end{aligned}
\tag{8.83}
$$

Let us now discuss weak solutions. First we introduce appropriate solution spaces. The fluid solution space will depend on the displacement η. If we denote

$$
V_F(t)=\left\{\boldsymbol{v}\in H^1(\Omega_\eta(t))\ :\ \operatorname{div}\boldsymbol{v}=0,\ \boldsymbol{v}|_{z=0}=0,\ \boldsymbol{v}\text{ is }\omega-\text{periodic in }x_1\right\},
$$

then the above energy estimate suggests that appropriate fluid solution space is

$$
\mathcal{V}_F(0,\infty;\Omega_\eta(t))=L^\infty(0,\infty;L^2(\Omega_\eta(t)))\cap L^2(0,\infty;V_F(t))
$$

while for the structure, again based on the energy estimate, we choose the solution space to be

$$
\mathcal{V}_S(0,\infty;\omega)=W^{1,\infty}(0,\infty;L^2(\omega))\cap L^\infty(0,\infty;H^2_{\mathrm{per}}(\omega))\cap H^1(0,\infty;H^1(\omega))\,.
$$

Definition 8.3.4. We call $(\boldsymbol{v}, \eta) \in \mathcal{V}_F(0,\infty;\Omega_\eta(t)) \times \mathcal{V}_S(0,\infty;\omega)$ a *weak solution* of the FSI problem (8.79)–(8.82) if for every $(\boldsymbol{\varphi}, \psi) \in C_c^1([0,\infty); \mathcal{V}_F(t) \times H^2_{\mathrm{per}}(\omega))$ satisfying $\boldsymbol{\varphi}(t,x,\eta(t,x')) = \partial_t \psi(t,x)\mathbf{e}_3$ it holds

$$
\begin{aligned}
-\varrho_f \int_0^\infty \int_{\Omega_\eta(t)} (\boldsymbol{v}\cdot\partial_t\boldsymbol{\varphi} + (\boldsymbol{v}\cdot\nabla)\boldsymbol{v}\cdot\boldsymbol{\varphi})\,\mathrm{d}\boldsymbol{x}\mathrm{d}t + 2\nu \int_0^\infty \int_{\Omega_\eta(t)} \operatorname{sym}\nabla\boldsymbol{v} : \operatorname{sym}\nabla\boldsymbol{\varphi}\mathrm{d}\boldsymbol{x}\mathrm{d}t \\
-\varrho_s \int_0^\infty \int_\omega \partial_t\eta\partial_t\psi\mathrm{d}x_1\mathrm{d}t + D\int_0^\infty\int_\omega \partial_x\partial_t\eta\cdot\nabla\psi\mathrm{d}x_1\mathrm{d}t + B\int_0^\infty\int_\omega \partial_x^2\eta\partial_x^2\psi\mathrm{d}x_1\mathrm{d}t \\
= \varrho_s\int_\omega \eta_0\psi(0)\mathrm{d}x_1 + \int_0^\infty\int_{\Omega_\eta(t)} f\cdot\boldsymbol{\varphi}\,\mathrm{d}\boldsymbol{x}\mathrm{d}t
\end{aligned}
\tag{8.84}
$$

and (8.82) is satisfied in the sense of traces. Moreover, the energy inequality (8.83) is satisfied.

The existence of weak solutions is by now well-established in the literature, see, e.g., [12, 38]. However, the existence results are not global and state that weak solutions exist as long as there is no contact between the elastic and rigid boundary, i.e., in our notation as long as $\eta > 0$. Even though there are results that contact will not occur in the case when the structure is rigid [28], to the best of our knowledge there are no global-in-time existence results for weak solution to problem (8.79)–(8.82). Since we are interested in the long time behavior, we rely on the following recent result on the existence of global-in-time strong solutions.

Theorem 8.3.5 ([24]). For every $\varepsilon > 0$, there exists global-in-time strong solution to problem (8.79)–(8.82) with initial conditions $\boldsymbol{v}_0 = 0$, $\eta_0 = \varepsilon\hat{\eta}_0$, $\eta_1 = 0$, and the right-hand side $f \in L^2(0,T; L^2_{\mathrm{per}}(\mathbb{R}^2))$. Solutions satisfy $\eta^\varepsilon > 0$ and additional

regularity properties that we briefly write

$$
\eta^\varepsilon \in H^2_t L^2_x \cap L^2_t H^4_x\,, \quad \boldsymbol{v}^\varepsilon \in L^2_t H^2_x\,, \quad \text{and} \quad p^\varepsilon \in L^2_t H^1_x\,. \tag{8.85}
$$

Motivated by the results from previous sections, we assume the following scaling ansatz in order to be in the thin-film regime:

(S1) $B = \hat{B}\varepsilon^{-1}$, $D = \hat{D}\varepsilon^{-2}$ and $\varrho_s = \hat{\varrho}_s\varepsilon^{-1}$ for some $\hat{B}$, $\hat{D}$, $\hat{\varrho}_s > 0$ independent of ε;

(S2) $\mathsf{T} = \varepsilon^{-2}$.

To the best of our knowledge, rigorous derivation of equation (8.15) as the reduced model for the FSI system (8.79)–(8.82) is still missing from the literature. Here we present the main steps of the derivation without proofs. The details of the proofs will be included in a forthcoming work [7]. The main steps are analogous to the main steps in linear case, but technically much more involved. The main difficulties are consequence of the fact that ε-problems are moving boundary problems, i.e., we have to deal with the geometrical nonlinearity at every step of the derivation.

Step 1: Uniform energy estimate. The first step is to quantify the energy estimate (8.83) in terms of the small parameter ε. Unlike in linear case, the domain depends on solution and therefore one has to carefully track dependence of functional inequalities (e.g., the Poincaré inequality) on the solution itself. By using the scaling ansatz (S1), we arrive to the following uniform (in small parameter ε) energy inequality: for a.e. $t \in (0,T)$

$$\frac{\varrho_f}{2}\|\boldsymbol{v}^\varepsilon(t)\|^2_{L^2(\Omega_\eta(t))}+\nu\int_0^t\int_{\Omega_\eta(t)}|\nabla\boldsymbol{v}^\varepsilon|^2+\frac{D}{\varepsilon^2}\int_0^t\|\partial_t\partial_x\eta^\varepsilon\|^2_{L^2(\omega)}+\frac{B}{\varepsilon}\|\partial_x^2\eta^\varepsilon\|^2_{L^2(\omega)}\le Ct\varepsilon^3. \tag{8.86}$$

By taking into account scaling ansatz (S2) and using the Sobolev embedding, we get an L^∞-estimate for the displacement

$$\|\eta^\varepsilon\|_{L^\infty(0,T;L^\infty(0,L))}\le C\varepsilon\,. \tag{8.87}$$

Step 2: Positivity of the limit displacement. Let us denote by η the weak limit of $\eta^\varepsilon/\varepsilon$. From Theorem 8.3.5, it is immediate that $\eta \ge 0$. However, in order to perform our analysis we need to prove the strict positivity $\eta > 0$. This can be done by adapting estimates used in the proof of Theorem 8.3.5 [24] to our case and combining them with the scaling ansatz.

Step 3: ALE formulation. In order to identify the limit model, we need to reformulate weak formulation (8.84) on the fixed reference domain. The main difference in comparison to the linear case is that now the change of variable depends on the solution itself. In numerical computation, this formulation is usually called Arbitrary Lagrangian–Eulerian (ALE) formulation. We use the following explicit form of the change of variables:

$$\begin{pmatrix}\hat{t}\\ y_1\\ y_2\end{pmatrix}=\begin{pmatrix}\varepsilon^2 t\\ x_1\\ \frac{x_2}{\eta^\varepsilon(t,x_1)}\end{pmatrix}. \tag{8.88}$$

Step 4: Identifying the limit model. In the last step we pass to the limit as $\varepsilon \downarrow 0$ and obtain the limit model (8.15). The limiting procedure follows heuristic described in the introduction and is similar as in the linear case. The main difference is that due to the nonlinearities in the system, we need to prove the strong convergence properties of sequence $(\boldsymbol{v}^\varepsilon, \eta^\varepsilon)$ in order to pass to the limit. For this, we need two main ingredients: Aubin–Lions lemma for strong convergence of the displacement sequence η^ε and the scaling ansatz for the convergence of the fluid convective term.

Acknowledgment

This work has been supported in part by the Croatian Science Foundation under projects 7249 (MANDphy) and 3706 (FSIApp).

References

1. Guy Bayada and Michéle Chambat. The transition between the Stokes equations and the Reynolds equation: a mathematical proof. *Appl. Math. Optim.*, 14(1):73–93, 1986.

2. J. Becker and G. Grün. The thin-film equation: Recent advances and some new perspectives. *J. Phys.: Condens. Matter* 17 (2005), 291–307.

3. F. Bernis and A. Friedman. Higher order nonlinear degenerate parabolic equations. *J. Diff. Eqs.* 83 (1990), 179–206.

4. A. Bertozzi. The mathematics of moving contact lines in thin liquid films. *Notices Amer. Math. Soc.*, 45 (1998), 689–697.

5. T. Bodnar, G. P. Galdi, Š. Nečasova. Fluid-Structure Interaction in Biomedical Applications. Springer/Birkhouser. 2014.

6. M. Bukal and B. Muha. Rigorous derivation of a linear sixth-order thin-film equation as a reduced model for thin fluid–thin structure interaction problems. *Appl. Math. Optim.* (2020). https://doi.org/10.1007/s00245-020-09709-9

7. M. Bukal and B. Muha. Justification of a nonlinear sixth-order thin-film equation as the reduced model for a fluid–structure interaction problem. *In preparation* (2021).

8. A. P. Bunger, and E. Detournay. Asymptotic solution for a penny-shaped near-surface hydraulic fracture. Engin. Fracture Mech. 72 (2005), 2468–2486.

9. M. Bukač, S. Čanić, B. Muha and R. Glowinski. An Operator Splitting Approach to the Solution of Fluid-Structure Interaction Problems in Hemodynamics, *in Splitting Methods in Communication and Imaging, Science and Engineering Eds. R. Glowinski, S. Osher, and W. Yin*, New York, Springer, 2016.

10. P. G. Ciarlet. Mathematical Elasticity. Vol. II: Theory of Plates. North-Holland Publishing Co, Amsterdam, 1997.

11. P. G. Ciarlet. Mathematical Elasticity. Vol. I: Three-dimensional elasticity. North-Holland Publishing Co, Amsterdam, 1988.

12. Antonin Chambolle, Benoît Desjardins, Maria J. Esteban, and Céline Grandmont. Existence of weak solutions for the unsteady interaction of a viscous fluid with an elastic plate. *J. Math. Fluid Mech.*, 7(3):368–404, 2005.

13. G. Cimatti. How the Reynolds equation is related to the Stokes equations. *Appl. Math. Optim.* 10 (1983), 267–274.

14. P. Constantin, T. Dupont, R. E. Goldstein, L. P. Kadanoff, M. J. Shelley, and S. M. Zhou. Droplet breakup in a model of the Hele-Shaw cell. *Phys. Rev. E* 47 (1993), 4169–4181.

15. S. Čanić and A. Mikelić. Effective equations modeling the flow of a viscous incompressible fluid through a long elastic tube arising in the study of blood flow through small arteries. *SIAM J. Appl. Dyn. Syst.*, 2(3):431–463, 2003.

16. A.Ćurković and E. Marušić-Paloka. Asymptotic analysis of a thin fluid layer-elastic plate interaction problem. *Applicable analysis* 98 (2019), 2118–2143.

17. R. Dal Passo, H. Garcke, and G. Grün. On a fourth order degenerate parabolic equation: global entropy estimates and qualitative behaviour of solutions. *SIAM J. Math. Anal.* 29 (1998), 321–342.

18. S. B. Das, I. Joughin, M. Behn, I. Howat, M. A. King, D. Lizarralde, M. P. Bhatia. Fracture propagation to the base of the Greenland ice sheet during supraglacial lake drainage. *Science* 320 (2008), 778–781.

19. R. Daw and J. Finkelstein. Lab on a chip. *Nature Insight* 442 (2006), 367–418.

20. P. Destuynder. Comparaison entre les modeles tridimensionnels et bidimensionnels de plaques en élasticité. *ESAIM: Mathematical Modelling and Numerical Analysis* 15 (1981), 331–369.

21. Earl H. Dowell. A modern course in aeroelasticity. Volume 217 of the Solid Mechanics and its Applications book series. Springer, 2015.

22. Q. Du, M. D. Gunzburger, L. S. Hou, and J. Lee. Analysis of a linear fluid-structure interaction problem. Discr. Cont. Dyn. Sys. 9 (2003), 633–650.

23. L. Giacomelli and F. Otto. Variational formulation for the lubrication approximation of the Hele-Shaw flow. *Calc. Var. PDEs*, 13 (2001), 377–403.

24. Céline Grandmont and Matthieu Hillairet. Existence of global strong solutions to a beam-fluid interaction system. *Arch. Ration. Mech. Anal.*, 220(3):1283–1333, 2016.

25. G. Griso. Asymptotic behavior of structures made of plates. *Anal. Appl.*, 3 (2005), 325–356.

26. I. J. Hewit, N. J. Balmforth, and J. R. de Bruyn. Elastic-plated gravity currents. *Euro. Jnl. of Applied Mathematics* 26 (2015), 1–31.

27. M. Heil, A. L. Hazel, and J. A. Smith. The mechanics of airway closure. *Respiratory Physiology & Neurobiology* 163 (2008), 214–221.

28. M. Hillairet and T. Takahashi Collisions in three-dimensional fluid structure interaction problems. *SIAM journal on mathematical analysis*, 40(6), pp.2451–2477, 2009.

29. A. E. Hosoi, and L. Mahadevan. Peeling, healing and bursting in a lubricated elastic sheet. *Phys. Rev. Lett.* 93 (2004).

30. R. Huang, and Z. Suo. Wrinkling of a compressed elastic film on a viscous layer. J. Appl. Phys. 91 (2002), 1135–1142.

31. J. R. King. The isolation oxidation of silicon the reaction-controlled case. SIAM J. Appl. Math. 49 (1989), 1064–1080.

32. E. Lauga, M. P. Brenner and H. A. Stone. Microfluidics: The No-Slip Boundary Condition. In *Handbook of Experimental Fluid Dynamics* Eds. J. Foss, C. Tropea and A. Yarin, Springer, New-York (2005).

33. Z. Li, A. M. Leshansky, L. M. Pismen, P. Tabelinga. Step-emulsification in a microfluidic device. *Lab Chip* 15 (2015), 1023–1031.

34. J. R. Lister, G. G. Peng, and J. A. Neufeld. Spread of a viscous fluid beneath an elastic sheet. *Phys. Rev. Lett.* 111 (15) (2013).

35. E. Marušić-Paloka. The effects of flexion and torsion on a fluid flow through a curved pipe. *Appl. Math. Optim.*, 44 (2001), 245–272.

36. C. Michaut. Dynamics of magmatic intrusions in the upper crust: Theory and applications to laccoliths on Earth and the Moon. *J. Geophys. Res.* 116 (2011).

37. Andro Mikelić, Giovanna Guidoboni, and Sunčica Čanić. Fluid-structure interaction in a pre-stressed tube with thick elastic walls. I. The stationary Stokes problem. *Netw. Heterog. Media*, 2(3):397–423, 2007.

38. Boris Muha and Sunčica Čanić. Existence of a weak solution to a nonlinear fluid-structure interaction problem modeling the flow of an incompressible, viscous fluid in a cylinder with deformable walls. *Arch. Ration. Mech. Anal.*, 207(3):919–968, 2013.

39. T. Myers. Thin films with high surface tension. *SIAM Rev.* 40 (1998), 441–462.

40. A. Oron, S. H. Davis, S. G. Bankoff. Long-scale evolution of thin liquid films. *Rev. Mod. Phys.* 69 (1997), 931–980.

41. G. P. Panasenko, R. Stavre. Asymptotic analysis of a periodic flow in a thin channel with visco-elastic wall. *J. Math. Pures Appl.* 85 (2006), 558–579.

42. G. P. Panasenko, R. Stavre. Asymptotic analysis of a viscous fluid-thin plate interaction: Periodic flow. *Mathematical Models and Methods in Applied Sciences* 24 (2014), 1781–1822.

43. D. Pihler-Puzović, P. Illien, M. Heil, and A. Juel. Suppression of complex fingerlike patterns at the interface between air and a viscous fluid by elastic membranes. *Phys. Rev. Lett.* 108 (2012).

44. D. Pihler-Puzović, A. Juel and M. Heil. The interaction between viscous fingering and wrinkling in elastic-walled Hele-Shaw cells. *Phys. Fluids* 26 (2014), 022102.

45. O. Reynolds. On the theory of lubrication and its application to M. Beauchamp Tower's experiments. *Phil. Trans. Roy. Soc. London A* 117 (1886), 157–234.

46. A. Z. Szeri. Fluid Film Lubrication. Cambridge University Press, Cambridge, 2012.

47. H. A. Stone, A. D. Stroock, A. Ajdari. Engineering Flows in Small Devices: Microfluidics Toward a Lab-on-a-Chip. *Annual Review of Fluid Mechanics* 36 (2004), 381–411.

48. J. Tambača, S. Čanić, and A. Mikelić. Effective model of the fluid flow through elastic tube with variable radius. In *XI. Mathematikertreffen Zagreb-Graz*, volume 348 of *Grazer Math. Ber.*, pages 91–112. Karl-Franzens-Univ. Graz, Graz, 2005.

49. M. Taroni, and D. Vella. Multiple equilibria in a simple elastocapillary system. *J. Fluid Mech.* 712 (2012), 273–294.

50. I. Titze. Principles of voice production. Prentice Hall, New York, 1994.

51. V. C. Tsai, and J. R. Rice. Modeling turbulent hydraulic fracture near a free surface. *J. App. Mech.* 79 (2012).

52. J. L. Vazquez. The Porous Medium Equation: Mathematical Theory. Oxford Science Publications, Oxford, 2007.

53. K. Yang, P. Sun, L. Wang, J. Xu, L. Zhang. Modeling and simulations for fluid and rotating structure interactions. Comp. Meth. App. Mech. Eng. 311 (2016), 788–814.

54. A. Yenduri, R. Ghoshal, and R. K. Jaiman. A new partitioned staggered scheme for flexible multibody interactions with strong inertial effects. *Computer Methods in Applied Mechanics and Engineering* 315 (2017), 316–347.

Chapter 9

Stability of a Steady Flow of an Incompressible Newtonian Fluid in an Exterior Domain

Jiří Neustupa (✉)
Czech Academy of Sciences, Institute of Mathematics, Prague, Czech Republic
e-mail: neustupa@math.cas.cz

This chapter provides a brief survey of studies on stability of a steady flow of an incompressible Newtonian fluid around a compact body $\mathcal{B}$. Results on the long-time behavior and stability under assumptions of "sufficient smallness" of some quantities are cited and briefly described in Sect. 9.2. Results, mainly based on assumptions on spectrum of a certain associated linear operator are presented in Sect. 9.3. Finally, Sect. 9.4 contains a short note on analogous results concerning the case when body $\mathcal{B}$ rotates.

9.1 Introduction

The Navier–Stokes exterior problem, existence of a steady solution. We deal with stability of a steady flow $\mathbf{V}$ of an incompressible Newtonian viscous fluid past a fixed compact body $\mathcal{B}$ in $\mathbb{R}^3$. We denote $\Omega := \mathbb{R}^3 \smallsetminus \mathcal{B}$ and we suppose

T. Bodnár et al. (eds.), *Waves in Flows*, Advances in Mathematical Fluid Mechanics, https://doi.org/10.1007/978-3-030-68144-9_9

that the boundary $\partial\Omega$ of domain Ω is of the class C^2. The flow $\mathbf{V}$ corresponds to a steady solution of the Navier–Stokes problem

$$\begin{aligned} \partial_t\mathbf{v}+\mathbf{v}\cdot\nabla\mathbf{v} &= -\nabla p+\nu\Delta\mathbf{v}+\mathbf{f} && \text{in } \Omega\times(0,\infty), && (9.1.1)\\ \operatorname{div}\mathbf{v} &= 0 && \text{in } \Omega\times(0,\infty), && (9.1.2)\\ \mathbf{v} &= \mathbf{0} && \text{in } \partial\Omega\times(0,\infty), && (9.1.3)\\ \mathbf{v}(\mathbf{x},t) \to \mathbf{v}_\infty &= (v_\infty,0,0) && \text{for } |\mathbf{x}|\to\infty, && (9.1.4)\end{aligned}$$

where $\mathbf{v}$ is the velocity, p is the pressure, $\mathbf{f}$ is the given external body force, $\nu>0$ is the coefficient of viscosity, and v_∞ is a real constant. The existence of a steady solution to the system (9.1.1)–(9.1.4) has been studied in many papers, let us cite e.g. [11] and [18], and it is described in detail in the book [28]. It follows from Theorem X.6.4 in [28] that provided $v_\infty\neq 0$ and $\mathbf{f}\in\mathbf{L}^q(\Omega)$ for all $q\in(1,q_0]$ (where $q_0>3$), every so-called generalized steady solution has the structure

$$\mathbf{V} = \mathbf{v}_\infty+\mathbf{U}, \tag{9.1.5}$$

where

$$\begin{aligned} &\mathbf{U}\in\mathbf{L}^r(\Omega) && \text{for all } r\in(2,\infty], && (9.1.6)\\ &\partial_j\mathbf{U}\in\mathbf{L}^s(\Omega) && \text{for } j=1,2,3 \text{ and for all } s\in(\tfrac{4}{3},\infty]. && (9.1.7)\end{aligned}$$

If, moreover, $\mathbf{f}$ has a compact support then $\mathbf{U}$ admits the representation

$$\mathbf{U}(\mathbf{x}) = \mathbf{m}\cdot\mathbb{E}(\mathbf{x})+\mathbf{U}'(\mathbf{x}), \tag{9.1.8}$$

where $\mathbf{m}$ is a certain constant vector, $\mathbb{E}$ is the Oseen fundamental tensor, and $\mathbf{U}'(\mathbf{x})$ is a perturbation which decays faster than $\mathbb{E}(\mathbf{x})$ for $|\mathbf{x}|\to\infty$. Concretely, $\mathbf{U}'(\mathbf{x})=O(|\mathbf{x}|^{-3/2+\delta})$ for $|\mathbf{x}|\to\infty$ for any $\delta>0$, see [28, Theorem X.8.1]. The gradient of $\mathbf{U}'$ satisfies

$$|\nabla\mathbf{U}'(\mathbf{x})| = \begin{cases} O(|\mathbf{x}|^{-2}) & \text{for } |\mathbf{x}|\to\infty\\ O(|\mathbf{x}|^{-2-2\sigma}) & \text{for } |\mathbf{x}|\to\infty,\ |\mathbf{x}|-x_1\geq|\mathbf{x}|^\sigma, \end{cases} \tag{9.1.9}$$

where $0<\sigma\leq\frac{1}{2}$, see [28, p. 717].

The problem for perturbations. Writing the solutions $\mathbf{v}$ of the problem (9.1.1) in the form $\mathbf{v}=\mathbf{V}+\mathbf{u}=\mathbf{v}_\infty+\mathbf{U}+\mathbf{u}$, we observe that stability of solution $\mathbf{V}$ of (9.1.1)–(9.1.4) is equivalent to stability of the zero solution of the problem

$$\begin{aligned} \partial_t\mathbf{u}+v_\infty\,\partial_1\mathbf{u}+\mathbf{U}\cdot\nabla\mathbf{u}+\mathbf{u}\cdot\nabla\mathbf{U}+\mathbf{u}\cdot\nabla\mathbf{u} & \\ = -\nabla q+\nu\Delta\mathbf{u} && \text{in } \Omega\times(0,\infty), && (9.1.10)\\ \operatorname{div}\mathbf{u} = 0 && \text{in } \Omega\times(0,\infty), && (9.1.11)\\ \mathbf{u} = \mathbf{0} && \text{in } \partial\Omega\times(0,\infty), && (9.1.12)\\ \mathbf{u}(\mathbf{x}) \longrightarrow \mathbf{0} && \text{for } |\mathbf{x}|\to\infty, && (9.1.13)\end{aligned}$$

for the velocity perturbation $\mathbf{u}$. (q denotes a perturbation of the pressure, associated with the velocity perturbation $\mathbf{u}$.)

Notation. We denote vector functions and spaces of vector functions by boldface letters. We also denote by c a generic constant, i.e. a constant whose value may change throughout the text. On the other hand, numbered constants have fixed values throughout the paper. Furthermore, we use this notation:

- Let $\rho_0 > 0$ be so large that $\mathcal{B} \subset B_{\rho_0}(\mathbf{0})$. For $\rho \geq \rho_0$ we denote $\Omega_\rho := \Omega \cap B_\rho(\mathbf{0})$ and $\Omega^\rho := \mathbb{R}^3 \smallsetminus \overline{B_\rho(\mathbf{0})}$.
- $\| \,.\, \|_q$ denotes the norm of a scalar- or vector- or tensor-valued function with the components in $L^q(\Omega)$.
- If Ω' differs from Ω, then $\| \,.\, \|_{q;\,\Omega'}$ denotes the norm in $L^q(\Omega')$.
- $\mathbf{C}^\infty_{0,\sigma}(\Omega)$ is the linear space of divergence-free vector functions from $\mathbf{C}^\infty_0(\Omega)$ and $\mathbf{L}^2_\sigma(\Omega)$ is the closure of $\mathbf{C}^\infty_{0,\sigma}(\Omega)$ in $\mathbf{L}^2(\Omega)$. The so called Helmholtz projection of $\mathbf{L}^2(\Omega)$ onto $\mathbf{L}^2_\sigma(\Omega)$ is denoted by P_σ.
- $\mathbf{D}^{1,2}_{0,\sigma}(\Omega)$ is the completion of $\mathbf{C}^\infty_{0,\sigma}(\Omega)$ in the norm $|.|_{1,2} := \|\nabla \,.\, \|_2$.
- $\mathbf{D}^{-1,2}_{0,\sigma}(\Omega)$ is the dual to $\mathbf{D}^{1,2}_{0,\sigma}(\Omega)$. The norm in $\mathbf{D}^{-1,2}_{0,\sigma}(\Omega)$ is denoted by $\| \,.\, \|_{-1,2}$.
- $\mathbf{D}^{2,2}(\Omega)$ is the linear space of functions $\mathbf{u} \in \mathbf{L}^1_{loc}(\Omega)$ such that $\partial_i\partial_j\mathbf{u} \in \mathbf{L}^2(\Omega)$ for all $i, j = 1, 2, 3$.
- $A\mathbf{u} := P_\sigma \Delta \mathbf{u}$ for $\mathbf{u} \in D(A) := \mathbf{W}^{2,2}(\Omega) \cap \mathbf{D}^{1,2}_{0,\sigma}(\Omega)$ is the so called Stokes operator in $\mathbf{L}^2_\sigma(\Omega)$. Its spectrum $\mathrm{Sp}(A)$ coincides with the interval $(-\infty, 0]$, see e.g. [13] or [17]. Operator A is self-adjoint in $\mathbf{L}^2_\sigma(\Omega)$ and $D(A)$ can be considered to be a Hilbert space with the graph norm $\big(\| \,.\, \|_2^2 + \|A \,.\, \|_2^2\big)^{1/2}$. Note that $(-A\mathbf{u}, \mathbf{u})_2 = \|\nabla \mathbf{u}\|_2^2$, where $(\,.\,,\,.\,)_2$ is the scalar product in $\mathbf{L}^2(\Omega)$.
- One can verify that if $\mathbf{U}$ satisfies (9.1.6), (9.1.7) and $\mathbf{u} \in D(A)$, then the functions $\mathbf{U} \cdot \nabla \mathbf{u}$ and $\mathbf{u} \cdot \nabla \mathbf{U}$ belong to $\mathbf{L}^2(\Omega)$. We denote

$$\begin{aligned} B^0\mathbf{u} &:= -P_\sigma\, \partial_1 \mathbf{u}, \\ B^1\mathbf{u} &:= -P_\sigma(\mathbf{U} \cdot \nabla \mathbf{u}) - P_\sigma(\mathbf{u} \cdot \nabla \mathbf{U}), \\ B^1_s\mathbf{u} &:= -P_\sigma\big[\mathbf{u} \cdot (\nabla \mathbf{U})_s\big], \\ B^1_a\mathbf{u} &:= -P_\sigma(\mathbf{U} \cdot \nabla \mathbf{u}) - P_\sigma\big[\mathbf{u} \cdot (\nabla \mathbf{U})_a\big]. \end{aligned}$$

  Here, $(\nabla \mathbf{U})_s$ (respectively $(\nabla \mathbf{U})_a$) denotes the symmetric (respectively the anti-symmetric = skew-symmetric) part of $\nabla \mathbf{U}$.
- For $\mathbf{u} \in D(\mathcal{L}_1) := D(A)$, we define

$$\mathcal{L}_1\mathbf{u} := \nu A\mathbf{u} + v_\infty B^0\mathbf{u} + B^1\mathbf{u} \tag{9.1.14}$$

$$= \nu A\mathbf{u} + v_\infty B^0\mathbf{u} + B^1_s\mathbf{u} + B^1_a\mathbf{u}, \tag{9.1.15}$$

$$\mathcal{N}\mathbf{u} := -P_\sigma(\mathbf{u}\cdot\nabla\mathbf{u}), \tag{9.1.16}$$

as the operators in $\mathbf{L}^2_\sigma(\Omega)$. The symmetric part of $\mathcal{L}_1$ is $(\mathcal{L}_1)_s := \nu A + B^1_s$ and the skew-symmetric part of $\mathcal{L}$ is $(\mathcal{L}_1)_a = v_\infty B^0 + B^1_a$.

The operator form (9.1.10)–(9.1.13). Treating equation (9.1.10) in $\mathbf{L}^2_\sigma(\Omega)$ and applying the projection P_σ, we exclude the pressure and we obtain the operator equation

$$\frac{\mathrm{d}\mathbf{u}}{\mathrm{d}t} = \mathcal{L}_1\mathbf{u} + \mathcal{N}\mathbf{u}, \tag{9.1.17}$$

which is equivalent to the system (9.1.10)–(9.1.13). Under solutions on a time interval $[0,T)$ (where $0 < T \leq \infty$), we understand functions $\mathbf{u}$ satisfying equation (9.1.17) a.e. in $(0,T)$ and such that

$$\mathbf{u} \in L^2(J;\, D(A)), \qquad \frac{\mathrm{d}\mathbf{u}}{\mathrm{d}t} \in L^2(J;\, \mathbf{L}^2_\sigma(\Omega))$$

for each bounded interval $J \subset [0,T)$. Each solution can be redefined on a set of measure zero in $[0,T)$ so that it becomes a continuous mapping from $[0,T)$ to $\mathbf{L}^2_\sigma(\Omega) \cap \mathbf{D}^{1,2}_{0,\sigma}(\Omega)$, see e.g. [46, Theorem I.3.1]. We further assume that each solution has been redefined in this sense.

Basic ideas in investigation of stability of the zero solution of Eq. (9.1.17). If the operators $v_\infty B^0$ and B^1 are in some sense "sufficiently small" in comparison to νA, then the "dissipative term" $\nu A\mathbf{u}$ dominates in Eq. (9.1.17), which leads to stability of the zero solution. The condition of sufficient smallness of $v_\infty B^0$ and B^1 in comparison to νA expresses the requirement that the basic flow $\mathbf{V}$ is in some sense small in comparison to ν. This idea, together with several other methods that use the assumption on smallness of some quantities, is explained in greater detail in Sect. 9.2.

Another approach to the question of stability of the zero solution to Eq. (9.1.17) or related equations is based on the spectral analysis of operator $\mathcal{L}_1$. It is well known that the zero solution is stable if

$$(\exists\,\delta > 0)\ (\forall\lambda \in \mathrm{Sp}(\mathcal{L}_1))\ :\ Re\,\lambda \leq -\delta, \tag{9.1.18}$$

see e.g. Sattinger [55] or Kielhöfer [42]. The authors show that the nonlinear operator $\mathcal{N}$ does not influence the stability of the zero solution, and the linear operator $\mathcal{L}_1$ need not be dissipative in order to satisfy the spectral condition (9.1.18). It means that the condition of "smallness" of the operators $v_\infty B^0$ and B^1 in comparison to νA is not needed. In this case, the stability is a consequence of an appropriate stabilizing influence of the skew-symmetric part of $\mathcal{L}_1$. However, condition (9.1.18) cannot be satisfied if Eq. (9.1.17) represents the problem in an exterior domain. The reason is that the spectrum of $\mathcal{L}_1$ consists of the essential part

$$\mathrm{Sp}_{\mathrm{ess}}(\mathcal{L}_1) = \left\{\lambda \in \mathbb{C};\ \mathrm{Re}\,\lambda \leq -\nu\,(\mathrm{Im}\,\lambda)^2/v_\infty^2\right\} \tag{9.1.19}$$

plus an at most countable family of eigenvalues which may possibly cluster only at points on the boundary of $\mathrm{Sp}_{\mathrm{ess}}(\mathcal{L}_1)$, see Babenko [1] or Farwig and Neustupa [14, 16]. The set in (9.1.19) coincides with a parabolic region in the half-plane $\{\lambda \in \mathbb{C};\ \mathrm{Re}\,\lambda \le 0\}$, symmetric about the real axis and touching the imaginary axis at point 0, whose shape is influenced by the basic steady flow $\mathbf{V}$ only through v_∞. (See Fig. 1 in Sect. 9.3.) An approach, which does not use any condition of "smallness" of $v_\infty B^0$ and B^1 in comparison to νA, and uses other conditions, based on spectral properties of operator $\mathcal{L}_1$ (which in fact imply a sufficiently strong stabilizing influence of the skew-symmetric part of $\mathcal{L}_1$), is described in Sect. 9.3.

9.2 Long-Time Behavior and Stability Under Assumptions of Smallness of Some Data

Recall that $\mathbf{V} \equiv \mathbf{v}_\infty + \mathbf{U}$ denotes a steady-state solution to problem (9.1.1)–(9.1.4). The associated perturbation $\mathbf{u}$ satisfies (9.1.10)–(9.1.13).

On the long-time behavior of solutions to (9.1.10)–(9.1.13). A number of results concern the long-time behavior of the unsteady perturbations $\mathbf{u}$ in the class of weak solutions. The first relevant contribution in this direction is due to K. Masuda [49], who assumed that $\mathbf{U}$ is continuously differentiable, $\nabla\mathbf{U} \in L^3(\Omega)$ and

$$\sup_{\mathbf{x}\in\Omega} |\mathbf{x}|\,|\mathbf{U}(\mathbf{x})| \;<\; \frac{1}{2}\,. \tag{9.2.1}$$

The perturbed unsteady solution is supposed to satisfy the momentum equation with a perturbed body force. Thus, the corresponding perturbation $\mathbf{u}$ satisfies equation (9.1.10) with an additional right-hand side $\mathbf{f}'$, representing the perturbation to the steady body force $\mathbf{f}$ in Eq. (9.1.1). The Helmholtz projection $P_\sigma\mathbf{f}'$ is assumed to be in $C^1\big([0,\infty);\, \mathbf{L}^2_\sigma(\Omega)\big) \cap L^1\big(0,\infty;\, \mathbf{L}^2_\sigma(\Omega)\big)$ and such that

$$\sup_{t>0} \int_t^{t+1} \Big\| \frac{\mathrm{d}}{\mathrm{d}s} P_\sigma\mathbf{f}'(s) \Big\|_2^2\, \mathrm{d}s + \int_0^\infty s^{1/2}\, \Big\| \frac{\mathrm{d}}{\mathrm{d}s} P_\sigma\mathbf{f}'(s) \Big\|_2\, \mathrm{d}s \;<\; \infty.$$

The perturbation $\mathbf{u}$ is supposed to be a weak solution to the problem (9.1.10)–(9.1.18) on the time interval $(0,\infty)$ with $\mathbf{f}'$ on the right-hand side of (9.1.10), satisfying the initial condition $\mathbf{u}(0) = \mathbf{u}_0 \in \mathbf{L}^2_\sigma(\Omega)$, and satisfying the so-called strong energy inequality

$$\begin{aligned} \frac{1}{2}\,\|\mathbf{u}(t)\|_2^2 \;&\le\; \frac{1}{2}\,\|\mathbf{u}(s)\|_2^2 \\ &\quad - \int_s^t \big[\big(\mathbf{u}(\tau)\cdot\nabla\mathbf{U}, \mathbf{u}(\tau)\big)_2 + \|\nabla\mathbf{u}(\tau)\|_2^2 + \big(\mathbf{f}'(\tau), \mathbf{u}(\tau)\big)_2 \big]\, \mathrm{d}\tau \end{aligned} \tag{9.2.2}$$

for a.a. $s > 0$ (including $s = 0$) and all $t \in [s, \infty)$. It is shown in [49] that there exists $T_* > 0$ such that $\mathbf{u}(t)$ becomes regular for $t > T_*$, and decays at this rate:

$$\|\nabla \mathbf{u}(t)\|_2 \le ct^{-1/4}, \quad \|\mathbf{u}(t)\|_\infty \le ct^{-1/8}, \quad \text{for all } t > T_*. \tag{9.2.3}$$

The proof uses (9.2.1) and the assumptions on the integrability of $\mathbf{f}'$ to deduce, first, that $\mathbf{u}(t)$ is "small" for large t. Then, combining this with the estimates of $A^{1/2}\mathbf{v}$ one shows that $\mathbf{u}(t)$ is regular and tends to zero for $t \to \infty$ in the norm $|\,.\,|_{1,2}$. The rate of decay is calculated from the energy-type inequality, satisfied by $\mathbf{u}(t)$. The author also generalizes these results to the case when the unperturbed solution $\mathbf{V}$ is time-dependent. It should be noted that in [49] no assumption on the size of the initial perturbation $\mathbf{u}_0$ is used: it can be arbitrarily large. However, nothing can be said about the behavior of $\mathbf{u}(t)$ for $t \in (0, T_*)$.

If $\mathbf{U} \equiv \mathbf{0}$, the decay rates (9.2.3) are sharpened by J. G. Heywood in [36].

These results have been further elaborated on by P. Maremonti in [47]. Maremonti studied the attractiveness of steady as well as unsteady solutions $\mathbf{V}$ to the problem (9.1.1)–(9.1.4) in the same class of weak solutions considered by Masuda with $\mathbf{f}' \equiv \mathbf{0}$. In particular, for the case $\mathbf{V}$ steady, he shows the following decay rates:

$$\|\mathbf{u}(t)\|_2 \le ct^{-1}, \quad |\mathbf{u}(t)|_{1,2} \le ct^{-1/2} \quad \|\mathbf{u}(t)\|_\infty \le ct^{-1/2},$$

thus improving and extending the analogous finding of [36] and [49]. Instead of condition (9.2.1), the author assumes that the maximum of certain variational problem involving $\mathbf{V}$ is not "too large." The latter condition is satisfied if $\mathbf{V}$ is sufficiently regular and obeys (9.2.1).

The somehow more complicated question of asymptotic stability of $\mathbf{V}$ in the L^2-norm was first addressed by P. Maremonti in [48]. In particular, he shows that all $\mathbf{u}$ in the class of weak solutions, with $\mathbf{u}_0 \in \mathbf{L}^2_\sigma(\Omega)$ and satisfying the strong energy inequality (9.2.1) with $\mathbf{f}' \equiv \mathbf{0}$, must decay to zero in the L^2-norm, provided that the magnitude of $\mathbf{V}$ is restricted in the same way as specified in [47] and discussed earlier on.

An important contribution to the studies of the asymptotic stability of the steady solution $\mathbf{V}$ was also made by T. Miyakawa and H. Sohr in [51]. The authors show that if the steady solution $\mathbf{V}$ of (9.1.1)–(9.1.4) is such that $\mathbf{V} \in L^\infty(\Omega)$, $\nabla\mathbf{V} \in L^3(\Omega)$ and the smallness condition (9.2.1) is satisfied, and if, in addition, the perturbation $\mathbf{f}'$ to the body force $\mathbf{f}$ is in $L^2\left([0,T);\, \mathbf{L}^2_\sigma(\Omega)\right)$ for all $T > 0$ and in $L^1\left([0,\infty);\, \mathbf{L}^2_\sigma(\Omega)\right)$, then the L^2-norm of each weak solution $\mathbf{u}$ to problem (9.1.10)–(9.1.13), satisfying the energy inequality (9.2.2), tends to zero for $t \to \infty$. In [51] it is also shown that the class of such weak solutions is not empty, thus solving a problem left open in [48] and partially solved in [22].

Further results concerning the L^2-decay of the perturbation $\mathbf{u}(t)$ (as a weak solution to (9.1.10)–(9.1.13)) for $t \to \infty$ are provided in the paper [3] by W. Borchers and T. Miyakawa: the authors assume that the steady solution $\mathbf{V}$ of (9.1.1)–(9.1.4) is in $L^3(\Omega)$, $\nabla\mathbf{V} \in L^3(\Omega)$, the smallness condition (9.2.1) is satisfied, and the perturbation $\mathbf{f}'$ to the body force $\mathbf{f}$ is in $L^2_{loc}\left([0,\infty);\, \mathbf{L}^2_\sigma(\Omega)\right) \cap L^1\left(0,\infty;\, \mathbf{L}^2_\sigma(\Omega)\right) \cap$

$L^1\big(0,\infty;\, \mathbf{D}_{0,\sigma}^{-1,2}(\Omega)\big)$. They show that then the L^2-norm of each weak solution $\mathbf{u}$ to problem (9.1.10)–(9.1.13), obeying (9.2.2) tends to zero for $t \to \infty$. Moreover, if $\|\mathrm{e}^{\mathcal{L}_1 t}\mathbf{u}_0\|_2 = O(t^{-\alpha})$ for some $\alpha > 0$, then $\|\mathbf{u}(t)\|_2 = O\big((\ln t)^{\epsilon - 1/2}\big)$ for any $\epsilon > 0$. (Here, $\mathrm{e}^{\mathcal{L}_1 t}$ denotes the semigroup generated by operator $\mathcal{L}_1$.) The results of [3] are generalized by the same authors to the case of n–space dimensions ($n \geq 3$) in [4].

Even sharper rates of decay of the norms $\|\mathbf{u}(t)\|_r$ ($2 \leq r \leq \infty$) and $\|\nabla\mathbf{u}(t)\|_r$ ($2 \leq r \leq 3$) were obtained by H. Kozono in [45], provided the perturbation $\mathbf{f}'$ to the body force is in $L^1\big(0,\infty;\, L^2(\Omega)\big) \cap C\big((0,\infty);\, L^2(\Omega)\big)$ and decays like t^{-1} for $t \to \infty$. Kozono does not use any condition of smallness of the basic flow $\mathbf{V}$ or its initial perturbation $\mathbf{u}_0$, but assumes that $\mathbf{V}$ lies in Serrin's class $L^r\,(0,\infty;\, L^s(\Omega))$ ($2/r + 3/s = 1$, $3 < s \leq \infty$). This implies that $\mathbf{V}$ is in fact a strong solution and it is in a suitable sense small for large t. Obviously, the only time-independent solution in the considered Serrin class is $\mathbf{V} = \mathbf{0}$.

Stability, obtained by energy-type methods. There exist a series of results on stability of solution $\mathbf{V}$ in the class of strong unsteady perturbations, which, unlike the cited papers [3, 4, 36, 47, 49] and [45], provide an information on the size of the perturbations at all times $t > 0$, and not just for "large" t. However, on the other hand, the initial value of the perturbation is always required to be "small" as well as $\mathbf{V}$ is also supposed to be "sufficiently small" in appropriate norms. The first results of this kind come from the early 1970s of the twentieth century and new results on this topic still appear.

The next paragraphs contain the sketch of the main steps to obtain a result of the above type. Assume that $\mathbf{U} \in \mathbf{L}^3(\Omega)$, $\nabla\mathbf{U} \in L^2(\Omega)^{3\times 3} \cap L^{3/2}(\Omega)^{3\times 3}$, and $\mathbf{u}$ is a strong solution to problem (9.1.10)–(9.1.13) in the time interval $(0, T_0)$, for some $T_0 > 0$. Multiplying equation (9.1.10) by $\mathbf{u}$ and integrating by parts over Ω, one obtains

$$\begin{aligned}\frac{1}{2}\,\frac{\mathrm{d}}{\mathrm{d}t}\,\|\mathbf{u}\|_2^2 + \nu\,\|\nabla\mathbf{u}\|_2^2 \;&=\; (\mathbf{u}\cdot\nabla\mathbf{U}, \mathbf{u}) \;\leq\; \|\nabla\mathbf{U}\|_{3/2}\,\|\mathbf{u}\|_6^2 \\ &\leq\; c_1^2\,\|\nabla\mathbf{U}\|_{3/2}\,\|\nabla\mathbf{u}\|_2^2. \end{aligned} \tag{9.2.4}$$

(The norm $\|\mathbf{u}\|_6$ is estimated by Sobolev's inequality: $\|\mathbf{u}\|_6 \leq c_1\,\|\nabla\mathbf{u}\|_2$, see e.g. [28, p. 54].) Multiplying equation (9.1.10) by $A\mathbf{u} \equiv P_\sigma\Delta\mathbf{u}$ and integrating over Ω, one obtains

$$\begin{aligned}&\frac{1}{2}\,\frac{\mathrm{d}}{\mathrm{d}t}\,\|\nabla\mathbf{u}\|_2^2 + \nu\,\|A\mathbf{u}\|_2^2 \\ &\quad= \int_\Omega \big[-v_\infty\,\partial_1\mathbf{u} + \mathbf{U}\cdot\nabla\mathbf{u} + \mathbf{u}\cdot\nabla\mathbf{U} + \mathbf{u}\cdot\nabla\mathbf{u}\big]\cdot A\mathbf{u}\,\mathrm{d}\mathbf{x} \\ &\quad\leq \frac{\nu}{4}\,\|A\mathbf{u}\|_2^2 + \frac{4}{\nu}\,|v_\infty|^2\,\|\partial_1\mathbf{u}\|_2^2 + \frac{4}{\nu}\,\|\mathbf{U}\cdot\nabla\mathbf{u}\|_2^2 + \frac{4}{\nu}\,\|\mathbf{u}\cdot\nabla\mathbf{U}\|_2^2 \\ &\qquad + \frac{4}{\nu}\,\|\mathbf{u}\cdot\nabla\mathbf{u}\|_2^2. \end{aligned} \tag{9.2.5}$$

The first term on the right-hand side can be absorbed by the left hand side. The other terms on the right-hand side can be estimated by means of the inequalities

$$\begin{aligned}\|\nabla\mathbf{u}\|_6 &\le c\,\|\nabla\mathbf{u}\|_{1,2} = c\left(\|\nabla\mathbf{u}\|_2^2+\|\nabla^2\mathbf{u}\|_2^2\right)^{1/2}\\ &\le c_2\left(\|\nabla\mathbf{u}\|_2^2+\|A\mathbf{u}\|_2^2\right)^{1/2},\end{aligned}$$

where the first inequality follows from the continuous imbedding $W^{1,2}(\Omega)\hookrightarrow L^6(\Omega)$ and the second one follows e.g. from [28, pp. 322–323]. Thus,

$$\begin{aligned}
\|\partial_1\mathbf{u}\|_2^2 &\le \|\nabla\mathbf{u}\|_2^2,\\
\|\mathbf{U}\cdot\nabla\mathbf{u}\|_2^2 &\le \|\mathbf{U}\|_3^2\,\|\nabla\mathbf{u}\|_6^2 \le c_2^2\,\|\mathbf{U}\|_3^2\left(\|\nabla\mathbf{u}\|_2^2+\|A\mathbf{u}\|_2^2\right)\\
\|\mathbf{u}\cdot\nabla\mathbf{U}\|_2^2 &\le \|\nabla\mathbf{U}\|_2^2\,\|\mathbf{u}\|_\infty^2 \le c\,\|\nabla\mathbf{U}\|_2^2\,\|\mathbf{u}\|_{1,6}^2\\
&\le c\,\|\nabla\mathbf{U}\|_2^2\left(\|\mathbf{u}\|_6^2+\|\nabla\mathbf{u}\|_6^2\right) \le c_3\,\|\nabla\mathbf{U}\|_2^2\left(\|\nabla\mathbf{u}\|_2^2+\|A\mathbf{u}\|_2^2\right),\\
\|\mathbf{u}\cdot\nabla\mathbf{u}\|_2^2 &\le \|\mathbf{u}\|_6^2\,\|\nabla\mathbf{u}\|_3^2 \le c_1^2\,\|\nabla\mathbf{u}\|_2^3\,\|\nabla\mathbf{u}\|_6\\
&\le c\,\|\nabla\mathbf{u}\|_2^3\left(\|\nabla\mathbf{u}\|_2^2+\|A\mathbf{u}\|_2^2\right)^{1/2}\\
&\le c_4\,\|\nabla\mathbf{u}\|_2^2\left(\|\nabla\mathbf{u}\|_2^2+\|A\mathbf{u}\|_2^2\right).
\end{aligned}$$

Substituting these inequalities to (9.2.5), we obtain

$$\begin{aligned}\frac{1}{2}\,\frac{\mathrm{d}}{\mathrm{d}t}\,\|\nabla\mathbf{u}\|_2^2+\frac{\nu}{2}\,\|A\mathbf{u}\|_2^2 &\le \frac{4}{\nu}\,|v_\infty|^2\,\|\nabla\mathbf{u}\|_2^2\\ &+\frac{4}{\nu}\left(c_2^2\,\|\mathbf{U}\|_3^2+c_3\,\|\nabla\mathbf{U}\|_2^2+c_4\,\|\nabla\mathbf{u}\|_2^2\right)\left(\|A\mathbf{u}\|_2^2+\|\nabla\mathbf{u}\|_2^2\right).\end{aligned}\tag{9.2.6}$$

Summing the inequalities (9.2.5) and (9.2.6) (multiplied by $\alpha>0$), we obtain

$$\begin{aligned}&\frac{1}{2}\,\frac{\mathrm{d}}{\mathrm{d}t}\left(\|\mathbf{u}\|_2^2+\alpha\,\|\nabla\mathbf{u}\|_2^2\right)+\nu\,\|\nabla\mathbf{u}\|_2^2+\frac{\alpha\nu}{2}\,\|A\mathbf{u}\|_2^2\\ &\le\left(c_1^2\,\|\nabla\mathbf{U}\|_{3/2}^2+\frac{4\alpha}{\nu}\,|v_\infty|^2\right)\|\nabla\mathbf{u}\|_2^2\\ &\quad+\frac{4\alpha}{\nu}\left(c_2^2\,\|\mathbf{U}\|_3^2+c_3\,\|\nabla\mathbf{U}\|_2^2+c_4\,\|\nabla\mathbf{u}\|_2^2\right)\left(\|A\mathbf{u}\|_2^2+\|\nabla\mathbf{u}\|_2^2\right).\end{aligned}$$

Provided that

$$c_1^2\,\|\nabla\mathbf{U}\|_{3/2}^2 < \frac{\nu}{2},\tag{9.2.7}$$

there exists $\alpha\in(0,1)$ so small that

$$c_1^2\,\|\nabla\mathbf{U}\|_{3/2}^2+\frac{4\alpha}{\nu}\,|v_\infty|^2 \le \nu\left(1-\frac{\alpha}{2}\right).\tag{9.2.8}$$

Then we get

$$\frac{1}{2}\,\frac{\mathrm{d}}{\mathrm{d}t}\left(\|\mathbf{u}\|_2^2+\alpha\,\|\nabla\mathbf{u}\|_2^2\right)+\frac{\alpha\nu}{2}\left(\|\nabla\mathbf{u}\|_2^2+\|A\mathbf{u}\|_2^2\right)$$

$$\leq \frac{4\alpha}{\nu}\left(c_2^2\|\mathbf{U}\|_3^2 + c_3\|\nabla\mathbf{U}\|_2^2 + c_4\|\nabla\mathbf{u}\|_2^2\right)\left(\|A\mathbf{u}\|_2^2 + \|\nabla\mathbf{u}\|_2^2\right).$$

Consequently,

$$\frac{1}{2}\frac{\mathrm{d}}{\mathrm{d}t}\left(\|\mathbf{u}\|_2^2 + \alpha\|\nabla\mathbf{u}\|_2^2\right) + \left[\frac{\alpha\nu}{2} - \frac{4\alpha}{\nu}c_2^2\|\mathbf{U}\|_3^2 - \frac{4\alpha}{\nu}c_3\|\nabla\mathbf{U}\|_2^2 \right.$$
$$\left. - \frac{4c_4}{\nu}\left(\|\mathbf{u}\|_2^2 + \alpha\|\nabla\mathbf{u}\|_2^2\right)\right]\left(\|A\mathbf{u}\|_2^2 + \|\nabla\mathbf{u}\|_2^2\right) \leq 0. \tag{9.2.9}$$

Now, we observe that if

$$\frac{\nu}{2} - \frac{4}{\nu}c_2^2\|\mathbf{U}\|_3^2 - \frac{4}{\nu}c_3\|\nabla\mathbf{U}\|_2^2 > 0 \tag{9.2.10}$$

and $\|\mathbf{u}_0\|_2^2 + \alpha\|\nabla\mathbf{u}_0\|_2$ (recall that $\mathbf{u}_0 = \mathbf{u}(0)$) is so small that

$$\frac{\alpha\nu}{2} - \frac{4\alpha}{\nu}c_2^2\|\mathbf{U}\|_3^2 - \frac{4\alpha}{\nu}c_3\|\nabla\mathbf{U}\|_2^2 - \frac{4c_4}{\nu}\left(\|\mathbf{u}_0\|_2^2 + \alpha\|\nabla\mathbf{u}_0\|_2^2\right) > 0, \tag{9.2.11}$$

then $\|\mathbf{u}(t)\|_2^2 + \alpha\|\nabla\mathbf{u}(t)\|_2^2$ is non-decreasing for t in some right neighborhood of 0. This consideration can be simply extended, by the bootstrapping argument, to the whole interval of existence of the strong solution $\mathbf{u}$ (let it be $(0, T_0)$) so that one obtains: $\|\mathbf{u}(t)\|_2^2 + \alpha\|\nabla\mathbf{u}(t)\|_2^2 \leq \|\mathbf{u}_0\|_2^2 + \alpha\|\nabla\mathbf{u}_0\|_2^2$ for all $t \in (0, T_0)$. This shows, among other things, that the norm $\|\mathbf{u}(t)\|_{1,2}$ cannot blow up when $t \to T_0-$. Consequently, $T_0 = \infty$ and the inequality

$$\|\mathbf{u}(t)\|_2^2 + \alpha\|\nabla\mathbf{u}(t)\|_2^2 < \|\mathbf{u}_0\|_2^2 + \alpha\|\nabla\mathbf{u}_0\|_2^2$$

holds for all $t \in (0, \infty)$.

Integrating inequality (9.2.9) with respect to t, one can also derive an information on the integrability of $\|A\mathbf{u}\|_2^2 + \|\nabla\mathbf{u}\|_2^2$ and on the asymptotic decay of $\|\nabla\mathbf{u}(t)\|_2$. Thus, also including the information on the uniqueness of strong solutions, and using the result of [48], one can formulate the theorem:

Theorem 9.2.1. Suppose the steady solution $\mathbf{V}$ to the problem (9.1.1)–(9.1.4) has the structure (9.1.5), where $\mathbf{U}$ belongs to $\mathbf{L}^3(\Omega)$ and $\nabla\mathbf{U} \in L^2(\Omega)^{3\times 3} \cap L^{3/2}(\Omega)^{3\times 3}$ satisfy conditions (9.2.7) and (9.2.10) and $\alpha > 0$ is chosen so that (9.2.8) holds. Then, if $\mathbf{u}_0 \in \mathbf{L}^2_\sigma(\Omega) \cap \mathbf{W}^{1,2}_0(\Omega)$ satisfies (9.2.11), the problem (9.1.10)–(9.1.13) with the initial condition $\mathbf{u}(0) = \mathbf{u}_0$ has a unique strong solution $\mathbf{u}$ on the time interval $(0, \infty)$. Furthermore, there exists $c_5 > 0$ such that this solution satisfies

$$\|\mathbf{u}(t)\|_2^2 + \alpha\|\nabla\mathbf{u}(t)\|_2^2 + c_5\int_0^t \left(\|\nabla\mathbf{u}(s)\|_2^2 + \alpha\|A\mathbf{v}(s)\|_2^2\right)\mathrm{d}s$$
$$\leq \|\mathbf{u}_0\|_2^2 + \alpha\|\nabla\mathbf{u}_0\|_2^2 \tag{9.2.12}$$

for all $t > 0$ and

$$\lim_{t\to\infty}\|\nabla\mathbf{u}(t)\|_2 = 0. \tag{9.2.13}$$

The ideas of the proof of Theorem 9.2.1 and similar energy-type considerations have been applied to many other studies of stability or instability of steady-state solutions to Navier–Stokes and related equations. Concerning flows in exterior domains, the readers are referred e.g. to [20, 21, 23, 34, 35].

Stability, obtained by means of semigroups. A different approach, based on a representation of a solution by means of semigroups generated by the operators A or $\mathcal{L}_1$ and on estimates of the semigroups has been used by H. Kozono and T. Ogawa [43], H. Kozono and M. Yamazaki [44], and Y. Shibata [57].

In particular, H. Kozono and M. Yamazaki [44] study the flow in an exterior "smooth" domain Ω in $\mathbb{R}^N$ $(N \geq 3)$, under the assumption that $\mathbf{v}_\infty = \mathbf{0}$. The steady-state solution $\mathbf{V}$ is supposed to belong to $\mathbf{L}^{N,\infty}_\sigma(\Omega) \cap \mathbf{L}^\infty(\Omega)$ and its gradient is supposed to be in $L^{r_*}(\Omega)^{N\times N}$ for some $r_* \in (N,\infty)$. (Here, $\mathbf{L}^{N,\infty}_\sigma(\Omega)$ is the Lorentz-type space.) The authors show that operator $\mathcal{L}_1$ generates a quasi-bounded analytic semigroup in $\mathbf{W}^{q,2}(\Omega)$. Furthermore, they prove that under some assumptions on the regularity of the basic steady flow $\mathbf{V}$ and the assumptions of smallness of $\mathbf{V}$ and the initial value $\mathbf{u}_0$ in the norm of $\mathbf{L}^{N,\infty}_\sigma(\Omega)$, there exists a strong solution $\mathbf{u}$ of the problem for perturbations, which, among other things, is in $BC\big((0,\infty);\, \mathbf{L}^{N,\infty}_\sigma(\Omega)\big) \cap C\big((0,\infty);\, D(A)\big) \cap C^1\big((0,\infty);\, \mathbf{W}^{1,r_*}(\Omega)\big)$ and satisfies

$$\|\mathbf{u}(t)\|_r \;\leq\; c\,t^{-\frac{N}{2}\left(N^{-1}-r^{-1}\right)}, \qquad N < r \leq r_*$$

for all $t > 0$, where $c = c(N, r, r_*)$. The authors also show that the better is the information on the spatial decay of the initial velocity $\mathbf{u}_0$, the sharper is the asymptotic behavior of $\mathbf{u}(t)$ for $t \to \infty$.

Note that the case $\mathbf{v}_\infty \neq \mathbf{0}$ is studied Y. Shibata [57]. The author strongly uses the L^q–L^r estimates of the semigroup, generated by the Oseen operator $\nu A + B_0$.

9.3 Spectral Methods

Some properties of operators A, B^0, B^1, and $\mathcal{N}$. Using the inclusions $\mathbf{U} \in \mathbf{L}^3(\Omega)$, $\nabla\mathbf{U} \in L^{3/2}(\Omega)^{3\times 3}$, Hölder's inequality and Sobolev's inequality (see [28, p. 54]), one can verify that the operators B^1_s and B^1_a satisfy the estimates

$$\|B^1_s\boldsymbol{\phi}\|_2 + \|B^1_a\boldsymbol{\phi}\|_2 \;\leq\; c_6\,\|\nabla\boldsymbol{\phi}\|_2, \tag{9.3.1}$$

$$\big|(B^1_s\boldsymbol{\phi},\boldsymbol{\psi})_2\big| + \big|(B^1_a\boldsymbol{\phi},\boldsymbol{\psi})_2\big| \;\leq\; c_7\,\|\nabla\boldsymbol{\phi}\|_2\,\|\nabla\boldsymbol{\psi}\|_2. \tag{9.3.2}$$

The operators B^1_s and B^1_a are relatively compact with respect to A and with respect to $\nu A + v_\infty B^0$ in $\mathbf{L}^2_\sigma(\Omega)$, see [52]. Thus, the operators $\mathcal{L}_1 \equiv \nu A + v_\infty B^0 + B^1_s + B^1_a$ and $\nu A + v_\infty B^0$ have the same essential spectrum, which is expressed by formula (9.1.19). The total spectrum of $\mathcal{L}_1$ therefore consists of $\mathrm{Sp}_{\mathrm{ess}}(\mathcal{L}_1)$ plus an at most countable set of isolated eigenvalues of $\mathcal{L}_1$, which can possibly cluster only on the boundary of $\mathrm{Sp}_{\mathrm{ess}}(\mathcal{L}_1)$. (See Fig. 1.)

The operator νA generates a bounded analytic semigroup in $\mathbf{L}^2_\sigma(\Omega)$, see e.g. [33]. Since $\|B^0\phi\|_2^2 \leq (-A\phi, \phi)_2$ for $\phi \in D(A)$, the operator B^0 is (νA)-bounded in $\mathbf{L}^2_\sigma(\Omega)$ with the relative (νA)-bound zero. Thus, the operator $\nu A + v_\infty B^0$ generates a quasi-bounded analytic semigroup in $\mathbf{L}^2_\sigma(\Omega)$, see [41, p. 498]. (The same statement is also proven in [50].) As B^1 is $(\nu A + v_\infty B^0)$-compact in $\mathbf{L}^2_\sigma(\Omega)$, the operator $\mathcal{L}_1 \equiv \nu A + v_\infty B^0 + B^1$ generates a quasi-bounded analytic semigroup in $\mathbf{L}^2_\sigma(\Omega)$ as well. We denote the semigroup by $\mathrm{e}^{\mathcal{L}_1 t}$.

Let $\kappa > 0$ be fixed. The operator $\nu A + (1+\kappa)B_s^1$ is self-adjoint in the space $\mathbf{L}^2_\sigma(\Omega)$. The spectrum of $\nu A + (1+\kappa)B_s^1$ consists of $\mathrm{Sp}_{\mathrm{ess}}(\nu A + (1+\kappa)B_s^1) = (-\infty, 0]$ and of at most a finite set of positive eigenvalues, each of whose has a finite multiplicity (see [52]). Let the positive eigenvalues be $\lambda_1 \leq \lambda_2 \leq \dots \leq \lambda_N$, each of them being counted as many times as is its multiplicity. Let $\phi_1, \dots, \phi_N$ be associated eigenfunctions. They can be chosen so that they constitute an orthonormal system in $\mathbf{L}^2_\sigma(\Omega)$. Each of the eigenfunctions $\phi_1, \dots, \phi_N$ belongs to $\mathbf{D}^{1,2}_{0,\sigma}(\Omega) \cap \mathbf{W}^{2,2}(\Omega)$ and to $\mathbf{D}^{-1,2}_{0,\sigma}(\Omega)$, see [53].

Note that the resolvent problem for the operator $\nu A\phi + (1+\kappa)B_s$ in $\mathbf{L}^2_\sigma(\Omega)$ is in main aspects analogous to the resolvent problem for the Schrödinger operator $\Delta - q(\mathbf{x})$ in $L^2(\mathbb{R}^3)$. An estimate of the number of non-negative eigenvalues of $\Delta - q(\mathbf{x})$, called the Cwiklel–Lieb–Rosenbljum bound, was obtained in late 1970s in the last century and it says that this number is less than or equal to $c\,\|q_-\|^{3/2}_{3/2;\,\mathbb{R}^3}$ where q_- denotes the negative part of q. (See e.g. [54, p. 101].) In our case, by analogy with [54], the inclusion $\nabla \mathbf{U} \in L^{3/2}(\Omega)^{3\times 3}$ also plays a fundamental role in the proof of finiteness of the set of positive eigenvalues of $\nu A\phi + (1+\kappa)B_s$ and their finite multiplicities, see [52].

The nonlinear operator $\mathcal{N}$ satisfies the estimate

$$\|\mathcal{N}\phi\|_2 \leq c_8\,\mathcal{Y}[\phi] \tag{9.3.3}$$

for $\phi \in \mathbf{W}^{2,2}(\Omega) \cap \mathbf{D}^{1,2}_{0,\sigma}(\Omega)$, where

$$\mathcal{Y}[\phi] := \|A\phi\|_2^2 + \|\nabla\phi\|_2^2.$$

An orthogonal decomposition of $\mathbf{L}^2_\sigma(\Omega)$. Denote by $\mathbf{L}^2_\sigma(\Omega)'$ the linear hull of $\phi_1, \dots, \phi_N$ and by P' the orthogonal projection of $\mathbf{L}^2_\sigma(\Omega)$ onto $\mathbf{L}^2_\sigma(\Omega)'$. Furthermore, denote by $\mathbf{L}^2_\sigma(\Omega)''$ the orthogonal complement to $\mathbf{L}^2_\sigma(\Omega)'$ in $\mathbf{L}^2_\sigma(\Omega)$ and by P'' the orthogonal projection of $\mathbf{L}^2_\sigma(\Omega)$ onto $\mathbf{L}^2_\sigma(\Omega)''$. Then $\mathbf{L}^2_\sigma(\Omega)$ admits the orthogonal decomposition

$$\mathbf{L}^2_\sigma(\Omega) = \mathbf{L}^2_\sigma(\Omega)' \oplus \mathbf{L}^2_\sigma(\Omega)'' \tag{9.3.4}$$

and the operator $\nu A + (1+\kappa)B_s^1$ is reduced on each of the orthogonal subspaces $\mathbf{L}^2_\sigma(\Omega)'$ and $\mathbf{L}^2_\sigma(\Omega)''$. Moreover, it is positive on $\mathbf{L}^2_\sigma(\Omega)'$ and non-positive on $\mathbf{L}^2_\sigma(\Omega)''$.

The essential dissipativity of $\nu A + B_s^1$ in $\mathbf{L}^2_\sigma(\Omega)''$. The operator $\nu A + (1+\kappa)B_s^1$ satisfies the inequality $\big(\nu A\phi + (1+\kappa)B_s\phi, \phi\big)_2 \leq 0$ for all $\phi \in \mathbf{L}^2_\sigma(\Omega)'' \cap D(A)$.

Hence

$$\begin{aligned} \big((\nu A + B_s^1)\phi, \phi\big)_2 &= \frac{\kappa}{1+\kappa}\,(\nu A\phi, \phi)_2 + \frac{1}{1+\kappa}\,\big((\nu A + B_s^1 + \kappa B_s^1)\phi, \phi\big)_2 \\ &\leq \frac{\kappa}{1+\kappa}\,(\nu A\phi, \phi)_2 \;=\; -c_9\,\|\nabla\phi\|_2^2, \end{aligned}$$

where $c_9 = \kappa\nu/(1+\kappa)$. This property is called the essential dissipativity of $\nu A + B_s^1$ in the subspace $\mathbf{L}^2_\sigma(\Omega)''$. It plays an important role in the proof of stability in [52]. Without having used $\kappa > 0$ in the definition of the space $\mathbf{L}^2_\sigma(\Omega)''$, one could not generally show that $\nu A + B_s^1$ is essentially dissipative in $\mathbf{L}^2_\sigma(\Omega)''$.

Stability due to an influence of the skew-symmetric part $(\mathcal{L}_1)_a$ of operator $\mathcal{L}_1$. Suppose that $\rho > 0$ is so large that

$$\|\nabla\mathbf{U}\|_{3/2;\,\Omega^\rho} \;\leq\; \frac{\nu}{8}. \tag{9.3.5}$$

Recall that $\mathbf{V} \equiv (v_\infty, 0, 0) + \mathbf{U}$ is a steady solution of the Navier–Stokes problem (9.1.1)–(9.1.4).

Theorem 9.3.1. Suppose that $\mathbf{U} \in \mathbf{L}^3(\Omega)$ and $\nabla\mathbf{U} \in L^{3/2}(\Omega)^{3\times 3}$. Suppose that

(i) there exists a function $\varphi \in L^1(0,\infty) \cap L^2(0,\infty)$ such that

$$\big\|\mathrm{e}^{\mathcal{L}_1 t}\phi_\iota\big\|_{2;\,\Omega_\rho} \;\leq\; \varphi(t) \tag{9.3.6}$$

for all $\phi_\iota \in \{\phi_1, \dots, \phi_N\}$ and $t > 0$.

Then there exist positive constants δ, c_{10}, c_{11} such that if $\mathbf{u}_0 \in \mathbf{L}^2_\sigma(\Omega) \cap \mathbf{D}^{1,2}_{0,\sigma}(\Omega)$ and

$$\|\mathbf{u}_0\|_2 + \|\nabla\mathbf{u}_0\|_2 \;\leq\; \delta, \tag{9.3.7}$$

then the problem defined by the Eq. (9.1.17) and the initial condition $\mathbf{u}(0) = \mathbf{u}_0$ has a unique solution $\mathbf{u}$ on the time interval $[0,\infty)$. The solution satisfies

$$\begin{aligned} \|\mathbf{u}(t)\|_2^2 + \|\nabla\mathbf{u}(t)\|_2^2 + c_{10}\int_0^t \big(\|\nabla\mathbf{u}(s)\|_2^2 + \|A\mathbf{u}(s)\|_2^2\big)\,\mathrm{d}s \\ \leq\; c_{11}\,\big(\|\mathbf{u}_0\|_2^2 + \|\nabla\mathbf{u}_0\|_2^2\big) \end{aligned} \tag{9.3.8}$$

(for all $t > 0$) and

$$\lim_{t\to\infty}\, \|\nabla\mathbf{u}(t)\|_2 \;=\; 0. \tag{9.3.9}$$

Remark (explaining the stabilizing role of $(\mathcal{L}_1)_a$). Assume that operator $\mathcal{L}_1$ is not dissipative. (The case when $\mathcal{L}_1$ is dissipative is treated in [23], etc.) Let us explain in greater detail that, in this case, condition (i) of Theorem 9.3.1 can be satisfied only due to an appropriate influence of the skew-symmetric part B_a of operator B (and also of operator $\mathcal{L}_1$). Suppose, hypothetically, that $\mathcal{L}_1$ equals only the sum $\nu A + B_s$. Then $\nu A + B_s$ has at least one positive eigenvalue ζ. Let

ψ be a corresponding eigenfunction. The eigenfunction ψ cannot be identically equal to zero in Ω_ρ, otherwise one can derive a contradiction with (9.3.5). Since

$$\begin{aligned}\left(\nu A\psi + (1+\kappa)B_s\psi,\, \psi\right)_2 &= (1+\kappa)\left(\nu A\psi + B_s\psi,\, \psi\right)_2 - \kappa\,(\nu A\psi, \psi)_2 \\ &= (1+\kappa)\,\zeta\,\|\psi\|_2^2 + \kappa\nu\,\|\nabla\psi\|_2^2 \;>\; \zeta\,\|\psi\|_2^2,\end{aligned}$$

the operator $\nu A + (1+\kappa)B_s$ must also have at least one positive eigenvalue λ with an associated eigenfunction ϕ such that $(\phi, \psi)_2 \neq 0$. The scalar product of $\mathrm{e}^{\mathcal{L}_1 t}\phi$ with ψ satisfies

$$\left(\mathrm{e}^{\mathcal{L}_1 t}\phi,\, \psi\right)_2 \;=\; \left(\mathrm{e}^{(\nu A+B_s)t}\phi,\, \psi\right)_2 \;=\; \left(\phi,\, \mathrm{e}^{(\nu A+B_s)t}\psi\right)_2 \;=\; \mathrm{e}^{\zeta t}\,(\phi, \psi)_2.$$

From this, we deduce that the norm $\|\mathrm{e}^{\mathcal{L}_1 t}\phi\|_{2;\,\Omega_\rho}$ is exponentially increasing. Thus, condition (i) could not be satisfied in the case of non-dissipative operator $\mathcal{L}_1$, without the influence of the skew-symmetric part B_a in $\mathcal{L}_1$. □

Theorem 9.3.1 is in detail proven in [52]. Nevertheless, to illustrate the importance of condition (i), we present below the main ideas of the proof.

Principles of the proof of Theorem 9.3.1. *Step 1.* Lemma 5 in [52] says that $\mathbf{u}$ is a solution of the Eq. (9.1.6) on the time interval $[0, T)$ if and only if $\mathbf{u} = \mathbf{v} + \mathbf{w}$ where the pair $\mathbf{v}$, $\mathbf{w}$ satisfies the system

$$\begin{aligned}\frac{\mathrm{d}\mathbf{v}}{\mathrm{d}t} &= \nu A\mathbf{v} + (1+\kappa)B_s^1\mathbf{v} - \kappa P'' B_s^1\mathbf{v} + P''(v_\infty B^0 + B_a^1)\mathbf{v} \\ &\quad + P''\mathcal{N}\mathbf{u},\end{aligned} \tag{9.3.10}$$

$$\begin{aligned}\frac{\mathrm{d}\mathbf{w}}{\mathrm{d}t} &= \nu A\mathbf{w} + v_\infty B^0\mathbf{w} + B_s^1\mathbf{w} + B_a^1\mathbf{w} - \kappa P' B_s^1\mathbf{v} \\ &\quad + P'(v_\infty B^0 + B_a^1)\mathbf{v} + P'\mathcal{N}\mathbf{u}\end{aligned} \tag{9.3.11}$$

on $[0, T)$ with the initial conditions

$$\mathbf{v}(0) = P''\mathbf{u}(0), \qquad \mathbf{w}(0) = P'\mathbf{u}(0). \tag{9.3.12}$$

Step 2. Multiplying equation (9.3.10) at first by $\mathbf{v}$ and then by $A\mathbf{v}$, and integrating over Ω and over the time interval $(0, t)$, we derive the inequalities

$$\begin{aligned}&\|\mathbf{v}(t)\|_2^2 + 2c_9\int_0^t \|\nabla\mathbf{v}(s)\|_2^2\,\mathrm{d}s \\ &\qquad\qquad \le\; \|\mathbf{v}(0)\|_2^2 + c\int_0^t \mathcal{Y}[\mathbf{u}(s)]\,\|\mathbf{v}(s)\|_2\,\mathrm{d}s,\end{aligned} \tag{9.3.13}$$

$$\begin{aligned}\|\nabla\mathbf{v}(t)\|_2^2 + \nu\int_0^t \|A\mathbf{v}(s)\|_2^2\,\mathrm{d}s \;&\le\; \|\nabla\mathbf{v}(0)\|_2^2 + c\int_0^t \|\nabla\mathbf{v}(s)\|_2^2\,\mathrm{d}s \\ &\quad + c\int_0^t \|\nabla\mathbf{u}(s)\|_2^2\,\mathcal{Y}[\mathbf{u}(s)]\,\mathrm{d}s.\end{aligned} \tag{9.3.14}$$

Step 3. Representing $\mathbf{w}$ by means of the semigroup $\mathrm{e}^{\mathcal{L}_1 t}$ in the form

$$\begin{aligned}\mathbf{w}(\vartheta) &= \mathrm{e}^{\mathcal{L}_1\vartheta}\mathbf{w}(0)\\ &\quad + \int_0^{\vartheta} \mathrm{e}^{\mathcal{L}_1(\vartheta-s)} \big[P'(-\kappa B_s^1 + v_\infty B^0 + B_a^1)\mathbf{v}(s) + P'\mathcal{N}\mathbf{u}(s)\big]\, \mathrm{d}s, \qquad (9.3.15)\end{aligned}$$

and applying condition (i), one can derive the estimate

$$\begin{aligned}&\int_0^t \|\mathbf{w}(\vartheta)\|_{2;\Omega_\rho}^2\, \mathrm{d}\vartheta\\ &\qquad \le\ c\,\|\mathbf{w}(0)\|_2^2 + c\int_0^t \mathcal{Y}[\mathbf{v}(s)]\, \mathrm{d}s + c\left(\int_0^t \mathcal{Y}[\mathbf{u}(s)]\, \mathrm{d}s\right)^2. \qquad (9.3.16)\end{aligned}$$

Multiplying equation (9.3.11) by $\mathbf{w}$, integrating over Ω and over the time interval $(0,t)$, and applying (9.3.16), we obtain

$$\begin{aligned}\|\mathbf{w}(t)\|_2^2 + \nu\int_0^t \|\nabla\mathbf{w}(s)\|_2^2\, \mathrm{d}s\ &\le\ c\,\|\mathbf{w}(0)\|_2^2 + c\int_0^t \mathcal{Y}[\mathbf{v}(s)]\, \mathrm{d}s\\ &\quad + c\left(\int_0^t \mathcal{Y}[\mathbf{u}(s)]\, \mathrm{d}s\right)^2 + c\int_0^t \mathcal{Y}[\mathbf{u}(s)]\, \|\mathbf{w}(s)\|_2\, \mathrm{d}s. \qquad (9.3.17)\end{aligned}$$

Step 4. Multiplying equation (9.3.11) by $A\mathbf{w}$, and integrating over Ω and the time interval $(0,t)$, we derive the inequality

$$\begin{aligned}\|\nabla\mathbf{w}(t)\|_2^2 + \nu\int_0^t \|A\mathbf{w}(s)\|_2^2\, \mathrm{d}s\ &\le\ \|\nabla\mathbf{w}(0)\|_2^2 + c\int_0^t \|\nabla\mathbf{w}(s)\|_2^2\, \mathrm{d}s\\ &\quad + c\int_0^t \mathcal{Y}[\mathbf{v}(s)]\, \mathrm{d}s + c\int_0^t \|\nabla\mathbf{u}(s)\|_2^2\, \mathcal{Y}[\mathbf{u}(s)]\, \mathrm{d}s. \qquad (9.3.18)\end{aligned}$$

Note that the generic constant c in the inequalities (9.3.14)–(9.3.18) is always independent of $\mathbf{v}$ and $\mathbf{w}$. The inequalities (9.3.14)–(9.3.18) hold for all t on the interval of existence of the solution $(\mathbf{v},\mathbf{w})$ of (9.3.10), (9.3.11)—let it be the interval $[0,T)$.

Step 5. Denoting $\mathcal{Z}[\mathbf{v},\mathbf{w}] := \|\mathbf{v}\|_2^2 + \alpha\,\|\nabla\mathbf{v}\|_2^2 + \beta\,\|\mathbf{w}\|_2^2 + \gamma\,\|\nabla\mathbf{w}\|_2^2$, and applying the estimates from Steps 1–4, one can derive that $\alpha > 0$, $\beta > 0$ and $\gamma > 0$ can be chosen so that provided that $\mathcal{Z}[\mathbf{v}(0),\mathbf{w}(0)]$ is "sufficiently small," one has

$$\int_0^t \mathcal{Y}[\mathbf{u}(s)]\, \mathrm{d}s + \mathcal{Z}[\mathbf{v}(t),\mathbf{w}(t)]\ \le\ c\,\mathcal{Z}[\mathbf{v}(0),\mathbf{w}(0)] \qquad (9.3.19)$$

for all $t \in [0,T)$. Using this inequality, one can further show that the interval of existence of the solution $(\mathbf{v},\mathbf{w})$ can be extended to $[0,\infty)$ and the sum $\mathbf{u} = \mathbf{v} + \mathbf{w}$ satisfies (9.3.8) and (9.3.9). □

Theorem 9.3.2. Theorem 9.3.1 remains valid if condition (i) is replaced by the condition

(ii) to given $\xi > 0$ there exist functions $\varphi \in L^1(0,\infty) \cap L^2(0,\infty)$ and $\boldsymbol{\psi}_1, \dots, \boldsymbol{\psi}_N \in \mathbf{L}^2_\sigma(\Omega) \cap \mathbf{D}^{-1,2}_{0,\sigma}(\Omega)$ such that

$$\|\boldsymbol{\phi}_j - \boldsymbol{\psi}_j\|_{-1,2} \leq \xi; \qquad \text{for } j = 1, \dots, N \tag{9.3.20}$$

$$\big|(\mathrm{e}^{\mathcal{L}_1 t}\boldsymbol{\phi}_i, \boldsymbol{\psi}_j)_2\big| \leq \varphi(t); \qquad \text{for } t > 0 \text{ and } i,j = 1, \dots, N. \tag{9.3.21}$$

Theorem 9.3.2 is proven in [53]. Note that (9.3.20) is automatically satisfied if we consider $\boldsymbol{\psi}_j = \boldsymbol{\phi}_j$ for $j = 1, \dots, N$. In this case, condition (ii) reduces to the requirement that

$$\big|(\mathrm{e}^{\mathcal{L}_1 t}\boldsymbol{\phi}_i, \boldsymbol{\phi}_j)_2\big| \leq \varphi(t); \qquad \text{for } t > 0 \text{ and } i,j = 1, \dots, N, \tag{9.3.22}$$

where $\varphi \in L^1(0,\infty) \cap L^2(0,\infty)$, which plays in the proof of stability the same role as (9.3.6). Thus, condition (ii) represents an alternative to (i), which is more general in the sense that the size of $\mathrm{e}^{\mathcal{L}_1 t}\boldsymbol{\phi}_i$ is not necessarily measured over Ω_ρ, as in (9.3.6), or from the projection onto the straight lines with directional vectors $\boldsymbol{\phi}_j$ ($j = 1, \dots, N$), as in (9.3.22). Condition (ii) provides more freedom in the choice of N straight lines, onto which is $\mathrm{e}^{\mathcal{L}_1 t}\boldsymbol{\phi}_i$ projected.

The case $\Omega = \mathbb{R}^3$. The first purely spectral theorem on stability of the zero solution of the equation (9.1.17) (and consequently, on stability of solution $\mathbf{V}$ to the problem (9.1.1)–(9.1.4)) was proven by P. Deuring and J. Neustupa in paper [6]. Here, the authors have shown that condition (i) of Theorem 9.3.1 is satisfied if

(iii) *there exists* $\delta > 0$ *such that all eigenvalues* λ *of operator* $\mathcal{L}_1$ *satisfy* $\operatorname{Re}\lambda < -\delta$,

(iv) $\lambda = 0$ *is not a generalized eigenvalue of operator* $\mathcal{L}_1$ *in the sense that the only function* $\mathbf{u} \in \mathbf{D}^{1,2}_{0,\sigma}(\Omega)$, *satisfying*

$$\int_\Omega \big[\nu\,\nabla\mathbf{u} : \nabla\mathbf{w} + v_\infty\,\partial_1\mathbf{u}\cdot\mathbf{w} + \mathbf{U}\cdot\nabla\mathbf{u}\cdot\mathbf{w} + \mathbf{u}\cdot\nabla\mathbf{U}\cdot\mathbf{w}\big]\,\mathrm{d}\mathbf{x} = 0$$

for all $\mathbf{w} \in \mathbf{L}^2_\sigma(\Omega) \cap \mathbf{D}^{1,2}_{0,\sigma}(\Omega)$, *is* $\mathbf{u} = \mathbf{0}$.

Note that condition (iv) says, in other words, that the only weak solution $\mathbf{u} \in \mathbf{D}^{1,2}_{0,\sigma}(\Omega)$ to the steady problem

$$\begin{aligned} -\nu\Delta\mathbf{u} + v_\infty\,\partial_1\mathbf{u} + \mathbf{U}\cdot\nabla\mathbf{u} + \mathbf{u}\cdot\nabla\mathbf{U} + \nabla p &= 0 && \text{in } \Omega,\\ \operatorname{div}\mathbf{u} &= 0, && \text{in } \Omega,\\ \mathbf{u} &= \mathbf{0} && \text{on } \partial\Omega \end{aligned}$$

is $\mathbf{u} = \mathbf{0}$. The decay rate of $\|\mathrm{e}^{\mathcal{L}_1 t}\boldsymbol{\phi}_\iota\|_{2;\Omega_\rho}$ (for $\boldsymbol{\phi}_\iota \in \{\boldsymbol{\phi}_1, \dots, \boldsymbol{\phi}_N\}$ and $t \to \infty$), required by condition (i), is obtained in [6] by means of relatively complicated and subtle resolvent estimates for the Oseen and perturbed Oseen operators.

Two spectral-stability results in the case $\Omega \neq \mathbb{R}^3$. In paper [53], the author deals with $\mathbf{U}$, satisfying (9.1.6)–(9.1.9) and considers the operator $(\mathcal{L}_1)_{\mathrm{ext}}$, which is an operator from $D((\mathcal{L}_1)_{\mathrm{ext}}) = \mathbf{D}^{2,2}(\Omega) \cap \mathbf{D}^{1,2}_{0,\sigma}(\Omega)$ to $\mathbf{L}^2_\sigma(\Omega)$, defined by the same formula as $\mathcal{L}_1$. (Thus, $(\mathcal{L}_1)_{\mathrm{ext}}$ is an extension of $\mathcal{L}_1$.) Further, he uses the conditions

(v) *there exist $\delta > 0$ and $a_0 > 0$ such that all eigenvalues λ of operator $\mathcal{L}_1$ satisfy* $\mathrm{Re}\,\lambda < \max\{-\delta;\ -a_0\,(\mathrm{Im}\,\lambda)^2\}$,

(vi) 0 *is not an eigenvalue of the extended operator* $(\mathcal{L}_1)_{\mathrm{ext}}$

and proves the same statement on stability of the zero solution of Eq. (9.1.17) as in Theorem 9.3.1.

It should be noted that conditions (iv) and (vi) are in fact equivalent. Indeed, if $\mathbf{u}$ is an eigenfunction of $(\mathcal{L}_1)_{\mathrm{ext}}$, corresponding to the eigenvalue $\lambda = 0$, then it is obviously a solution of the weak problem from condition (iv). On the other hand, if $\mathbf{u} \in \mathbf{D}^{1,2}_{0,\sigma}(\Omega)$ is a function from condition (iv) then, applying especially Theorem IV.5.1 and Theorem V.5.3 from [28], one can deduce that $\mathbf{u}$ also belongs to $\mathbf{D}^{2,2}(\Omega)$, which means that $\mathbf{u}$ is in $D((\mathcal{L}_1)_{\mathrm{ext}})$ and satisfies $(\mathcal{L}_1)_{\mathrm{ext}}\mathbf{u} = \mathbf{0}$.

In order to prove the statement on stability, the author applies Theorem 9.3.2. Showing that $\mathcal{L}_1^*$ (the adjoint operator to $\mathcal{L}_1$) maps $\mathbf{C}^\infty_{0,\sigma}(\Omega)$ onto a dense subset of $\mathbf{D}^{-1,2}_{0,\sigma}(\Omega)$, he considers functions $\boldsymbol{\psi}_j$ in the form $\boldsymbol{\psi}_j = \mathcal{L}_1^*\boldsymbol{\zeta}_j$ for appropriate $\boldsymbol{\zeta}_j \in \mathbf{C}^\infty_{0,\sigma}(\Omega)$ $(j = 1,\dots,N)$. The semigroup $\mathrm{e}^{\mathcal{L}_1 t}$ can be represented by the line integral over an appropriate infinite curve Γ, lying essentially in the left half-plane of the complex plane $\mathbb{C}$, touching the imaginary axis and the spectrum of $\mathcal{L}_1$ only at point 0 (see Fig. 9.1), so that

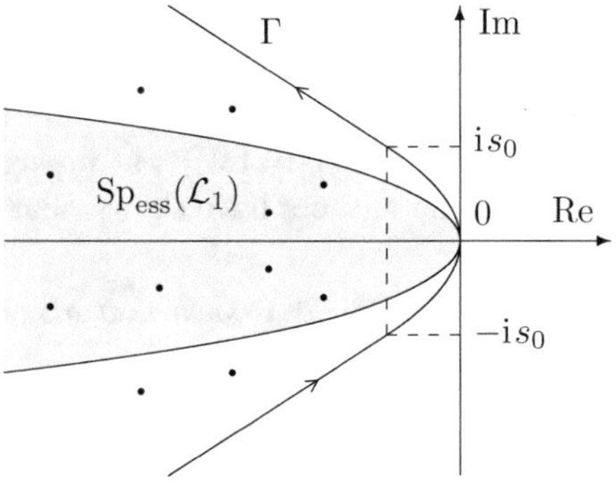

Figure 9.1: Spectrum of operator $\mathcal{L}_1$ and curve Γ

$$\mathrm{e}^{\mathcal{L}_1 t}\boldsymbol{\phi}_i = \frac{1}{2\pi\mathrm{i}}\int_\Gamma \mathrm{e}^{\lambda t}\,(\lambda I - \mathcal{L}_1)^{-1}\boldsymbol{\phi}_i \text{ for } i=1,\dots,N$$

and

$$\begin{aligned}
(\mathrm{e}^{\mathcal{L}_1 t}\boldsymbol{\phi}_i, \boldsymbol{\psi}_j)_2 &= (\mathrm{e}^{\mathcal{L}_1 t}\boldsymbol{\phi}_i, \mathcal{L}_1^*\boldsymbol{\zeta}_j)_2 = \frac{1}{2\pi\mathrm{i}}\int_\Gamma \mathrm{e}^{\lambda t}\,\big((\lambda - \mathcal{L}_1)^{-1}\boldsymbol{\phi}_i, \mathcal{L}_1^*\boldsymbol{\zeta}_j\big)_2\,\mathrm{d}\lambda \\
&= \frac{1}{2\pi\mathrm{i}}\int_\Gamma \mathrm{e}^{\lambda t}\,\big((\lambda - \mathcal{L}_1)^{-1}\boldsymbol{\phi}_i, (\mathcal{L}_1^* - \overline{\lambda})\boldsymbol{\zeta}_j\big)_2\,\mathrm{d}\lambda \\
&\quad + \frac{1}{2\pi\mathrm{i}}\int_\Gamma \mathrm{e}^{\lambda t}\,\big((\lambda - \mathcal{L}_1)^{-1}\boldsymbol{\phi}_i, \overline{\lambda}\boldsymbol{\zeta}_j\big)_2\,\mathrm{d}\lambda \\
&= -\frac{1}{2\pi\mathrm{i}}\int_\Gamma \mathrm{e}^{\lambda t}\,(\boldsymbol{\phi}_i, \boldsymbol{\zeta}_j)_2\,\mathrm{d}\lambda + \frac{1}{2\pi\mathrm{i}}\int_\Gamma \mathrm{e}^{\lambda t}\,\lambda\,\big((\lambda I - \mathcal{L}_1)^{-1}\boldsymbol{\phi}_i, \boldsymbol{\zeta}_j\big)_2\,\mathrm{d}\lambda.
\end{aligned}$$

for $i, j = 1, \dots, N$. As, the integrand in the first term on the right-hand side depends on λ only through $e^{\lambda t}$, one can show (applying an appropriate parametrization of line Γ) that this term decays exponentially for $t \to \infty$. It is shown in [53] that the second term on the right-hand side is, as a function of t, in $L^1(1,\infty) \cap L^2(1,\infty)$. The proof is quite complicated and one has to use the resolvent estimates for operator $\mathcal{L}_1$, the fact that $\boldsymbol{\zeta}_j$ has a compact support, and also the factor λ in the integrand plays an important role, because it enables one to control the integral for λ in the neighborhood of point $\lambda = 0$.

Later on, P. Deuring [9] succeeded in verification of condition (i) from Theorem 9.3.1 just using the assumptions (iv) and

(vii) *all eigenvalues* λ *of operator* $\mathcal{L}_1$ *satisfy the inequality* $\operatorname{Re}\lambda < 0$.

The used method differs from the one used in [53], and it is again (as in [6]) based on subtle and technical resolvent estimates for the Oseen and perturbed Oseen operators in Ω_ρ, where the bounds for velocity depend on the resolvent parameter λ in a specific and explicit way. (See Section 6 in [9].) In order to obtain the estimates for small $|\lambda|$, the author applies the results from his former paper [10].

The relation between conditions (v) and (vii). Obviously, condition (vii) is weaker than condition (v) in the sense that (v) $\Longrightarrow$ (vii). However, although it is not clear at the first sight, provided condition (iv) holds, the opposite implication (vii) $\Longrightarrow$ (v) is also true.

Let us sketch the proof. Assume, by contradiction, that conditions (iv), (vii) hold and (v) does not hold. Then, since the eigenvalues of $\mathcal{L}_1$ may possibly cluster only at points of the essential spectrum of $\mathcal{L}_1$, and the only point of $\operatorname{Sp}_{\text{ess}}(\mathcal{L}_1)$ that does not satisfy the inequality in condition (v) for any $a_0 > 0$ is $\lambda = 0$, there exists a sequence $\{\lambda_n\}$ of eigenvalues of $\mathcal{L}_1$ (with corresponding eigenfunctions $\mathbf{u}_n$, such that $\operatorname{Re}\lambda_n < 0$ and $\lambda_n \to 0$ for $n \to \infty$. The eigenfunctions satisfy the equation

$$\nu A\mathbf{u}_n + v_\infty B^0\mathbf{u}_n + B^1\mathbf{u}_n - \lambda_n\mathbf{u}_n = \mathbf{0} \tag{9.3.23}$$

and they can be normalized so that $\|\mathbf{u}_n\|_2 = 1$. Multiplying equation (9.3.23) by $\mathbf{u}_n$ and integrating in Ω, multiplying afterwards equation (9.3.23) by $A\mathbf{u}_n$ and integrating in Ω, and using the properties of function $\mathbf{U}$ stated in (9.1.6)–(9.1.9), one can derive that the norms $\|\nabla\mathbf{u}_n\|_2$ and $\|A\mathbf{u}_n\|_2$ are bounded independently of n. (See analogous estimates in [53, pp. 149–151], the proof of Lemma 4.4.) Put $k_n := \|\nabla\mathbf{u}_n\|_2 + \|A\mathbf{u}_n\|_2$ and $\mathbf{v}_n := \mathbf{u}_n/k_n$. Functions $\mathbf{v}_n$ satisfy the same equation as functions $\mathbf{u}_n$, i.e.

$$\nu A\mathbf{v}_n + v_\infty B^0\mathbf{v}_n + B^1\mathbf{v}_n - \lambda_n\mathbf{v}_n = \mathbf{0}. \tag{9.3.24}$$

The norms of $\mathbf{v}_n$ in $\mathbf{D}^{1,2}_{0,\sigma}(\Omega) \cap \mathbf{D}^{2,2}(\Omega)$ are equal to one (for all $n \in \mathbb{N}$). Hence there exists a subsequence (which is also denoted by $\{\mathbf{v}_n\}$), weakly convergent to some function $\mathbf{v}$ in $\mathbf{D}^{1,2}_{0,\sigma}(\Omega) \cap \mathbf{D}^{2,2}(\Omega)$.

Suppose at first that $\mathbf{v} \neq \mathbf{0}$. Then, considering the limit transition in the weak form of (9.3.24), we can show that $\mathbf{v}$ is an eigenfunction of operator $(\mathcal{L}_1)_{\text{ext}}$, corresponding to the eigenvalue 0. However, this contradicts condition (vi) and therefore also condition (iv).

Assume now that $\mathbf{v} = \mathbf{0}$. The weak convergence of $\{\mathbf{v}_n\}$ to $\mathbf{0}$ implies the strong convergence to $\mathbf{0}$ in $\mathbf{W}^{1,q}(\Omega)_\rho)$ for $1 < q < 6$ and ρ arbitrarily large. Then, applying the estimates (4.14) and (4.18) of $\|\nabla\mathbf{v}_n\|_2$ and $\|A\mathbf{v}_n\|_2$ from [53], one can deduce that

$$\lim_{n\to\infty} \big(\|\nabla\mathbf{v}_n\|_2 + \|A\mathbf{v}_n\|_2\big) \;=\; 0.$$

This is, however, also impossible, because $\|\nabla\mathbf{v}_n\|_2 + \|A\mathbf{v}_n\|_2 = 1$.

Consequently, we observe that under the validity of condition (iv), conditions (v) and (vii) are equivalent.

The final spectral-stability theorem. Summarizing the aforementioned results, we can formulate the next theorem. Recall that the theorem directly follows from [9] and, employing the equivalence [(iv),(v)] $\Longleftrightarrow$ [(iv), (vii)], also from [53].

Theorem 9.3.3. Let function $\mathbf{U}$ have the properties (9.1.6)–(9.1.9). Let operator $\mathcal{L}_1$ satisfy conditions (iv) and (vii). Then the zero solution of Eq. (9.1.17) is stable in the sense, stated in Theorem 9.3.1.

Sufficient conditions for instability. In addition to sufficient conditions for stability of the zero solution of Eq. (9.1.17), one can also be naturally interested in conditions that imply instability of the zero solution of Eq. (9.1.17), which is equivalent to instability of the steady solution $\mathbf{V}$ of the problem (9.1.1)–(9.1.4). In this context, we can cite two papers, where the authors deal with flows in exterior domains: L. I. Sazonov [56] proved the instability of a steady-state flow $\mathbf{V}(\mathbf{x}) \equiv \mathbf{v}_\infty + \mathbf{U}(\mathbf{x})$ in the norm of $\mathbf{L}^3(\Omega)$, assuming that at least one eigenvalue of the associated linear operator $\mathcal{L}_1$ has positive real part and $\mathbf{U}(x)$ is suitably integrable. S. Friedlander et al. [19] considered a steady flow $\mathbf{V} \in \mathbf{C}^\infty(\Omega)$, assuming that operator $\mathcal{L}_1$ has a part of the spectrum in the right half-plane of $\mathbb{C}$. The instability is proven in the norm of $\mathbf{L}^r(\Omega)$ for any $r \in (1,\infty)$. Note that it is important in both the papers [56] and [19] that operator $\mathcal{L}_1$ generates an analytic semigroup.

9.4 A Note on Stability of a Steady Flow Past a Rotating Body

The transformed problem, existence of a steady solution. In this section, we suppose that the compact body $\mathcal{B}$ is rotating about the x_1–axis with a constant angular velocity ω. In this case, the region of flow $\mathbb{R}^3 \smallsetminus \mathcal{B}$ is generally time-dependent. In order to obtain a problem in a fixed domain, it is convenient to transform the problem (9.1.1)–(9.1.4) to a coordinate system attached to the body. In such a system, the Eqs. (9.1.1), (9.1.2) take the form

$$\partial_t\mathbf{v} - (\omega\mathbf{e}_1 \times \mathbf{x})\cdot\nabla\mathbf{v} + \omega\mathbf{e}_1 \times \mathbf{v} + \mathbf{v}\cdot\nabla\mathbf{v} \;=\; -\nabla p + \nu\Delta\mathbf{v} + \mathbf{f}, \qquad (9.4.1)$$

$$\operatorname{div} \mathbf{v} = 0, \tag{9.4.2}$$

where $\mathbf{e}_1$ denotes the unit vector in the direction of the x_1–axis, $\mathbf{v}$, p, $\mathbf{f}$, and $\mathbf{x}$ are the transformed velocity, pressure, external body force, and spatial variable, respectively. The system (9.4.1), (9.4.2) is considered for $(\mathbf{x}, t) \in \Omega \times (0, \infty)$, with Ω a fixed domain. The no-slip boundary condition (9.1.3) for the velocity on the surface of the body transforms to

$$\mathbf{v}(\mathbf{x}, t) = \omega \mathbf{e}_1 \times \mathbf{x} \qquad \text{for } \mathbf{x} \in \partial\Omega, \tag{9.4.3}$$

and the condition in infinity (9.1.4) is preserved:

$$\mathbf{v}(\mathbf{x}, t) \to \mathbf{v}_\infty = (v_\infty, 0, 0) \qquad \text{for } |\mathbf{x}| \to \infty. \tag{9.4.4}$$

(See e.g. the papers [13, 37, 38], or [29] for more details concerning the transformation.)

For basic qualitative properties of the problem (9.4.1)–(9.4.4), we refer to [2, 5, 7, 8, 12, 15, 24–27, 30, 32, 37–39], and [40].

A steady-state solution. We further suppose that Ω is an exterior domain in $\mathbb{R}^3$ with the boundary of the class $C^{2+\mu}$ for some $\mu \in (0, 1)$. It follows e.g. from [28, Sec. XI.6] (for $v_\infty \neq 0$) and [28, Sec. XI.7] or [25] (for $v_\infty = 0$ and ω "sufficiently small") that the problem (9.4.1)–(9.4.4) has a steady solution $\mathbf{V} = \mathbf{v}_\infty + \mathbf{U}$ (corresponding to a steady-state velocity field), such that

$$\nabla \mathbf{U} \in L^{r_0}(\Omega)^{3\times 3} \cap L^\infty(\Omega)^{3\times 3} \tag{9.4.5}$$

for some $r_0 \in (1, 3)$ and there exist c_{12}, $c_{13} > 0$ such that

$$|\mathbf{U}(\mathbf{x})| \le c_{12}, \qquad |\nabla \mathbf{U}(\mathbf{x})| \le \frac{c_{13}}{1 + |\mathbf{x}|} \qquad \text{for } \mathbf{x} \in \Omega. \tag{9.4.6}$$

Recall that $\mathbf{V}$ is a steady solution in the "rotating frame," i.e. in the coordinate system attached to the rotating body. However, this solution is $(2\pi/\omega)$-periodic in the frame in which body $\mathcal{B}$ rotates about the x_1-axis with the constant angular velocity ω.

On stability of solution V. Of works, where the stability of a steady-state solution $\mathbf{V}$ has been studied , we cite the papers [26] (by G. P. Galdi and A. Silvestre), [40] (by T. Hishida and Y. Shibata), [58] (by Y. Shibata), and [29] (by G. P. Galdi and J. Neustupa).

In paper [26], the authors consider $\mathbf{f} = \mathbf{0}$ and $v_\infty = 0$. The steady solution $\mathbf{V}$ is supposed to satisfy

$$\|\nabla^2 \mathbf{V}\|_2 + \sup_{\mathbf{x}\in\Omega} (1 + |\mathbf{x}|^2)\, |\nabla \mathbf{V}(\mathbf{x})| + \sup_{\mathbf{x}\in\Omega} (1 + |\mathbf{x}|)\, |\mathbf{V}(\mathbf{x})| < \infty$$

and the initial value of the perturbed solution $\mathbf{v}$, together with $\mathbf{V}$, are supposed to satisfy $(\omega \mathbf{e}_1 \times \mathbf{x}) \cdot \nabla[\mathbf{v}(0) - \mathbf{V}] \in \mathbf{L}^2(\Omega)$. The authors prove that there exists $\delta = \delta(\Omega, \nu)$ such that if

$$\|\mathbf{v}(0) - \mathbf{V}\|_{1,2} + |\omega| < \delta,$$

then the problem (9.4.1)–(9.4.4) has a unique strong solution $\mathbf{v}$ on the time interval $(0,\infty)$, such that, among other properties,

$$(\mathbf{v}-\mathbf{b}),\ \partial_t\mathbf{v} \in L^\infty(0,\infty;\ \mathbf{L}^2(\Omega)), \quad \nabla\mathbf{v} \in L^2(0,\infty;\ L^2(\Omega)^{3\times3}),$$
$$\nabla^2\mathbf{v} \in L^2(0,\infty;\ L^2(\Omega)^{3\times3\times3}),$$

where $\mathbf{b}$ is a suitable extension of the boundary value $\omega\mathbf{e}_1\times\mathbf{x}$ from $\partial\Omega$ to Ω. Moreover,

$$\lim_{t\to\infty} \big\|\nabla[\mathbf{v}(t)-\mathbf{V}]\big\|_2 = 0.$$

In [40], the authors also deal with the case $v_\infty = 0$. They assume that $\mathbf{V}\in\mathbf{L}^{3,\infty}(\Omega)$ (the Lorentz space) and $\mathbf{v}(0)\in\mathbf{L}^{3,\infty}_\sigma(\Omega)$, where the subscript σ has an analogous meaning as in $\mathbf{L}^2_\sigma(\Omega)$. They show that there exists $\delta=\delta(\nu,\omega,\Omega)>0$ such that if $\|\mathbf{V}\|_{\mathbf{L}^{3,\infty}(\Omega)}+\|\mathbf{v}(0)\|_{\mathbf{L}^{3,\infty}(\Omega)}<\delta$, then $\mathbf{v}\in BC((0,\infty);\ \mathbf{L}^{3,\infty}(\Omega))$. Moreover, if $q\in(3,\infty)$, then there exists $\widetilde{\delta}=\widetilde{\delta}(\nu,\omega,\Omega,q)>0$ such that if $\|\mathbf{V}\|_{\mathbf{L}^{3,\infty}(\Omega)}+\|\mathbf{v}(0)\|_{\mathbf{L}^{3,\infty}(\Omega)}<\widetilde{\delta}$, then

$$\|\mathbf{v}(t)\|_r = O\big(t^{-\frac{1}{2}+\frac{3}{2r}}\big) \qquad \text{as } t\to\infty$$

for each $r\in(3,q)$.

The case $v_\infty\neq 0$ has been studied in paper [58]. Considering $\mathbf{V}$ in the form $\mathbf{V}=\mathbf{v}_\infty+\mathbf{U}$ and solutions $\mathbf{v}$ of (9.4.1)–(9.4.4) in the form $\mathbf{v}=\mathbf{V}+\mathbf{u}\equiv\mathbf{v}_\infty+\mathbf{U}+\mathbf{u}$, one obtains the equations for perturbations

$$\partial_t\mathbf{u}-(\omega\mathbf{e}_1\times\mathbf{x}+u_\infty\mathbf{e}_1)\cdot\nabla\mathbf{u}+\omega\mathbf{e}_1\times\mathbf{u}+\mathbf{U}\cdot\nabla\mathbf{u}+\mathbf{u}\cdot\nabla\mathbf{U}+\mathbf{u}\cdot\nabla\mathbf{u}$$
$$= -\nabla q+\Delta\mathbf{u}, \tag{9.4.7}$$
$$\operatorname{div}\mathbf{u} = 0 \tag{9.4.8}$$

and the conditions

$$\mathbf{u}(\mathbf{x},t) = \mathbf{0} \qquad \text{for } \mathbf{x}\in\partial\Omega, \tag{9.4.9}$$
$$\mathbf{u}(\mathbf{x},t) \to \mathbf{0} \qquad \text{for } |\mathbf{x}|\to\infty. \tag{9.4.10}$$

(Function q in Eq. 9.4.7) again represents a perturbation of the pressure.) In [58], the author assumes that $3<q<\infty$ and $\epsilon>0$ is "small," and shows that there exists $\delta>0$ such that if

$$\|\mathbf{U}\|_{3-\epsilon}+\|\mathbf{U}\|_{3+\epsilon}+\|\nabla\mathbf{U}\|_{3/2-\epsilon}+\|\nabla\mathbf{U}\|_{3/2+\epsilon}+\|\mathbf{u}(0)\|_3 < \delta,$$

then

$$[\mathbf{u}]_{3,0,t}+[\mathbf{u}]_{q,\mu(q),t}+[\nabla\mathbf{u}]_{3,1/2,t} \le \delta$$

for any $t>0$, where $[\mathbf{u}]_{r,\rho,t}:=\sup_{0<s<t}\ s^\rho\,\|\mathbf{u}(s)\|_r$.

Sufficient conditions for stability of the steady solution $\mathbf{V}$, without any condition of smallness of $\mathbf{V}$, have been derived in [29]. Here, in addition to operators A, B^0, and B^1 (defined in Sect. 9.1), the authors also use the operator

$$B^2\mathbf{u} := (\mathbf{e}_1 \times \mathbf{x}) \cdot \nabla\mathbf{u} - \mathbf{e}_1 \times \mathbf{u}$$

and define

$$\mathcal{L}_2\mathbf{u} := \nu A\mathbf{u} + v_\infty B^0\mathbf{u} + B^1\mathbf{u} + \omega B^2\mathbf{u}$$

for $\mathbf{u} \in D(\mathcal{L}_2) := \{\mathbf{u} \in D(A);\ (\mathbf{e}_1 \times \mathbf{x}) \cdot \nabla\mathbf{u} \in \mathbf{L}^2(\Omega)\}$. As $B^2\mathbf{u} \in \mathbf{L}^2_\sigma(\Omega)$ for $\mathbf{u} \in D(\mathcal{L}_2)$ (due to [14, Lemma 2.3]), $\mathcal{L}_2$ is an operator in $\mathbf{L}^2_\sigma(\Omega)$. In contrast to $\mathcal{L}_1$, operator $\mathcal{L}_2$ generates only a C_0-semigroup in $\mathbf{L}^2_\sigma(\Omega)$. The spectrum of $\mathcal{L}_2$ is studied in [13], [14], [16], and [17]: it consists of the essential part

$$\begin{aligned}
\mathrm{Sp}_{\mathrm{ess}}(\mathcal{L}_2) &= \{\lambda = \alpha + \mathrm{i}k\omega \in \mathbb{C};\ \alpha \le 0,\ k \in \mathbb{Z}\} \quad (\text{if } u_\infty = 0),\\
\mathrm{Sp}_{\mathrm{ess}}(\mathcal{L}_2) &= \Big\{\lambda = \alpha + \mathrm{i}\beta + \mathrm{i}k\omega \in \mathbb{C};\ \alpha, \beta \in \mathbb{R},\ k \in \mathbb{Z},\ \alpha \le -\frac{\nu\beta^2}{u_\infty^2}\Big\}\\
&\quad (\text{if } u_\infty \ne 0)
\end{aligned}$$

plus an at most countable family of eigenvalues which may possibly cluster only at points on the boundary of $\mathrm{Sp}_{\mathrm{ess}}(\mathcal{L}_2)$. By analogy with (9.1.17), the system (9.4.7)–(9.4.10) is equivalent to the operator equation

$$\frac{\mathrm{d}\mathbf{u}}{\mathrm{d}t} = \mathcal{L}_2\mathbf{u} + \mathcal{N}\mathbf{u}. \tag{9.4.11}$$

Note that $\mathcal{L}_1$ and $\mathcal{L}_2$ differ by the operator ωB^2, which is skew-symmetric. Using especially this fact, we can define the subspaces $\mathbf{L}^2_\sigma(\Omega)'$ and $\mathbf{L}^2_\sigma(\Omega)''$ of $\mathbf{L}^2_\sigma(\Omega)$ in the same way as in Sect. 9.3. Again, the subspace $\mathbf{L}^2_\sigma(\Omega)'$ is finite-dimensional and its orthonormal basis is constituted by the eigenfunctions $\boldsymbol{\phi}_1, \dots, \boldsymbol{\phi}_N$ of the operator $\nu A + (1+\kappa)B^1_s$. Although the proof is subtler and more complicated, one can derive the same theorem as Theorem 9.3.1, where one must only consider operator $\mathcal{L}_2$ instead of $\mathcal{L}_1$, see [29] for more details.

It is a challenging open problem to prove a theorem, analogous to Theorem 9.3.3, with the operator $\mathcal{L}_2$ instead of $\mathcal{L}_1$ and Eq. (9.4.11) instead of (9.1.17). The main reason, why the same procedure as in the proof of Theorem 9.3.3 does not work in the case of operator $\mathcal{L}_2$ and Eq. (9.4.11), lies in the facts that the semigroup $\mathrm{e}^{\mathcal{L}_2 t}$ is only of class C_0 (and not analytic), and the spectrum of $\mathcal{L}_2$ cannot be encircled by a curve similar to Γ as on Fig. 1.

A sufficient condition for instability. It is shown in paper [31] that the steady solution $\mathbf{V}$ of (9.4.1)–(9.4.4) is unstable in the norm of $\mathbf{L}^2(\Omega)$ if the linear operator $\mathcal{L}_2$ has a part of the spectrum in the half-plane $\{\lambda \in \mathbb{C};\ \mathrm{Re}\,\lambda > 0\}$. As an auxiliary result of independent interest, the authors have also proved that the uniform growth bound of the semigroup $\mathrm{e}^{\mathcal{L}_2 t}$ is equal to the spectral bound of operator $\mathcal{L}_2$. (This equality is obvious for analytic semigroups, but generally does not hold for C_0-semigroups.)

Acknowledgments

The research has been supported by the Grant Agency of the Czech Republic (grant No. 19-04243S) and by the Academy of Sciences of the Czech Republic (RVO 67985840).

References

1. K. I. Babenko: Spectrum of the linearized problem of flow of a viscous incompressible liquid round a body. *Sov. Phys. Dokl.* **27** (1982), 25–27.
2. W. Borchers: *Zur Stabilität und Faktorisierungsmethode für die Navier–Stokes Gleichungen inkompressibler viskoser Flüssigkeiten.* Habilitation thesis, University of Paderborn, 1992.
3. W. Borchers, T. Miyakawa: L^2–decay for Navier-Stokes flows in unbounded domains, with application to exterior stationary flows. *Arch. Rat. Mech. Anal.* **118** (1992), 273–295.
4. W. Borchers, T. Miyakawa: On stability of exterior stationary Navier–Stokes flows. *Acta Math.* **174** (1995), 311–382.
5. P. Cumsille, M. Tucsnak: Wellpossedness for the Navier–Stokes flow in the exterior of a rotating obstacle. *Math. Meth. Appl. Sci* **29** (2006), 595–623.
6. P. Deuring, J. Neustupa: An eigenvalue criterion for stability of a steady Navier–Stokes flow in $\mathbb{R}^3$. *J. Math. Fluid Mech.* **12** (2010), 202–242.
7. P. Deuring, S. Kračmar, Š. Nečasová: Pointwise decay of stationary rotational viscous incompressible flows with nonzero velocity at infinity. *J. Diff. Equations* **255** (2013), No. 7, 1576–1606.
8. P. Deuring, S. Kračmar, Š. Nečasová: A leading term for the velocity of stationary viscous incompressible flow around a rigid body performing a rotation and a translation. *Discr. Cont. Dyn. Syst. A* **37** (2017), No. 3, 1389–1409.
9. P. Deuring: Stability of stationary viscous incompressible flow around a rigid body performing a translation. *J. Math. Fluid Mech.* **20** (2018), 3, 937–967.
10. P. Deuring: Oseen resolvent estimates with small resolvent parameter. *J. Diff. Equations* **265** (2018), 1, 280–311.
11. R. Farwig: The stationary Navier–Stokes equations in a 3D–exterior domain. *Lecture Notes in Num. Appl. Anal.* **16** (1998), 53–115.
12. R. Farwig: An L^q-analysis of viscous fluid flow past a rotating obstacle. *Tohoku Math. J.* **58** (2005), 129–147.
13. R. Farwig, J. Neustupa: On the spectrum of a Stokes-type operator arising from flow around a rotating body. *Manuscripta Mathematica* **122** (2007), 419–437.

14. R. Farwig, J. Neustupa: On the spectrum of an Oseen–type operator arising from flow past a rotating body. *Integral Equations and Operator Theory* **62** (2008), 169–189.

15. R. Farwig, M. Krbec, Š. Nečasová: L^q–approach to Oseen flow around a rotating body. *Math. Meth. Appl. Sci.* **31** (2008), 5, 551–574.

16. R. Farwig, J. Neustupa: On the spectrum of an Oseen–type operator arising from fluid flow past a rotating body in $L^q_\sigma(\Omega)$. *Tohoku Math. J.* **62** (2010), 2, 287–309.

17. R. Farwig, Š. Nečasová, J. Neustupa: Spectral analysis of a Stokes–type operator arising from flow around a rotating body. J. Math. Soc. Japan **63** (2011), 1, 163–194.

18. R. Finn: On the exterior stationary problem for the Navier–Stokes equations and associated perturbation problems. Arch. Rat. Mech. Anal. **19** (1965), 363–406.

19. S. Friedlander, N. Pavlovich, R. Shvydkoy: Nonlinear instability for the Navier–Stokes equations. *Comm. Math. Phys.* **204** (2006), 335–347.

20. G. P. Galdi, S. Rionero: Local estimates and stability of viscous flows in an exterior domain. *Arch. Rat. Mech. Anal.* **81** (1983), 333–347.

21. G. P. Galdi, S. Rionero: *Weighted Energy Methods in Fluid Dynamics and Elasticity.* Springer–Verlag, Lecture Notes in Mathematics 1134, Berlin–Heidelberg–New York–Tokyo 1985.

22. G. P. Galdi, P. Maremonti: Monotonic decreasing and asymptotic behavior of the kinetic energy for weak solutions of the Navier-Stokes equations in exterior domains. *Arch. Rat. Mech. Anal.* **94** (1986), 253–266.

23. G. P. Galdi, M. Padula: A new approach to energy theory in the stability of fluid motion. *Arch. Rat. Mech. Anal.* **110** (1990), 187–286.

24. G. P. Galdi: On the motion of a rigid body in a viscous liquid: a mathematical analysis with applications. In *Handbook of Mathematical Fluid Dynamics,* eds. S. Friedlander and D. Serre, Vol. **1**, Elsevier 2002.

25. G. P. Galdi: Steady flow of a Navier-Stokes fluid around a rotating obstacle. *J. Elasticity* **71** (2003), 1–31.

26. G. P. Galdi, A. L. Silvestre: Strong solutions to the Navier–Stokes equations around a rotating obstacle. *Arch. Rat. Mech. Anal.* **176** (2005), 331–350.

27. G. P. Galdi, A. L. Silvestre: The steady motion of a Navier–Stokes liquid around a rigid body. *Arch. Rat. Mech. Anal.* **184** (2007), 371–400.

28. G. P. Galdi: *An Introduction to the Mathematical Theory of the Navier–Stokes Equations. Steady–State Problems.* 2nd Edition, Springer 2011.

29. G. P. Galdi, J. Neustupa: Stability of steady flow past a rotating body. In *Mathematical Fluid Dynamics, Present and Future,* Series: *Springer Proceedings in Mathematics & Statistics,* Vol. 183, eds. Y. Suzuki and Y. Shibata, Springer Japan, Tokyo 2016, pp. 71–94.

30. G. P. Galdi, J. Neustupa: Steady flows around moving bodies. In *Handbook of Mathematical Analysis in Mechanics of Viscous Fluids,* ed. Y. Giga and A. Novotný, Springer Int. Publishing 2018, 341–417.

31. G. P. Galdi, J. Neustupa: Nonlinear spectral instability of steady-state flow of a viscous liquid past a rotating obstacle. *Math. Annal.* (2020), https://doi.org/10.1007/s00208-020-02045-x.

32. M. Geissert, H. Heck, M. Hieber: L^p–theory of the Navier–Stokes flow in the exterior of a moving or rotating obstacle. *J. Reine Angew. Math.* **596** (2006), 45–62.

33. Y. Giga, H. Sohr: On the Stokes operator in exterior domains. *J. Fac. Sci. Univ. Tokyo,* Sec. IA **36** (1989), 103–130.

34. J. G. Heywood: On stationary solutions of the Navier–Stokes equations as limits of nonstationary solutions. *Arch. Rat. Mech. Anal.* **37** (1970), 48–60.

35. J. G. Heywood: The exterior nonstationary problem for the Navier–Stokes equations. *Acta Math.* **129** (1972), 11–34.

36. J. G. Heywood: The Navier–Stokes equations: On the existence, regularity and decay of solutions. *Indiana Univ. Math. J.* **29** (1980), 639–681.

37. T. Hishida: The Stokes operator with rotation effect in exterior domains. *Analysis* **19** (1999), 51–67.

38. T. Hishida: An existence theorem for the Navier–Stokes flow in the exterior of a rotating obstacle. *Arch. Rat. Mech. Anal.* **150** (1999), 307–348.

39. T. Hishida: L^q estimates of weak solutions to the stationary Stokes equations around a rotating body. *J. Math. Soc. Japan.* **58** (2006), 743–767.

40. T. Hishida, Y. Shibata: L_p–L_q estimate of the Stokes operator and Navier-Stokes flows in the exterior of a rotating obstacle. *Arch. Rat. Mech. Anal.* **193** (2009), 339–421.

41. T. Kato: *Perturbation Theory for Linear Operators.* Springer–Verlag, Berlin–Heidelberg–New York 1966.

42. H. Kielhöfer: On the Lyapunov stability of stationary solutions of semilinear parabolic differential equations, *J. Diff. Equations* **22** (1976), 193–208.

43. H. Kozono, T. Ogawa: On stability of Navier–Stokes flows in exterior domains. *Arch. Rat. Mech. Anal.* **128** (1994), 1–31.

44. H. Kozono, M. Yamazaki: On a large class of stable solutions to the Navier–Stokes equations in exterior domains. *Math. Z.* **228** (1998), 751–785.

45. H. Kozono: Asymptotic stability of large solutions with large perturbation to the Navier-Stokes equations. *J. Func. Anal.* **176** (2000), 153–197.

46. J. L. Lions, E. Magenes: *Nonhomogeneous Boundary Value Problems and Applications I.* Springer–Verlag, New York 1972.

47. P. Maremonti: Asymptotic stability theorems for viscous fluid motions in exterior domains. *Rend. Sem. Mat. Univ. Padova* **71** (1984), 35–72.

48. P. Maremonti: Stabilità asintotica in media per moti fluidi viscosi in domini esterni. *Ann. Mat. Pura Appl.* **142** (1985), 4, 57–75.

49. K. Masuda: On the stability of incompressible viscous fluid motions past objects. *J. Math. Soc. Japan* **27** (1975), 294–327.

50. T. Miyakawa: On nonstationary solutions of the Navier–Stokes equations in an exterior domain. *Hiroshima Math. J.* **12** (1982), 115–140.

51. T. Miyakawa, H. Sohr: On energy inequality, smoothness and large time behavior in L^2 for weak solutions of the Navier–Stokes equations in exterior domains, *Math. Z.* **199** (1988), 455–478.

52. J. Neustupa: Stability of a steady viscous incompressible flow past an obstacle. *J. Math. Fluid Mech.* **11** (2009), 22–45.

53. J. Neustupa: A spectral criterion for stability of a steady viscous incompressible flow past an obstacle. *J. of Math. Fluid Mech.* **18** (2016), 133–156.

54. M. Reed, B. Simon: *Methods of Modern Mathematical Physics,* Vol. IV, Academic Press, New York–San Francisco–London 1978.

55. D. H. Sattinger: The mathematical problem of hydrodynamic stability. *J. of Math. and Mech.* **18** (1970), 797–817.

56. L. I. Sazonov: Justification of the linearization method in the flow problem. *Izv. Ross. Akad. Nauk Ser. Mat.* **58** (1994), 85–109 (Russian).

57. Y. Shibata: On an exterior initial boundary value problem for Navier–Stokes equations. *Quarterly of Appl. Math.* **LVII** (1999), 1, 117–155.

58. Y. Shibata: A stability theorem of the Navier-Stokes flow past a rotating body. *Proc. of the conf. "Parabolic and Navier-Stokes equations",* Banach Center Publications Vol. 81, Institute of Mathematics, Polish Academy of Sciences, Warsaw 2008, 441–455.

Printed in the United States
by Baker & Taylor Publisher Services